Michael Abbott's Hydroinformatics

IWA Digital Water Book Series No. 1

Michael Abbott's Hydroinformatics

Poiesis of New Relationships with Water

Edited by
Andreja Jonoski

IWA
PUBLISHING

Published by **IWA Publishing**
Unit 104–105, Export Building
1 Clove Crescent
London E14 2BA, UK
Telephone: +44 (0)20 7654 5500
Fax: +44 (0)20 7654 5555
Email: publications@iwap.co.uk
Web: www.iwapublishing.com

First published 2022

British Library Cataloguing in Publication Data
A CIP catalogue record for this book is available from the British Library

ISBN: 9781789062649 (Paperback)
ISBN: 9781789062656 (eBook)
ISBN: 9781789062922 (ePUB)

Doi: 10.2166/9781789062656

This eBook was made Open Access in December 2022.

Contents

Articles of Michael B. Abbott

Editorial

I first heard of Mike Abbott in the 1980s, during my studies of civil engineering in the 'hydrotechnics' specialization at the Faculty of Civil Engineering, University St. Cyril and Methodius – Skopje, Macedonia. My Professor of hydraulics, Angel Angelovski, himself participant of post-graduate courses at IHE Delft in the Netherlands, introduced Abbott to us as the leading authority in the field of computational hydraulics. Computers were just becoming popular in what was then Yugoslavia, and we were instructed to start learning how to use them for solving water-related problems. At that time, Michael Abbott was just a name of a famous scientist, working somewhere very far from my reality, and I never dreamt that I would actually meet him.

That changed in 1994 when I came to IHE Delft for my post-graduate studies and actually saw the famous Professor Abbott. During my first two years of study, I specialized in groundwater hydrology, and I only heard stories about Mike Abbott's captivating lectures (both inside and outside of IHE classrooms) from my friends and colleagues who were enrolled in the studies of hydroinformatics. I first met Mike in 1996, when he decided to consider me as a candidate for carrying out PhD research under his supervision. On the first day that we met, he told me: "*Look Andreja, if you want to do a PhD with me you should know the following: others may be interested in things that are; I am only interested in things that are not.*" I learned the meaning of these words over the subsequent years of my PhD research, which thoroughly transformed me professionally and as a person. Prof. Abbott became 'Mike Abbott', and then simply 'Mike'. His supervision primarily consisted of two methods. Firstly, he would provide various books to me, all already read by him, often with his extensive written commentaries on the blank sides of the printed pages. I needed to understand both the original author, and the arguments that Mike had with that author. Secondly, there were many long talks, during which research aspects were discussed together with our personal experiences. Mike Abbott was an educator who aimed at creating new persons, not only PhD graduates.

Mike Abbott[1] was already widely recognized as a pioneering figure in the field of computational hydraulics at the time that he established himself as the founding father of the new discipline of hydroinformatics. For him, the evolution from computational hydraulics to hydroinformatics was both natural and necessary. His judgement was that it was time to move from things 'that are' to things 'that are not'. During the years after my PhD graduation, I have been trying to bring (and keep) together all the different aspects of hydroinformatics that Mike Abbott was promoting. His interweaving of water engineering with information and communication technology, philosophy, sociology, semiotics and religious literature was both puzzling and challenging. Understanding his vision required time and dedication, but for me this journey has been one of immense intellectual satisfaction and I consider myself truly privileged to have become one of his followers.

This book is a tribute to Mike and to his vision of hydroinformatics. It consists of seven original chapters, all written by his colleagues, students and close collaborators. Chapter 1 attempts to revisit the different motives for hydroinformatics, as originally conceived and further elaborated by Mike over the last 30 years. These have merged to form one comprehensive hydroinformatics approach that is highly relevant and necessary for dealing with our current water challenges. Chapter 2 presents the early efforts and later developments of computational hydraulics, as a precursor

---

[1] In the first stages of preparation of this book the group of authors had some discussion regarding reference to Mike Abbott. To most of us, his friends, colleagues, students and collaborators he was simply 'Mike'. On the other hand, in scientific citations he was 'M.B. Abbott', or 'Michael Abbott'. We have finally decided to use a 'middle way' option and throughout the book we refer to him as 'Mike Abbott'.

to hydroinformatics. The authors, who all worked closely with Mike in this field, revise the scientific and technological advances in the field, together with their integration in the new businesses centred around water-modelling software products. Chapter 3 introduces the role of 'modern' artificial intelligence (primarily associated with data-driven modelling and machine learning), recognized by Mike Abbott as indispensable to hydroinformatics for the analysis and modelling of water systems. Written by authors who have been pioneers in introducing these methods into applications for hydroinformatics, the chapter presents historical and current developments, together with exciting future opportunities. One widely recognized contribution of Mike Abbott to the field of hydrology has been the development and promotion of physically-based hydrological modelling, through the SHE system (Système Hydrologique Européen). In Chapter 4, some of the collaborators who actively participated in this effort present the SHE origins and its subsequent progress. They highlight the ongoing debates among hydrologists regarding the value of this modelling approach, and the current practices of its adoption and further development as necessary for dealing with contemporary hydrological challenges. Chapter 5 presents the business aspects of hydroinformatics in more detail and from the perspective of water engineers actively engaged in developing and using water modelling software in their consulting businesses. The original vision of Mike is presented, together with past experiences and future prospects of hydroinformatics-related water businesses and consultancy. The role of modelling, information technology and the electronic networks of communication in transforming the water management in China is presented in Chapter 6. Some showcasing examples of using such technologies for managing the formidable Yangtze (Chiangjiang) river basin are also presented. Although many of these developments took place independently of hydroinformatics, the authors point out that they did occur in many ways as envisioned and promoted by Mike Abbott. Moreover, with some of his students and followers taking up leading positions in Chinese water management (as is the case of the lead author of Chapter 6), some of these developments were indeed along the lines that Mike promoted. Finally, Chapter 7 presents a revision of developments in hydroinformatics education, as it has been experienced at IHE Delft, in the Netherlands. This chapter is written by co-authors belonging to the hydroinformatics group at IHE Delft, where Mike Abbott initiated the postgraduate course that later evolved into the MSc specialization in hydroinformatics. The chapter also presents developments of hydroinformatics education in Europe and globally, and some future challenges for hydroinformatics educators.

Following the content of the chapters, the book concludes with seven published articles of Mike Abbott, some written with his collaborators. The selection of these articles was made to cover as broadly as possible the wide spectrum of Mike's contributions to computational hydraulics and hydroinformatics. These include: numerical schemes and modelling of real-world water flows (Abbott and Ionescu, 1966, Abbott et al., 1978); modelling systems (Abbott et al., 1991), introduction of hydroinformatics (Abbott, 1999), role of hydraulics knowledge, its representation and usage in postmodern context (Abbott, 1999), impact of hydroinformatics on the field of hydraulics (Abbott et. al., 2001) and hydroinformatics for water-related social justice (Abbott and Vojinovic, 2014). Readers can identify and trace many of the issues and aspects addressed in the seven chapters of this book in these previously published articles.

This book was made possible by joint efforts of many friends and colleagues. First and foremost, I am most grateful to the authors and co-authors of the seven chapters. During the last two years, we have gone through a series of interactions and revisions of the chapters, trying to meet deadlines, sometimes with long evenings of writing, reviewing and editing. I am really grateful to all for their collaboration and patience for the production of this book. I am most grateful to Vladan Babovic, Editor in Chief of IWA Digital Water Programme for trusting me with the responsibility to be the editor of this volume, and for his support during its preparation. My sincere thanks to IWA Publishing, especially Mark Hammond and Niall Cunniffe, who provided guidance and assistance whenever it was needed. I am also grateful to all my colleagues from our IHE hydroinformatics group for their support in this project, including my colleagues from the IHE library who helped me in the final stages of the book production. I have benefited greatly from the leadership of Roland Price and Dimitri Solomatine, who succeeded Mike as leaders of the Hydroinformatics Chair Group. My very special thanks go to Ioana Popescu and Schalk Jan van Andel for our long and inspiring talks regarding hydroinformatics and our roles and responsibilities in it. Finally, I am grateful to Louise Abbott for encouraging me to take on this project, and for her continuing support.

Public access to the electronic version of this book has been made available thanks to the generous support provided by IHE Delft, Institute for water education, the Netherlands, and Hydroinformatics Institute, Singapore. This support is immensely appreciated. Such public access can surely enable many more readers, especially younger professionals, to become aware and appreciate the field of hydroinformatics and the immense contribution of Mike Abbott for its establishment and development.

Andreja Jonoski,
November, 2022

Preface

Over the last decade, almost every aspect of our lives has become digital – from how we deal with finances to the ways in which we are entertained. Pervasive availability of powerful and interconnected technologies, including the Cloud and AI, are leading many business and governmental organisations on a journey of digital transformation.

The water sector is also undergoing process of being digitised. While "digital water" may appear to be the 21st century development, for true visionaries, such as Mike Abbott, this evolution was anticipated already in 1980s.

Some of us were his students following a graduate programme on Computational Hydraulics at the time. Mike was considered as one of the pioneers and leading figures in the field. In the spring of 1990, Mike invited a number of leading scientists and technologists specialising in hydraulics and hydrology for an informal workshop to discuss his outline of a new subject area that he was preparing over past few years. The subject he called hydroinformatics.

The energy at the meeting was extraordinary. We, the students, were invited to attend the sessions. Like flies on the wall, we were able to witness speeches of intellectual giants of the time, sharing their views to support and supplement Mike's vision. As students we could not possibly foresee meteoric growth and the impact hydroinformatics would have on our field, and eventually, for many of us, on our own careers. Nevertheless, it was obvious that something significant was developing right in front of our own eyes.

Hydroinformatics eventually turned into a precursor of today's digital water. I congratulate Andreja Jonoski for bringing together some of the Mike's closest colleagues and collaborators to contribute to this volume – the volume that should serve as a tribute to Mike Abbott's remarkable career, intellectual legacy and leadership.

As digital disruption continues to reshape business domains, individuals and organisations remain under pressure to adapt to change. Doing so will allow them to take full advantage of the opportunities ahead. This book, hopefully, will provide perspectives on digital strategies for water management. It is meant to support you as you become more proactive in the digital domain, help turn digital threats into opportunities, and allow to leverage digital to create competitive advantages and to enhance performance.

Professor Vladan Babovic
Book Series Editor
National University of Singapore

doi: 10.2166/9781789062656_0001

Chapter 1

Revisiting the motives for hydroinformatics

A. Jonoski

IHE Delft Institute for Water Education, Hydroinformatics and Socio-technical Innovation, Delft, the Netherlands

ABSTRACT

Since Mike Abbott established hydroinformatics about 30 years ago, the field has been characterized by tremendous growth in both research and practice. With its natural dependence on developments in information and communication technologies, the expansion and diversity of hydroinformatics concepts and applications have hardly been surprising. Such developments have brought some issues of concern within the hydroinformatics community, mainly associated with apparent loss of some foundational principles. This chapter attempts to clarify the current situation by revisiting the main motivations for hydroinformatics, as originally proposed and later developed by Abbott. It also re-introduces and re-interprets the main aspects of the field, with the hope that such a re-interpretation will be useful for the future growth and development of the subject. Hydroinformatics originated as an engineering discipline, specifically linked to hydraulic engineering and to the development and use of numerical models from within the field of computational hydraulics. This initiated the central idea that modelling in hydroinformatics is motivated by applications in the real world. The union of fourth-generation modelling systems with artificial intelligence, was originally regarded as being the essence of hydroinformatics. The initial emphasis of the subject was on developing models of our natural aquatic environment and man-made water infrastructure that were more reliable, leading to improved support for decision making. As a consequence, improving model generation and application for decision support became central, and the field of hydroinformatics was established as a technology for enabling transparent knowledge creation, mobilization and sharing among different stakeholders participating in water- and environmental decision making. Aiming at transforming decision-making processes (and not only supporting existing ones), hydroinformatics identified the Internet as the key platform for introducing environments that would enable the conjunctive use of different 'knowledges' (scientific, as well as social/cultural/narrative knowledge). These last aspects are ongoing motivations for hydroinformatics to continue its development as a technology with the responsibilities of caring for the aquatic environment and improving human well-being.

Keywords: Hydroinformatics, physically based modelling, data-driven modelling, decision support, technology, sociotechnology, narrative knowledge

1.1 INTRODUCTION

The hydroinformatics community widely recognizes Mike Abbott as the founding father of the field: others also contributed to the process of establishing hydroinformatics as a distinct field of research and practice. Mike Abbott notes on the very first page of his seminal book titled: 'Hydroinformatics: Information Technology and the Aquatic Environment', that hydroinformatics has emerged as a 'place' where a 'considerable band of individuals … have assembled' (Abbott, 1991a, p. vii). Still, it was he who introduced and coined the term 'hydroinformatics' and took upon himself the responsibility to provide the foundations and directions of future developments.

In fact, the term, 'founding *father*' is fitting to describe the relationship between Mike Abbott and hydroinformatics. Like a loving and considerate parent, Mike Abbott was working and writing extensively to provide proper 'upbringing' to this new-born 'child'. Within a decade, however, the 'child' started developing its own characteristics and idiosyncrasies, not always in line with the directions and wishes of its father. Mike Abbott's foundational ideas did establish hydroinformatics as an all-encompassing field regarding the use of information technology when dealing with the aquatic environment. However, as different researchers and practitioners adopted hydroinformatics, they started focusing on particular aspects of their fields of interest, and quickly started emphasizing them as essential for hydroinformatics. Consequently, the *comprehensive idea of hydroinformatics* moved gradually into the background, leaving space for the multiple aspects of the field. This is perhaps natural when growing from 'childhood' to 'adolescence', but in the case of hydroinformatics this growth has been spectacular, both in terms of addressing different aspects of the relationship between human beings and the aquatic environment mediated by information technology, and by the diversity of proposed, designed and implemented hydroinformatics applications.

Two of the reasons that brought about this situation can be identified easily. First, the past 30 years have seen radically new developments in information and communication technologies (ICTs). Beyond the explosive growth of computational speed and available memory, we have witnessed the establishment, growth and current all-pervasive presence of the Internet and the World Wide Web. This has been further extended by the use of wireless communications and fast penetration of mobile telephony throughout the world. New data sources coming from intelligent sensor networks, remote sensing, social media and citizens' observatories led the transition from data-poor to data-rich environments for hydroinformatics applications. Each of these developments brought new challenges and opportunities for further advancements in hydroinformatics. In fact, the response of the field has been to embrace all these ICT developments almost without question. In the myriad of hydroinformatics applications that followed we see such a diversity of subjects that sometimes it becomes almost impossible to see any link among them. Except for the fact that they have something to do with water, and that they implement some sort of ICT, it seems there is nothing else that keeps them together under the common umbrella term of 'hydroinformatics applications'.

Second, many water- and environment-related fields (such as hydrology, hydraulics, water resources development and management, aquatic ecology or environmental engineering), which were somewhat slower in recognizing and utilizing the ICT potential (some of them even doubtful and suspicious about the way in which hydroinformatics was foreseeing and actively embracing the new ICTs), soon started to include various ICTs in their mainstream research and practice. Most notably, computer-based modelling became quite central activity, including both physically based (or process-based) modelling, and, soon afterwards, data-driven (machine-learning) modelling as well. This expansion of modelling brought about different views and approaches, not always aligned with those originating from hydroinformatics. Related to this, most of these fields put comparatively less emphasis on developing *modelling systems* (generic modelling tools for instantiating water-based models), and the associated need for the integration of models with data from different data sources for model development, calibration/validation, data assimilation and so on. The full integration of models and data in decision support tools and systems as proposed and promoted in hydroinformatics was also coming more gradually in these related fields, again associated with different ideas and approaches regarding the purpose and composition of such tools and systems. These developments also contributed to the ambiguity in discerning hydroinformatics applications from those rapidly emerging in other fields.

Mike Abbott, while content with such progress, also saw difficulties and even dangers in these processes, and he continued writing and working to maintain the main idea of hydroinformatics, while still dealing with all the transformations that the field was experiencing. One main concern shared by him and many of his close collaborators was that hydroinformatics would become an *eclectic* field where 'anything goes'. This concern can be explained by looking at the introduction of hydroinformatics in the first issue of *Journal of Hydroinformatics*, where Mike Abbott clearly associated hydroinformatics with the '*irreversible change from knowers to consumers of knowledge [as] cornerstone of postmodernity*' (Abbott, 1999, see originally Lyotard, 1979/1984). He then further elaborates as follows:

> This general tendency within societies that enter into the postmodern condition then leads to the formation of a new way of employing knowledge, which in turn necessitates a new kind of knowledge, which is a knowledge of how to access, absorb and apply an electronically encapsulated knowledge, ubiquitously called information, which itself truly becomes knowledge precisely to the extent that is genuinely accessed, authentically absorbed and properly applied.....within the postmodern context, hydroinformatics is the name of this new kind of knowledge as it is applied to the worlds of the waters.

Thus, as knowledge is compartmentalized into products it is challenging and transforming knowledge/power relations and the ways in which knowledge is legitimized in our societies. It impacts everything, including our relationships with the aquatic environment. Hydroinformatics then needs to take *responsibility* for shaping and guiding these relationships in the world of knowledge consumerism, when it comes to dealing with our water systems and aquatic environments. In taking this position, the alternative view of postmodernity, characterized with complete epistemological relativism, where 'any truth is possible', facts and 'alternative facts' (or, news and 'fake news') can coexist, captured in the 'anything goes' slogan, would be the most dangerous association for hydroinformatics. One aspect of this has been the continuous struggle for proper understanding of our natural and man-made water systems and environments and how this understanding is adequately translated/represented in our computer models. Another, potentially more serious aspect, has been the way in which 'anything goes' knowledge in fact masks all other legitimizing criteria except those of *performance* and *efficiency*, and the only true criterion of knowledge becomes its market value (Jameson, 1991; Lyotard, 1979/1984).

Already in 1991 Mike Abbott stated (Abbott, 1991a, p. viii):

> hydroinformatics has to do primarily with the social dimensions of hydraulics and environmental engineering… It is for the most part concerned with how science is applied, rather than science itself… Indeed, the very fact that it proceeds in social dimensions means that it has to take account of non-scientific aspects.

From its very beginning, hydroinformatics was established as a field that would respect and employ scientific knowledge with all its rigour, but it would also recognize its limits, as well as the necessity to combine it with non-scientific knowledge. One could say that this recognition also contributed to some sort of 'vulnerability' of the field of hydroinformatics, to be perceived (and discarded) by some as 'non-scientific', and by others who would ignore such 'uncomfortable' aspects in scientific research and in some related institutional contexts.

It is for these reasons that this chapter aims to revisit the motivations for hydroinformatics, as introduced by Mike Abbott and taken further by some of his collaborators. The hope is that such revisiting will bring forward again the comprehensive idea of hydroinformatics, which future researchers and practitioners will recognize, maintain and advance.

1.2 ENGINEERING ROOTS OF HYDROINFORMATICS

When Mike Abbott introduced hydroinformatics as a 'place' where a 'considerable band of individuals … have assembled' (p. vii, Abbott, 1991a), he mentioned that such individuals arrived at this 'place' from different origins: some from simulation (numerical) modelling, like himself, others from data modelling, or from artificial intelligence (AI) (mostly knowledge-based or expert rule-based systems, which were very popular at that time). Yet, the reality is that hydroinformatics emerged from the field of numerical modelling, particularly with applications in hydraulics, namely the field of computational hydraulics. It is in fact difficult to imagine 'paths' through which other disciplines would have arrived at hydroinformatics, without computational hydraulics being in their remit. The obvious reason for this is that computational hydraulics had become an integral part of *hydraulic engineering*.

The evolution of computational hydraulics, from its struggles with the 'higher reaches' of hydraulics, which were completely dominated by analytical approaches, through its five generations of modelling, till its establishment as the dominant field that provides tools for hydraulic engineering tasks, is well known (see again Abbott, 1991a, pp. 16–29). Chapter 2 of this volume is devoted to this most important condition for the emergence of hydroinformatics. The emphasis here is on the fact that engineers, particularly *hydraulic engineers*, were those who needed easy-to-use and efficient tools for their tasks of design, operation and control of hydraulic engineering structures and overall water-related infrastructure. Such tasks were then executed in the *real world*, and the quality of performance of the developed tools could be tested in actual engineering interventions. The results of the testing had a very high impact on the computational hydraulics tools (both models and their interconnections with relevant data). Physically based models which could represent accurately the real-world systems (their elements, structure and interconnections) were vital, especially for tasks related to engineering *design*, where changes, or completely new engineering interventions, were to be introduced. The power of data-driven models, especially for forecasting, operational and control tasks was recognized later and such models were added and demonstrated in real-world systems. One could say that engineering led computational hydraulics, and later hydroinformatics, to adopt *models for action* in the real world.

It is therefore not surprising that when Mike Abbott discusses the future market realization of hydroinformatics, he identifies *consultants* and *contractors* as primary agents – users of hydroinformatics systems (Abbott, 1991a, pp. 77–84). Consultants would be more concerned with advanced simulation modelling, whereas contractors would focus more on data flows and data modelling to gain efficiencies. '*The legal/political*' agent was identified as the third category of users of hydroinformatics, where contributions would mostly come from artificial intelligence for legal and political decision making. Behind this tripartite categorization of hydroinformatics users is the established practice of engineering interventions in the real world. To be clear, the needed transformations of these agents with hydroinformatics contributions were presented from within the context of the environmental movement, and the *ecological dimension of hydroinformatics* was introduced as '*mobilising a new solicitude towards the world of the waters*' (idem, p. 60). However, it would be the engineers of the consulting and contracting firms who would realize this 'new solicitude' by adopting hydroinformatics.

Attribution of dominant roles and responsibilities to engineering is perhaps best seen towards the end of the section on market realization of hydroinformatics (idem, p. 84), where in relation to the need for a new 'social contract' to deal with environmental challenges, Mike Abbott introduces the following:

> .. 'pure' scientists (e.g., most biologists and ecologists, many, if not most, hydrologists) are rather suddenly called upon 'to serve society' as 'applied' scientists. Such individuals are, however, intrinsically incorrigible hobbyists: who follow an own bent, for whom a study is properly made only 'for its own sake – for whom, for the most part, the very notion of business enterprise is anathema.

'Hobbyist' is perhaps a somewhat unfortunate term to describe 'most biologists, ecologists and most, if not all, hydrologists', but the intended point was that these professionals were mostly excluded from participation in the *engineering* interventions carried out by the main consultants – contractors – legal/political arrangements. Mike Abbott immediately introduces this as a 'deep problem', the solution of which would be that: '*hydroinformatics must accommodate itself to hobbyism*'. However, the main role and responsibility would still be with the hydraulician as '*best prepared… to take up this challenge*' (idem, p. 84).

This 'accommodation' was to be radically new, via the hydroinformatics systems, and transformational for all disciplines involved, including computational hydraulics, as explained above (idem, pp. 68–69):

> Very generally, the need here is to introduce systems that are 'open' to their users: these will be much less specific 'tools' and much more 'environments' in which chemists, biologists, sedimentologists and others can themselves build and modify tools....
>
> It should further be mentioned that a movement in the ecological dimension of hydroinformatics must have considerable repercussions on hydraulics, and in particular, on computational hydraulics. This is because many interactions between organisms are sensitive to a variety of fine scale flows, such as are currently grouped together under the rubric of 'turbulence'. The study of ecosystems will necessitate a more detailed description of fine flow structures, of processes of turbidity and of mechanical and thermodynamic interactions between solids, gases, and waters generally. Thus, the challenge of marine and aquatic-ecological modelling has to be passed on, at least to some extent to computational hydraulic modelling

This is a tall order for computational hydraulic modelling, but it also presents an openness and readiness for full collaboration with others to introduce and realize the *ecological dimension*. Nevertheless, the hydroinformatics system is the main locus of all such interactions, and this was going to be designed, implemented and run by hydraulicians and hydraulic engineers.

Thirty years ago these were indeed lofty goals for hydroinformatics, and which have only been partially realized. Environmental movements have become much stronger, driven by continuing environmental destruction, climate change impacts and significant loss of biodiversity. Moreover, engineers and 'engineering mindset' have often been blamed for many of these continuing problems. Over the last 200 years, attitudes towards large water engineering projects have been radically transformed. From being emancipating, liberating from poverty and glorious, they have become associated with elitist decision making, technocratic, undemocratic and causing unforeseen damage. This has been exposed in the literature in different ways (e.g., Feldman, 1991, from the time when hydroinformatics emerged). Mike Abbott himself wrote extensively about this major concern, especially regarding poor decision making for large water infrastructure projects in the 'Third world' and the need for the realization of social justice in such projects through the introduction of hydroinformatics systems that would facilitate transparent and participatory decision making (e.g., Abbott, 2001, 2007; Abbott & Jonoski, 2001; Abbott & Vojinovic, 2010; Vojinovic & Abbott, 2012, pp. 55–77).

However, the direction taken in the last few decades has been one in which many 'hobbyists' are not at all interested in being 'accommodated' by any engineering-driven framework (including hydroinformatics), no matter how open and inviting it would appear. Instead, they are demanding, and increasingly managing to set the agendas, gradually transforming all three 'agents' where hydroinformatics was to make major contributions. Rather than classical engineering solutions, interdisciplinary teams are being set up to identify holistic and sustainable 'green solutions', such as: water harvesting, water conservancy and natural water storage for droughts, non-structural measures and 'making room for rivers' for floods, nature-based solutions for both droughts and floods, as well as for water quality improvements. Similar solutions are sought in urban water supply and drainage systems, with classical cost-efficiency criteria for water and energy savings being augmented with criteria for the reduction of emissions, and contribution of water and wastewater facilities to the circular economy.

Such developments are not occurring under the umbrella of hydroinformatics, yet some 'open environments' and platforms for such interdisciplinary collaboration have emerged, more driven by necessity and the urgency of our environmental problems. 'Hobbyists' and engineers have started working together, even if they encounter frequent difficulties. They are developing a more shared understanding about the challenges they face together, the details of their individual approaches and the potential for developing joint solutions.

In the meantime, large infrastructural projects have certainly not stopped. Many countries in the 'Global South' see in such projects key opportunities for their economic development, with similar drives as the West experienced several decades ago (An Ethiopian would be as proud with the Grand Ethiopian Renaissance Dam – GERD in 2022, as an American with the Hoover Dam in 1935, for mostly similar reasons). Chapter 6 of this volume is devoted to the contribution of hydroinformatics to managing such large water infrastructure development in China, which has been spectacular over the last few decades. In the West, stakeholders in such projects are still present and active, even when facing 'sustainability commitments' (Crow-Miller *et al.*, 2017).

It is notable, however, that even in these new settings, when it comes to the implementation of different 'green solutions', or new water infrastructure projects *in the real world*, the engineering consultants and contractors are still being called upon. Whereas, they have been transformed in their attitudes and perspectives, it is their engineering rigour and approach that seem indispensable for the actual implementation of the solutions. To achieve these solutions, these agents turn to their tools and systems, which are still very much centred around the *models for action* introduced above, and now made even more powerful through their integration with new and more reliable data from different sources, and with support of other novel ICT technologies.

1.3 MODELLING AT THE CENTRE OF HYDROINFORMATICS

At the time of the introduction of hydroinformatics, the well-established *models for action* (mostly computational hydraulics models for engineering applications) had both commercial and free modelling systems available as products. Their value in engineering projects had already been demonstrated, so that hydroinformatics focused on introducing new goals for their future development. First, these proven and reliable models were to become the central components of newly proposed *localized* hydroinformatics systems. In these systems the main challenges were model-data integration, the inclusion of all relevant knowledge (including regulatory/legal knowledge) for effective and efficient use of such models, where 'traditional' AI (based on explicit knowledge representation in knowledge-based and expert systems) was to provide a significant contribution (see again Abbott, 1991a, pp. 29–46).

Systems with the proposed level of sophistication are yet to be fully realized, although partial aspects have been utilized in some dedicated projects (Thorkilsen & Dynesen, 2001). Apart from other challenges, one significant reason for this situation was the already established market and business of *generic* modelling systems as *products*, offered to a variety of consultants, contractors and other water-related government institutions and agencies (see Chapter 5 in this volume devoted to the hydroinformatics business developments). Such products then became the main attractors for acquiring consulting projects, with little attention to the establishment of long-term dedicated *localized* hydroinformatics systems that would be put in place during planning and design, and then used and maintained during operations of a given aquatic system. This has been especially challenging for large natural systems with man-made infrastructure (catchments, large rivers, aquifers, lakes), where competing interests and overlapping and unclear management structures did not allow the emergence of agents that would champion the development of such localized hydroinformatics system. The situation is different for urban water systems where such agents with clear responsibilities do exist, namely the water utilities, and a number of integrated systems have been developed and implemented (see Price & Vojinović, 2011). In any case, the main point here is that hydroinformatics is addressing more the challenges of developing such systems, leaving further improvements of the computational hydraulics models to take their natural course with inputs from research that would be brought in more advanced software implementations.

Second, during the same period, under the leadership of Mike Abbott, the new 'sub-symbolic' modelling paradigm emerged in hydraulics and other water-related sciences, which later acquired many different labels such as: 'data-driven modelling', 'machine learning', 'computational intelligence', 'data mining for knowledge discovery' (Abbott, 1991a, 1994; Abbott *et al.*, 2001). In this paradigm models are built without an explicit representation of the 'system'. Instead, they use available data (data are being 'mined'), to 'learn' and 'discover' useful, sometimes even unknown classification patterns or relationships between inputs and outputs. The 'learning' happens with different mathematical constructs such as neural networks, genetic algorithms, support vector machines and others. Most recently, such modelling approaches have been re-labelled as 'modern' AI, because they have become so successful and dominant in many areas of research and practice. Undoubtedly, Mike Abbott and his collaborators were pioneers in introducing this modelling paradigm for water-related applications (see, e.g., Babovic, 1996; Babovic & Abbott, 1997; Dibike *et al.*, 1999; Minns, 1998; Minns & Hall, 1996; Solomatine & Torres, 1996), later on to be successfully expanded to a broad range of problems in water sciences (see examples in Abrahart *et al.*, 2008; Babovic, 2009; Bhattacharya *et al.*, 2003; Solomatine & Ostfeld, 2008), leading to current 'modern AI' water-related applications explored in many different communities (Huang *et al.*, 2021; Kapelan *et al.*, 2020; Sit *et al.*, 2020). This expansion has led many to associate the whole field of hydroinformatics with this modelling paradigm only. (For a more thorough overview of these developments see Chapter 3 in this volume devoted to this topic.)

Driven by such success, some quickly presented the sub-symbolic paradigm as a potentially superior alternative to physically (process)-based models. Mike Abbott presented this paradigm in 1991 as an alternative to the traditional AI associated with explicit (symbolic) knowledge representation (Abbott, 1991a, pp. 94–106). Furthermore, he predicted that the most valuable contributions from the sub-symbolic paradigm would be for dealing with social knowledge to be used in hydroinformatics systems aimed at supporting collaborative decision making of multiple stakeholders (Abbott, 1999). However, he never put the sub-symbolic modelling *in competition* with physically based modelling, and was always insisting on the complementarity of these two paradigms. As stated in Abbott *et al.* (2001):

> We strongly believe that the most appropriate way forward is to combine the best of the two approaches: theory-driven, understanding-rich processes with data-driven discovery processes.

It took some years for members of the community engaged in this discussion to discover that indeed this would be the 'most appropriate way forward'. There are clear advantages of data-driven models for operational (including forecasting) applications, for discovering previously unknown patterns in complex data sets, and even for evolving new functional relationships of hidden and unknown processes. Physically based models remain superior for design and planning applications, where system representations are absolutely necessary, and they offer a more immediate understanding of causal relationships to their users. Important and fascinating combined uses have been identified in which fast-running data-driven models serve as emulators of physically based models (needed when these models are required in optimization or sensitivity/uncertainty frameworks requiring multiple model runs), data-driven models that make classifications so that an appropriate process-based model can be applied, and configurations in which data-driven models replace sub-components of physically based models.

One important notion in the quote above is that 'understanding rich' sits in between 'theory-driven' and 'data-driven', which means that it must be applicable to both approaches. As is frequently stated, in the 'theory-driven', physically based modelling, this need for understanding comes more at the beginning of the modelling process, and in data-driven approaches it must come at the end. This in fact remains a challenge for most data-driven (machine learning) models. Post-modelling analysis of results obtained from such models is required, which it is not always done. The problem has been recognized to such an extent that an entirely new sub-field has emerged named 'Explainable AI' attempting to deal with these problems in a generic way, across many different domains where 'modern' AI is being applied (Arrieta *et al.*, 2020). While some progress is expected to come from these efforts, domain knowledge will remain critically important for exploiting the full potential of this exciting modelling paradigm (see Jiang *et al.* 2020).

The third direction that needs to be mentioned here is the goal of introducing spatially distributed physically (process)-based modelling in water and environmental domains beyond hydraulics. One important example is the introduction of such an approach in hydrological (catchment) modelling, primarily through the Système Hydrologique Européen (SHE) effort (see, Abbott *et al.*, 1986a, 1986b). The initial development has later been taken up by products such as MIKE SHE (Refsgaard & Storm, 1995) and SHETRAN (Ewen *et al.*, 2000). These efforts provoked heated debates in the hydrological modelling community, with many objecting to this approach with the argument that there would always be sub-grid processes that need to be parametrized, necessarily leading to parameter values that cannot be measured (or independently assessed) and to the unavoidable need for calibration, always leading to the possibility of 'equifinality' (Beven, 1989). Objections also came from the rejection of the possibility of having sufficiently accurate *general* theories that can be applied to 'unique places' (Beven, 2000, 2001). Some of these objections led to very bold and completely wrong predictions that physically based modelling in hydrology will be abandoned (Beven, 2002). The alternatives proposed and promoted were in allowing different conceptualizations (and parametrizations) of the hydrological system, and through frameworks that enable searches through these conceptualizations – the 'best' model would be selected. Because these conceptualizations could be different, with different parameters, and with structure and parameter values that cannot be assessed independently of the model, uncertainty-based frameworks to search for the best model became unavoidable. Others in the hydrological community brought additional arguments against physically based modelling using a new vocabulary labelling this kind of modelling 'mechanistic' and 'Newtonian', and emphasizing that catchments need to be perceived more holistically as 'organisms' with a 'Darwinian' approach (Savenije & Hrachowitz, 2017; Sivapalan *et al.*, 2011).

Mike Abbott and his collaborators hardly participated directly in these discussions, especially as they developed within the hydrological community. (Some engagement with critics regarding modelling approaches promoted in hydroinformatics as applied in hydraulics can be found in Abbott *et al.*, 2001, and the discussion of Beven & Pappenberger, 2003, followed by the authors' response; related to this – see also Cunge, 2003). They continued the development of hydroinformatics – advancing both physically (process)-based and data-driven modelling in different water and environment fields, including hydrology. When Mike Abbott was introducing hydroinformatics in 1999 he did refer to '*mindless calibration*' in some of the modelling approaches proposed (related to the alternative approaches to hydrological modelling as discussed above), and this was put forward as a further drive towards data-driven modelling approaches of the sub-symbolic paradigm (Abbott, 1999). (Indeed, after three decades of research and applications of data-driven modelling in hydrology it is now clear that with such approaches one can put aside the issues of inaccessible parameter values altogether and still have models that provide reasonably reliable input–output (e.g., rainfall–runoff) relationship.) A fairly comprehensive presentation of the hydroinformatics position regarding these issues is given in Abbott *et al.* (2005), in which the need for the collaboration of many different disciplines and the mobilization of all modern data-acquisition and communication possibilities are discussed. The motivation comes again from the engineering interest to improve the *models for action*. If we aim to develop such models (now obviously on a larger scale than traditional engineering applications) the path followed should always be towards a sufficiently faithful representation of system elements where such interventions are proposed. This inevitably demands much more data about different hydrological processes, associated with substantially larger investments to acquire the necessary data and to mobilize these large datasets in modelling activities (model setup, calibration, validation, application). As Abbott *et al.* (2005) state:

> It had … become clear that only distributed physically based models like the SHE systems could make use of the rapid advances in instrumentation, supervisory control and data acquisition (SCADA) systems with real-time data transmission, data assimilation techniques, remotely sensed data and its interpretation systems, seamless geographical information systems (GIS) interfacing, advances in geodetic surveying incorporating GPS-based position-fixing equipment, cartographic transformation packages, intranetted and extranetted communication systems, and the many other developments that hydroinformatics had already woven together ….

The argument that we can never measure appropriate parameter values at appropriate scales may lead to *non-action* in this regard, that is, not investing in new measuring techniques, abandoning and not upgrading observation networks, ignoring new sources of data (remote sensing, data from citizens). One could even argue that these alternative modelling approaches are to some extent driven by cheap and widely available computing, while data acquisition remains expensive. The discussions in the hydrological community regarding such issues have continued (see Refsgaard *et al.*, 2010 and the follow-up discussion), with some convergence of opinion that a physical basis for hydrological modelling is certainly needed, and calls are being made to 'effectively use our diversity of modelling approaches in order to advance our collective quest for a physically realistic hydrologic model' (Clark *et al.*, 2017, see also Hrachowitz & Clark, 2017). As this is an important aspect for hydroinformatics in general, a separate Chapter 4 is devoted to it in this volume.

It should also be noted that such discussions were notably absent when advances in hydroinformatics to deal with urban water systems (water supply or urban drainage) were being introduced. One can imagine that for such systems similar points about poor knowledge of 'parameters' can be made (many parameters are not easily identifiable, large systems may have many pipes and other elements, sometimes we don't even know where exactly these elements are located, they can have a varying performance due to ageing, clogging with material, pipes can leak, groundwater can intrude, etc.). Yet, when it comes to operational management or the design of urban water systems it is hard to imagine that the approaches promoted in hydrology and discussed above would be at all considered as viable alternatives to physically based distributed models. The nature of the problems is such that the calculations of variables (flows, heads, water quality conditions) need to be known everywhere in the network in order to make appropriate decisions. Alternative model conceptualizations that could reproduce the 'general' system behaviour would be rather limiting for operational management or design decisions that are regularly needed. Admittedly, such man-made systems are still 'simpler', or at least 'more identifiable', than the large natural systems like catchments. Also, when dealing with urban water systems there is an identifiable 'decision maker', that is, the utility, with clear modelling demands and objectives that in most cases only physically based models can meet. The lack of such a clear decision-making context in river basins, has an impact on the modelling approaches proposed and used. The position of hydroinformatics is that for serious 'informed decision-making' in these natural systems, physically based models would need to be continuously improved, but would still be the most viable alternative, and consequently, the challenges of poor data availability and the difficulties in 'identifiability' need to be faced head-on with support of the latest available technologies.

One clear indicator about this direction of the full representation of urban water systems with physically based simulation models integrated with data from different sensors is in the application of the concept of *digital twins* (Bartos & Kerkez, 2021; Pedersen *et al.*, 2021; Valverde-Pérez *et al.*, 2021). The aim is to provide as faithful as possible virtual representations of the real system, to better support design and operational decisions. There is clearly some marketing value in this term, as it becomes more frequently used by many consultants and commercial software companies operating in the water sector. The overall approach however, started to be adopted also in hydrology, especially in view of the possibilities to exploit remote sensing data for detailed physically based hydrological modelling (see, e.g., the Digital Twin Hydrology project as part of the Digital Earth initiative of the European Space Agency, CNR-IRPI, 2021).

To be clear, the general advancement of modelling in many different domains related to water- and the environment is certainly to be valued and supported. Many models, even within the water fields are *not* aimed to be used as *models for action* (engineering-like decisions). Models are being built for 'assessment' (in the broadest possible sense, which may include many different aspects, like drought/flood potential, climate change impact, environmental impact of pollution, etc.). Such models may have very high value to inform policy decisions. Often, the scale of such modelling is beyond conceivable engineering actions. Large-scale catchments, whole countries, continents and even the whole planet are being modelled. On the other hand, the integrated modelling of environments made of natural and man-made systems and the socio-economic actors operating in these systems have brought into play modelling paradigms such as system dynamics and agent-based models (Abebe *et al.*, 2019; Sušnik *et al.*, 2021). Again, these can be highly valuable modelling approaches for developing better insights and understanding of such complex systems, next to informing policy. Here we enter the realm of modelling as a fundamental activity for improved human understanding, where many different modelling concepts are indeed welcomed. This very broad, effectively limitless domain of developing models raises questions of model usefulness, validity, verification and applicability (see e.g. in the domain of earth system modelling the much-cited work of Oreskes *et al.*, 1994).

Although hydroinformatics has *primarily been concerned with models for action in the real world*, its developments also proceed through engagement in modelling of the kind just discussed above. In fact, it often advances and learns lessons from the broader modelling approaches developed elsewhere. One demonstrable example of this is realizing the need to include a proper uncertainty analysis in model development and application. As much as Mike Abbott and some of his close collaborators were clearly arguing against substituting poor model representation with uncertainty analyses frameworks (see again Abbott *et al.*, 2001 and the follow-up discussion), the overall importance of uncertainty analysis was gradually recognized. In forecasting applications, for example, ensemble forecasts became important, next to methods such as data assimilation (Ramos *et al.*, 2013; van Andel *et al.*, 2008). New uncertainty assessment frameworks were also proposed (Solomatine & Shrestha, 2009). Contribution to such awareness came from modellers outside of the hydroinformatics community, who developed comprehensive methods for assessing the sensitivity and uncertainty of models (often the same hydrologists and broader environmental modellers were engaged in debates about the role of physically based models) (e.g., Pianosi *et al.*, 2016; Refsgaard *et al.*, 2007). Here again however, hydroinformatics needs to focus on the relevance of such analyses when used with *models for actions*, and work towards developing methods for *using* uncertainty analysis directly in decision making, or for the identification of further data requirements (additional measurements) that would reduce uncertainties and therefore lead to better predictions and decisions.

1.4 HYDROINFORMATICS: FROM TECHNOLOGY TO SOCIOTECHNOLOGY

Throughout most of the writings of Mike Abbott on hydroinformatics, one can identify the key statement that *hydroinformatics is essentially a technology*. This simple statement is in fact of great importance for understanding hydroinformatics, primarily because it defines the role of hydroinformatics as a technology that can mobilize and unite scientific and non-scientific knowledge for more responsible (and, consequently, more sustainable) management and use of our water systems and the aquatic environment. Behind this affirmative statement in fact lies a hidden negation (expressing what hydroinformatics 'is not'), primarily aiming to distinguish hydroinformatics from its characterization as *science*, or as *modern technology*. Both of these characterizations have been considered as incomplete, and could actually lead to erroneous understanding of hydroinformatics, and, subsequently to applications rather different than intended.

The relationship between hydroinformatics and science has been introduced already in Abbott (1991a). It shows that hydroinformatics must research how science is applied, first because of reliance on the best science for producing its reliable models, but, also, because of its concerns for applications of science beyond its limits. Essentially, because hydroinformatics must be concerned with the world of *values* in its ecological and social dimensions, the limits of a value-neutral, or value-free science must be recognized. Moreover, the dangers have been pointed out of science being mis-used to become *scientism* (nowadays understood as an ideology that all problems in the world can be reduced to scientific problems and then solved using scientific methods). The process by which science was brought in position of such supremacy in relation to all other sources of knowledge is now fully recognized as a product of the modern era (with motivations of power and progress, variably defined), and elaborated in a vast body of post-modern literature (for detailed elaborations in relation to hydroinformatics, see, Jonoski, 2002; Vojinovic & Abbott, 2012). This supremacy is in fact closely intertwined with modern technology, in this context primarily understood as 'applied modern science'.

Technology as 'applied modern science' brought spectacular transformations of human societies over the last few centuries. However, it has also been realized that the technology-driven progress led by technocratic governance has been achieved at very high costs to the natural environment, and to a large portion of the human population, who have effectively been excluded from such progress. Furthermore, serious concerns have been raised regarding technological neutrality (is it really neutral, if it is applied *science*?), technological autonomy (is it running out of control?) and overall technological determinism (is technology conditioning, shaping and controlling all social arrangements and culture?) (Bauman, 1996; Ellul, 1964; Lyotard, 1979/1984). (Such concerns, somewhat less full of doom in general, are still present, although we are also witnessing concerns for a 'runaway technology' associated with the 'singularity' moment when machines will take over from humans as AI progresses.). Lyotard (1979/1984) in his highly influential 'The Post-modern Condition' gave a very clear presentation on how the driving force for such developments is in fact in the 'Grand Narratives' of modernity, with their performance and efficiency criteria as sole legitimizers of knowledge, led to the present position of both science and modern technology in our societies. The simple quote that 'the 'organic' connection between technology and profit preceded its union with science' (idem, p. 45), points clearly to the false characterization of technology as 'applied science'.

As a contrast to such modern technology Mike Abbott (following Heidegger, 1963//1977) brought in the understanding that the essence of technology is in *revealing*, or *unconcealment*. Technology may be considered as an assembly of means for achieving ends, but, beyond that, its essence resides in the way that revealing and unconcealment lead us to experience *alētheia*, or *truth*. The origin of the word technology is in the Greek *technē*, denoting the unity of all creative, or *poietic*, human activities (including both the arts and the work of the craftsman). Technology is thus a human poietic action in which revealing takes place, resulting in *truth*.

Two aspects should be noted about this understanding of technology. First, the truth from revealing is very different than scientific truth (correspondence, or logical truth). It may be constrained by 'laws of nature' (which can be captured by science), but ultimately it comes as a result of the creative process in which humans are engaged as individuals or groups, where a different kind of knowledge and experience is mobilized. Related to this, the second aspect is that such an essence of technology appears as 'free' from being only 'applied science', and, more importantly, free from 'enslavement' to the legitimizing criteria of performance and efficiency imposed by the modern state and market institutions. This aspect is of crucial importance for understanding Mike Abbott's vision of hydroinformatics as a liberating and almost anarchistic discipline with respect to such impositions, which becomes necessary for its *autonomous* development. It is then not surprising that the very first sentences of the Hydroinformatics book of Mike Abbott are Heidegger's quotes (Abbott, 1999):

> Thus the coming-to-presence of technology harbours in itself what we least suspect, the possible arising of the saving power.
>
> ...The closer we come to the danger, the more brightly do the ways in the saving power begin to shine and the more questioning we come. For questioning is the piety of thought
>
> Martin Heidegger, The Question Concerning Technology

Following Heidegger then, technology has a dual essence. On one side is the 'Enframing' – converting virtually everything in the world into a 'resource' or 'standing reserve' by 'ordering', 'counting' and 'computing', which is the 'greatest danger' as it blocks the other essential side of human-technological action – the revelation of truth (Heidegger, 1963//1977, p. 33).

> The essence of technology is in a lofty sense ambiguous. Such ambiguity points to the mystery of all revealing, i.e., of truth.
>
> On the one hand, Enframing challenges forth the frenziedness of ordering that blocks every view into the coming-to-pass of revealing and so radically endangers the relation to the essence of truth.
>
> On the other hand, Enframing comes to pass for its part in the granting that lets man endure – as yet unexperienced, but perhaps more experienced in the future – that he may be the one who is needed and used for the safekeeping of the coming to presence of truth. Thus does the arising of the saving power appear.

Mike Abbott clearly associates hydroinformatics with the second essence of technology, namely 'the saving power', as is obvious from most of his writings. Two questions do arise, though:

To some, this may still appear as some sort of 'grand' technological fix, where the cure for all technological maladies will come from other, different, better technologies. This would be a misunderstanding, because the fundamentally different thinking about the technology involved here does allow for the choice of approaches when dealing with our natural environment, including non-technological ones. As will be discussed in the next sections, hydroinformatics is about developing systems that would allow different perspectives and combinations of different knowledges (denoted in the plural deliberately) in environmental management and decision making. These would still be ICT-based systems, so an objection could be raised that there is too much faith put in that kind of technology as 'saving power', given the impact of digital technologies across our societies. However, when we are concerned with most immediate problems of environmental destruction, hydroinformatics indeed proposes such ICT-based systems for an alternative, less technocratic and more democratic and inclusive governance of the natural environment.

The second question relates more to the way in which the realization of such ICT systems could in fact emerge, so that the 'saving power' could be harnessed and used. As mentioned above, hydroinformatics was introduced from the very beginning as having ecological and social dimensions, with full awareness that all environmental problems are 'hybrid' problems that involve intertwined social and technical aspects, and are therefore sociotechnical problems (Latour, 1993). The notion of sociotechnology was then introduced as a potentially better characterization of hydroinformatics as well, so that it would become clear that development of the intended hydroinformatics systems and environments would also depend critically on both available ICTs and enabling social/institutional arrangements. Characterizing hydroinformatics as sociotechnology then leads to conceptualizing, designing and implementing 'new decision support tools', capable of harnessing, combining and using knowledge from different domains, perspectives and stakeholders (see again Abbott, 1999, 2002; Jonoski, 2002; Vojinovic & Abbott, 2012).

1.5 DECISION SUPPORT AND HYDROINFORMATICS

Mike Abbott introduced the concept of hydroinformatics systems as 'decision support systems'. From the very beginning this was about combining scientific knowledge with other kinds of knowledge, or bringing together scientific truth with yet another kind of truth, namely the so-called *common-sense* truths. Following Heidegger again, common-sense truths are those that originate from experience, and are always associated with the *intentionalities* of different stakeholders, for example the truth about a river for a fisherman, or a truth of a river for a nature conservationist, or a truth of a river for a captain of a towboat that pushes barges for transporting goods. Such truths may be associated with particular efficiency or performance criteria for these different stakeholders, corresponding to their interests, but other criteria may also enter the subsequent evaluation process that will eventually lead to some decisions. In any case the 'pragmatic task of a hydroinformatics system is' (Abbott, 1991a, p. 29):

> to combine and integrate all scientific information and, further, all common-sense truths to the extent that these truths can be given a standard-scientific informational representation

The 'standard-scientific informational representation' of common-sense truths is necessary for internal operation of any such system, which essentially means representation in some quantifiable, numerical representation. In fact, the common-sense truths need to be somehow translated to 'objectives' or 'criteria', associated with particular indicators so that any intervention in the real system in question could be evaluated with respect to these. Here we see a clear link with the notion of decision support system as defined in the broad field of 'systems analysis for water resources' (Loucks, 1995; Loucks & van Beek, 2017).

When hydroinformatics emerged, the field of system analysis in water resources was already established with approaches that enable scientific knowledge from the water sciences into the domain of decision making. It introduced various decision support methods, such as simulation, optimization, multi-criteria analysis (MCA), decision making under uncertainty, control theory and risk theory that could be pulled together for the goals of decision support. The overlaps with hydroinformatics were significant, and many developments of hydroinformatics could easily be recognized as belonging to systems analysis. This is particularly the case for the different optimization methods and techniques researched and promoted by hydroinformatics, especially those of 'model-based optimization', where simulation models are directly coupled with optimization algorithms, mostly using evolutionary approaches (for overviews of these approaches, see Maier *et al.*, 2014; Reed *et al.*, 2013 and numerous particular applications reported in the *Journal of Hydroinformatics*).

The difficulty faced by both the field of systems analysis and hydroinformatics was the opening of such systems to *actual* participation of stakeholders in water- and environmental decision making. Most reported successful applications still assumed that the 'representation of the common-sense truths', that is, the expressions of various criteria with relevant indicators would already be carried out, or, that the indicators involved would still be rather easily identifiable, as, for example, in decisions on urban water systems with clear decision-making entities – the utilities (see Price & Vojinović, 2011). In such cases one could still assume that there is one decision-making body that needs to be 'supported', even in multi-criteria settings. One can easily recognize how this approach could still be perceived as overly technocratic and top-down, especially as our societies increasingly strongly demanded participatory environmental decision making.

One attempt to address this problem was to introduce models as central tools around which stakeholders could gather to jointly present, discuss and hopefully recommend commonly agreed solutions. The 'Shared Vision Planning' approaches from the early 1990s in the USA (see for later developments, Palmer *et al.*, 2013), were later expanded to 'modelling with stakeholders' (Basco-Carrera *et al.*, 2017; Voinov *et al.*, 2016). Such approaches do not necessarily focus on detailed process-based models, instead so called 'system dynamics' models or other more conceptual models are used for various environmental problems. Moreover, until recently, the Internet has not been considered as the platform on which such collaborative modelling efforts could be realized, and most collaboration activities around selected models have been organized via dedicated workshops.

Hydroinformatics introduced some radically new proposals in this area. First, on a very fundamental level, it introduced the idea that the term 'decision *support*' could be misleading and misused, as it could lead to tools that would in fact support the established technocratic decision-making processes. This concern has actually been shared across many different fields engaged in decision support. For example, from the field of policy analysis (Thissen & Twaalfhoven, 2001):

> From the perspective of policy network theory, policy analytic activities have a role as an intermediator, supporting communicative debates among multiple parties at interest, developing shared strategies, creating win-win situations, and breaking through cognitive fixations...A logical consequence of the network perspective is that policy analytic activities should not aim at rational advice given to a single authoritative public decision maker, but should start from, and end with, the interactions between relevant stakeholders.

As an alternative idea to the traditional approach of 'science provides the facts – decision makers need to use them and make decisions', in which the role of associated technology is still mainly *to inform*, hydroinformatics proposed new tools and environments that should enable *changing of attitudes* of all participants in the decision-making process (traditional decision makers from state and market institutions, diverse stakeholders and the general public – the citizens), to lead towards more caring, commonly accepted and sustainable solutions for the natural environment. Here the role of technology goes beyond being informational to become *transformational*.

Second, the main challenge for such systems and tools would mainly be in enabling the *individual and social learning* of the involved participants. For this purpose, the system design, with all its components should be primarily focused on capturing and representing in a *transparent* manner the 'inference chains' of the form (Abbott, 1999):

'(beliefs, facts (data))->attitudes->positions->judgements-> decisions->actions'

This is, of course, a formidable challenge. The role of accurate and reliable models (with possibilities for presenting results across different spatial and temporal scales, and in many different forms) remains central and necessary, but they are only associated with the part of 'facts(data)' of the above chain. The whole process depends on enabling the making of judgements (leading to decisions and actions), and, at the next levels, revising one's own beliefs, attitudes and positions in view of the knowledge learned about such chains used by other participants in the process. Thus, learning should not be restricted to isolated and apparently quantifiable aspects of the problems in question, but needs to be extended with learning about the overall state of the problem, the possible solutions and accompanying consequences, and, most importantly, with learning about the values of other persons and interest groups, so that one's own values and interests could be understood in the context of this 'social landscape'. The overall goal of this process would be the *emergence* of holistic thinking and integral solutions, which would take into account diversity of beliefs, attitudes and interests, while restricting the opportunities for the creation of dominance patterns in which some of the stakeholders would be severely disadvantaged.

The question now arises: How is this to be achieved? Of course, the goal would not be to have explicit representation of the items in the 'inference chains' mentioned above. In order to have the possibility to realize all such processes, Mike Abbott proposes to use all the knowledge available from the branch of Philosophy known as *Phenomenology*, most elaborately established by Husserl, and developed further by Heidegger, Gadamer and other thinkers. Phenomenology is introduced as a strict science that studies how objects (real or imaginary) create human experiences. The concern here is the design of systems in virtual environments. From this broad perspective the focus shifts to studying the functioning and use of different objects as signs or symbols, towards particular human experiences. Already at the level of numerical modelling Mike Abbott introduced these notions as critically important. His famous definition of a 'model as a collection of indicative signs that serves as an expressive sign' (e.g., Abbott, 2003) was frequently misunderstood, but it was already clearly presenting how different indicative signs (e.g., the colour red along a river stretch) could create overall expressions (e.g., severe water quality degradation), which could eventually create *symbolic knowledge* to a particular stakeholder (loss of fishing) (for a clearer elaboration, see Vojinovic & Abbott, 2012, pp. 184–187). Such a use of signs and symbols becomes critical for bridging the 'gap' between 'neutral' scientific truths and value-laden common-sense truths. Obviously, these semiotic devices would become even more important when different stakeholders need to interact as subjects in the envisaged virtual environments for 'decision support'.

While being fundamental, Phenomenology and Semiotics are fields where hydroinformaticians have very limited expertise, and further progress in this area requires linkages with other experts from these fields. Next to these challenges, however, the design and implementation of such novel 'systems' has faced many other difficulties. If they are introduced to allow very flexible approaches, including 'non-systemic' approaches to governance of our natural environment (see O'Kane, 2015), perhaps these virtual tools should not be designed and implemented as systems themselves, but as being more like 'environments'? How is this to be achieved? What should be the locus for such an environment? In an early study of such possibilities, Jonoski (2002, pp. 227–230) envisaged three options: governmental or non-governmental institutions, multiplayer role-play games, or via business enterprises. Although some progress has been made with such options, development of such environments as originally envisaged by Mike Abbott is still rather limited. It appears that the sociotechnical challenge of merging scientific and non-scientific knowledge is much harder to realize, especially in the era of our digitized societies, where the boundaries between our virtual and real worlds become blurred and masked.

1.6 HYDROINFORMATICS AND THE VIRTUAL WORLD

Although hydroinformatics is concerned with the real world of the aquatic environment and our relationships with it, its constructs belong to the 'virtual world'. The growth of hydroinformatics in fact coincides in time, and is conditioned by, the radical transformations that this virtual world has undergone during the last three decades. Of course, the most important condition here is the emergence of the electronic networks such as the Internet, followed by the wireless mobile telephony networks. In fact, the fourth-generation modelling systems, were already widely available in the early 1990s as *products* before the widespread deployment of Internet services such as the World Wide Web. Therefore, the transition from the established 'product-based' business to the new reality of the Internet and introduction of 'modelling as a service' has been relatively slow. With spectacular advances in virtualization, where users can choose infrastructure, platform, software or container services, new possibilities emerge for advancing hydroinformatics businesses centred on traditional modelling products. On the other hand, the same electronic networks enable the unprecedented integration of data from many different sources (smart sensors, remote sensing, citizens' data) with both physically based and machine-learning models for different objectives, such as water systems design and rehabilitation, operational management, forecasting and warning, and so on. Again, practical successes were mostly for urban water systems, or similar systems where the decision-making context was clear (normally with the utility or the water agency as a single decision-making entity) and performance objectives were relatively straightforward to identify. Future developments on this front are expected to be in the further integration of distributed models (or modelling components) with distributed data, all via frameworks of integrated web services, and increasingly equipped with semantic support (taxonomies and ontologies) for their easy identification and re-use (e.g., see Choi *et al.* 2021, for some recent developments).

Mike Abbott and his collaborators saw greatest potential in electronic networks for development and deployment of the new kinds of 'decision support systems' (discussed in the previous section) for engagement of multiple stakeholders, even the general public and the citizens (Abbott & Jonoski, 1998; Jonoski, 2002; Jonoski & Abbott 1998; Yan *et al.*, 1999). They emphasized the opportunities offered, such as speed of communication, access to both corporate and individual stakeholders (even those traditionally excluded for any participation in environmental decision making), possibilities for the development of transparent systems, and, very importantly – 'customized' knowledge delivery and use (using advanced semiotic devices) that would ensure meaningful engagement. Systems that utilized some of these advantages have been developed (e.g., Almoradie *et al.*, 2015; Kumar *et al.*, 2015), though they are still relatively far from having the capabilities and scale as originally envisioned.

An attempt to understand the reason for this situation must begin with the most fundamental challenge that such systems face, namely the integration of *narrative* knowledge with scientific knowledge, as already introduced earlier. The term 'narrative knowledge', introduced by Lyotard (1979/1984), is fitting because it encompasses all non-scientific social and individual knowledge that may be targeted to be mobilized in such systems, and because this knowledge is predominantly in narrative form. 'Stories' that guide our existence, provide our obligatory frameworks and are in fact the source of our values, which come from different sources. They can indeed be the 'Grand Narratives' of Lyotard (idem), such as 'human progress', 'human emancipation', 'emancipation of the working classes', or religions, nations, families, all the way to favourite sport clubs, or local bird-watching society. Even current mobilization for 'reducing emissions of greenhouse gasses' or 'sustainability goals' are primarily due to narrative knowledge (even when some are informed by science). In fact, individuals and communities maintain-and are guided – by powerful and competing 'stories', and isolating 'common sense truths that can be given a standard-scientific informational representation' becomes a formidable, almost impossible challenge. Yet, at the same time, engaging narrative knowledge remains an absolute necessity if any progress in this area is to be made. Projects with attempts to deal with some water- and environmental problems (increasing water or energy efficiency, decreasing pollution, etc.) aimed at changing attitudes and behaviours of particular communities, are destined for failure if they do not engage the narrative knowledge of that same community in relation to the environmental unit in question (city, river valley, coastal zone, etc.).

One area that emerged in recent years with potential for engaging narrative knowledge in environmental decision making is 'citizen science', and its more focused sub-area of 'citizens observatories', where citizens and citizen groups are engaged in collecting environmental data (mostly using mobile phones, sometimes connected to portable devices) that can then be used for a better analysis of environmental systems, eventually leading to improved management (Alfonso *et al.*, 2022; Assumpção *et al.*, 2018; Zheng *et al.*, 2018). Justification for this approach has often been given in 'lack of data from existing monitoring networks', or, 'excessive costs in additional data gathering activities'. However, the fundamental advantage of such activities, is in the opportunity to engage citizens' narrative knowledge in the environmental management through the data collection process (Wehn *et al.*, 2021). Only when 'data collection' or even 'overall environmental management' activities are with full connection to the existing narrative knowledge of the engaged community they become meaningful. *Observations* made by citizens (that could be done by anybody) need to become their *testimonies* (unique to their experience and relationship with that environment) (Marcel, 1956/1995).

As scientists, technologists and other experts have been struggling to realize this engagement of narrative knowledge in their systems for supporting environmental management or citizens' science campaigns, *social media platforms have emerged as focal points for sharing and creating narrative knowledge.* Their spectacular growth and attraction to billions of people have been due to the fact that they are 'social' platforms, with opportunities for sharing personal stories as well as stories of very different communities and groups. The dominance of few commercial social media platforms in this area has become so strong that envisaging *separate* systems in which narrative knowledge could be engaged together with scientific knowledge for any purpose, including environmental management and governance, becomes increasingly difficult. Similarly to other government and market organizations, such separate systems, if possible at all, would need to maintain strong links and presence on these social media platforms. In some way, these platforms have taken much of the virtual space that hydroinformatics (as well as other field with similar ideas) planned to utilize for its creations.

Few large corporations that own and manage these platforms (as well as many of their users) are quick to promote the benefits coming from the content generated there, such as mobilizing and maintaining communities, enabling knowledge creation and sharing and development and even giving voices to the traditionally disadvantaged in our societies. On the other hand, whenever disturbing content is being generated on these networks, often originating from narratives of hatred or ignorance, the same organizations and acolytes are bringing in the defensive argument that these are 'just platforms' that companies are maintaining, and only users are responsible for the actual content. At the same time, it has been recently revealed that some social media companies have in fact used hate as part of their business models (centred around advertising to large number of users, where 'customization' of content has turned into 'profiling' of users). Further concerns regarding privacy and dangers coming from surveillance (not only from state actors, but also regarding 'surveillance capitalism') are still unresolved.

One clear challenge for future hydroinformatics endeavours is in understanding and defining its relations with large social media platforms, especially as their power continues to grow. If hydroinformatics is indeed a field that aims 'to take *responsibility* for shaping and guiding our relationships with the aquatic environment in the world of knowledge consumerism', as introduced in the beginning of this chapter, can it ignore the processes of knowledge production and consumption occurring on these platforms? One response could be in 'monitoring' these platforms (to the extent possible), collecting data and then using them for own analysis and in own systems with different purposes. Developing tools and environments inside these social media platforms (as most of these are offering such possibilities) could also be pursued, although issues such as ownership of content and potentially conflicting purposes of content development could arise. Yet another consideration would be to still focus more on development of platforms for local social networks, dedicated to environmental issues of particular concern, and deal with the many challenges in achieving this, already mentioned earlier. In this respect, Mike Abbott foresaw that hydroinformatics would need to explore and use the so-called 'technologies of persuasion' (including studies how paralogical and *aporiaic* constructs are used in advertising, currently widespread on social media platforms), going beyond usage of 'standard' phenomenological and semiotic devices (Abbott, 1999).

1.7 ROLE OF HYDROINFORMATICIANS

Hydroinformatics is exciting but a challenging field. The vision of Mike Abbott has been that a hydroinformatician (researcher or practitioner) needs to attain diverse competences such as: hydraulics, hydrology and other water sciences, theory and practice of numerical modelling, software development and integration, as well as strands of social sciences and philosophy. Such competences are required so that the hydroinformatician can fulfil the role of 'integrator', or 'the person in the middle who weaves everything together' (see again Vojinovic & Abbott, 2012). This 'hybrid' expertise is in fact very difficult to be obtained solely via training and education. Hydroinformatics educational programmes exist only at post-graduate level (MSc and PhD), so that candidates already educated in particular 'basic' field (e.g., civil/hydraulic engineering, computer science, or field of relevance), could enter as more mature learners of different aspects of hydroinformatics. Chapter 7 of this volume presents such hydroinformatics education, as it has been originally developed at IHE Delft Institute for water education in the Netherlands, which demonstrates the challenges of maintaining relevant curriculum, in the face of rapid technological and social changes.

However, career opportunities for hydroinformatics graduates do not always require the full 'integrating' role. Many in fact continue with various sub-specializations in the fields of physically based modelling, artificial intelligence for water systems, or software development. This challenge of balancing the 'hybrid' and 'integrative' study of hydroinformatics with developing expertise in particular sub-fields has always persisted throughout the history of hydroinformatics education. Most graduates fully appreciate such breath of hydroinformatics education, and feel better prepared for future challenges, although some are left with the impression that hydroinformatics studies are with insufficient depth in the sub-field of their particular interest. Educators in hydroinformatics, especially those who follow the vision of Mike Abbott, remain confident that the broad and 'hybrid' education in this field is necessary, and may become even more important as environmental challenges require interdisciplinary approaches.

Beyond education, the important question that Mike Abbott posed for any future hydroinformatician was the one of *motivation*. From the very beginning he introduced the *European possibility and responsibility of hydroinformatics*, because Europe was a primary locus of science-driven modern developments, which next to their successes brought unprecedented segregation and conflict among human populations and destruction of the natural environment. As a person of deep faith, he clearly associated this condition with the modern secularization, allowing the most destructive threat of the 'kingdom of nothingness' (Abbott, 1991a, pp. 73–76). Throughout his writings he therefore insisted that the true motivation for any hydroinformatician must be discovered by deep self-reflection at the Kierkegaardian 'level of the religious' (Abbott, 1999, see also Abbott, 1991a, pp. 129–135, see also Abbott, 1991b). He presents the same quote from Heidegger in both Abbott (1991a, pp. 78–79) and in Abbott (1999):

> The flight from the world of the suprasensory is replaced by historical progress. The otherworldly goal of everlasting bliss is transformed into the earthly happiness of the greatest number.
>
> The careful maintenance of the cult of religion is relaxed through enthusiasm for creating a culture or the spreading of civilisation. Creativity, previously the unique property of the biblical God, becomes the distinctive mark of human activity. Human creativity finally passes over into business enterprise.

In this condition we can turn to 'technology as the continuation of creation with creaturely means', and hydroinformatics in particular becomes a technology that 'strengthens the covenant between the Creator and its Creature in the realms of the arteries and the veins of the biosphere' (Abbott, 1991b).

Furthermore, in Abbott (1991a, p. 133) he states:

> In the first instance, technology can only properly serve this world through a spirit of dedication, but dedication is itself only provided by devotion.

..but also (idem, p. 134):

> It remains further to add that although technology is properly sprung between science and religious consciousness, it can in no way be associated with any particular social religion.

Abbott recognizes that the 'level of the religious', where devotion is to be found, may be very different for different individual hydroinformaticians, and some would not even use that same term to describe their motivations. Indeed, there are certainly many hydroinformaticians who do find their motivation in 'historical progress' or 'spreading of civilization', and adhere to many other secular values. Here again we come back to the power of narrative knowledge, which may be very diverse, and still provide the motivations for individual and social actions.

The European responsibility of hydroinformatics introduced by Mike Abbott, should certainly not be mis-understood as anti-scientific. Rather, the objection here is to the dominant *narrative of science itself* as 'modern', 'superior', ('European' and 'Western'), that has been 'discovered' and needs to be 'applied' so that others would be 'civilized' and 'developed'. After many decades of post-modern analysis and objection to this stance, we are currently witnessing new narratives calling for 'decolonization of science' and 'decolonization of knowledge', precisely for the same objections that Mike Abbott raised when he introduced hydroinformatics.

Finally, even though current manifestations of hydroinformatics successes are mainly in its technological dimension, further development of the ecological and social dimensions that Mike Abbott introduced need to be pursued by current and future generations of hydroinformaticians, if they are to fulfil their responsibility introduced above. Mike Abbott's holistic vision of hydroinformatics opened many avenues for further development of this exciting field, which need to be explored further. As usual, the role of the hydroinformatics community is of great importance here. Without such vision, hydroinformatics may degenerate into a purely technological adventure, hardly distinguishable from many similar applications of ICTs in the fields of water and the environment and meet a paradoxical destiny of *oblivion through ubiquity*. Mike Abbott's extensive writings remain with us, as guardians against such a prospect, and as reminders of the true nature and purpose of hydroinformatics.

REFERENCES

Abbott M. B. (1991a). Hydroinformatics: Information Technology and the Aquatic Environment. Ashgate, Aldershot, UK/Brookfield, USA.

Abbott M. B. (1991b). A central issue of teaching within the hydroinformatics paradigm. *La Houille Blanche*, **3–4**, 257–261, https://doi.org/10.1051/lhb/1991024

Abbott M. B. (1994). Hydroinformatics, a Copernican revolution in hydraulics. *Journal of Hydraulic Research*, **32**, 3–14, Extra Issue, https://doi.org/10.1080/00221689409498800

Abbott M. B. (1999). Introducing hydroinformatics. *Journal of Hydroinformatics*, **01**(1), 3–19, IWA Publishing, https://doi.org/10.2166/hydro.1999.0002

Abbott M. B. (2001). The democratisation of decision-making processes in the water sector I. *Journal of Hydroinformatics*, **03**(1), 23–34, https://doi.org/10.2166/hydro.2001.0004

Abbott M. B. (2002). On definitions. *Journal of Hydroinformatics*, **04**(2), i–xxvii, IWA Publishing, electronic version only, https://doi.org/10.2166/hydro.2002.0008

Abbott M. B. (2003). The conversion of data into information for public participation in decision making processes. In: Integrated Technologies for Environmental Monitoring and Information Production, N. B. Harmancioglu, S. D. Ozkul, O. Fistikoglu and P. Geerders (eds), Nato Science Series, Springer, Dordrecht, **23**, 17–24, https://doi.org/10.1007/978-94-010-0231-8_2

Abbott M. B. (2007). Stakeholder participation in creating infrastructure. *New Civil Engineer*, **160**(1), 26–32.

Abbott M. B. and Jonoski A. (1998). Promoting collaborative decision-making through electronic networking. In: Hydroinformatics '98, Babovic & Larsen (eds), Balkema, Rotterdam, pp. 911–918.

Abbott M. B. and Jonoski A. (2001). The democratisation of decision-making processes in the water sector II. *Journal of Hydroinformatics*, **03**(1), 35–48, https://doi.org/10.2166/hydro.2001.0005

Abbott M. B. and Vojinovic Z. (2010). Realizing social justice, I and II. *Journal of Hydroinformatics*, **12.1/12.2**, 97–117/225–239, IWA Publishing.

Abbott M. B., Babovic V. and Cunge J. (2001). Towards the hydraulics of the hydroinformatics era. *Journal of Hydraulic Research*, **38**(4), 339–349, https://doi.org/10.1080/00221680109499839

Abbott M. B., Bathurst J. C., Cunge J. A., O'Connell P. E. and Rasmussen J. (1986a). An Introduction to the European hydrological system – système hydrologique européen 'SHE', 1: history and philosophy of a physically based distributed modelling system. *Journal of Hydrology*, **87**, 45–59, https://doi.org/10.1016/0022-1694(86)90114-9

Abbott M. B., Bathurst J. C., Cunge J. A., O'Connell P. E. and Rasmussen J. (1986b). An Introduction to the European Hydrological System – Système Hydrologique Européen 'SHE', 2: structure of a physically based distributed modelling system. *Journal of Hydrology*, **87**, 61–77, https://doi.org/10.1016/0022-1694(86)90115-0

Abbott M. B., Mynett E. A. and O'Kane J. P. (2005). Hydroinformatics and its contributions to hydrology: from computation to communication. In: Encyclopedia of Hydrological Sciences, G. M. Anderson and J. J. McDonnell (eds), John Wiley & Sons Ltd, Chichester, UK, pp. 223–238.

Abebe Y. A., Ghorbani A., Nikolic I., Vojinovic Z. and Sanchez A. (2019). A coupled flood-agent-institution modelling (CLAIM) framework for urban flood risk management. *Environmental Modelling & Software*, **111**, 483–492, https://doi.org/10.1016/j.envsoft.2018.10.015

Abrahart R. J., See M. L. and Solomatine D. P. (eds) (2008). Practical Hydroinformatics: Computational Intelligence and Technological Developments in Water Applications. Springer-Verlag, Berlin Heidelberg, Germany.

Alfonso L., Gharesifard M. and Wehn U. (2022). Analysing the value of environmental citizen-generated data: complementarity and cost per observation. *Journal of Environmental Management*, **303**, 114157, https://doi.org/10.1016/j.jenvman.2021.114157

Almoradie A., Cortes V. J. and Jonoski A. (2015). Web-based stakeholder collaboration in flood risk management. *Journal of Flood Risk Management*, **8**(1), 19–38, John Wiley & Sons Ltd and The Chartered Institution of Water and Environmental Management (CIWEM), first published online on 31 October 2013, https://doi.org/10.1111/jfr3.12076

Arrieta A. B., Díaz-Rodríguez N., Del Ser J., Bennetot A., Tabik S., Barbado A., Garcia S., Gil-Lopez S., Molina D., Benjamins R., Chatila R. and Herrera F. (2020). Explainable artificial intelligence (XAI): concepts, taxonomies, opportunities and challenges toward responsible AI. *Information Fusion*, **58**, 82–115, https://doi.org/10.1016/j.inffus.2019.12.012

Assumpção T. H., Popescu I., Jonoski A. and Solomatine D. P. (2018). Citizen observations contributing to flood modelling: opportunities and challenges, *Hydrology and Earth System Sciences*, **22**(2), 1473–1489, https://doi.org/10.5194/hess-22-1473-2018

Babovic V. (1996). Emergence, Evolution, Intelligence: Hydroinformatics. PhD thesis, Balkema Publishers.

Babovic V. (2009). Introducing knowledge into learning based on genetic programming. *Journal of Hydroinformatics*, **11**(3–4), 181–193, https://doi.org/10.2166/hydro.2009.041

Babovic V. and Abbott M. B. (1997). The evolution of equations from hydraulic data part I: theory and part II: applications. *Journal of Hydraulic Research*, **35**(3), 397–410/411–430, https://doi.org/10.1080/00221689709498420

Bartos M. and Kerkez B. (2021). Pipedream: an interactive digital twin model for natural and urban drainage systems. *Environmental Modelling & Software*, **144**, 105120, https://doi.org/10.1016/j.envsoft.2021.105120

Basco-Carrera L., Warren A., van Beek E., Jonoski A. and Giardino A. (2017). Collaborative modelling or participatory modelling? A framework for water resources management. *Environmental Modelling & Software*, **91**, 95–110, ISSN 1364-8152, https://doi.org/10.1016/j.envsoft.2017.01.014

Bauman Z. (1996). Postmodern Ethics. Blackwell Publishers Inc., Oxford.

Bhattacharya B., Lobbrecht A. H. and Solomatine D. P. (2003). Neural networks and reinforcement learning in control of water systems. *ASCE Journal of Water Resources Planning and Management*, **129**(6), 458–465, https://doi.org/10.1061/(ASCE)0733-9496(2003)129:6(458)

Beven K. (1989). Changing ideas in hydrology – the case of physically based models. *Journal of Hydrology*, **105**, 157–172, https://doi.org/10.1016/0022-1694(89)90101-7

Beven K. J. (2000). Uniqueness of place and process representations in hydrological modelling. *Hydrology and Earth System Sciences*, **4**(2), 203–213, https://doi.org/10.5194/hess-4-203-2000

Beven K. J. (2001). How far can we go in distributed hydrological modelling?. *Hydrology and Earth System Sciences*, **5**(1), 1–12, https://doi.org/10.5194/hess-5-1-2001

Beven K. J. (2002). Towards an alternative blueprint for a physically-based digitally simulated hydrologic response modelling system. *Hydrological Processes*, **16**(2), 189–206, https://doi.org/10.1002/hyp.343

Beven K. J. and Pappenberger F. (2003). Discussion. *Journal of Hydraulic Research*, **41**(3), 331–336, https://doi.org/10.1080/00221680309499978

Choi Y.-D., Goodall J. L., Sadler J. M., Castronova A. M., Bennett A., Li Z., Nijssen B., Wang S., Clark M. P., Ames D. P., Horsburgh J. S., Yi H., Bandaragoda C., Seul M., Hooper R. and Tarboton D. G. (2021). Toward open and reproducible environmental modeling by integrating online data repositories, computational environments, and model application programming interfaces. *Environmental Modelling & Software*, **135**, 104888, ISSN 1364-8152, https://doi.org/10.1016/j.envsoft.2020.104888

Clark M. P., Bierkens M. F. P., Samaniego L., Woods R. A., Uijlenhoet R., Bennett K. E., Pauwels V. R. N., Cai X., Wood A. W. and Peters-Lidard C. D. (2017). The evolution of process-based hydrologic models: historical challenges and the collective quest for physical realism. Hydrology and Earth System Sciences, **21**, 3427–3440, https://doi.org/10.5194/hess-21-3427-2017

CNR-IRPI. (2021). DTE Hydrology, http://hydrology.irpi.cnr.it/projects/dte-hydrology/ [last accessed 15 July 2022].

Crow-Miller B., Webber M. and Molle F. (2017). The (re)turn to infrastructure for water management? *Water Alternatives*, **10**(2), 195–207.

Cunge J. A. (2003). Of data and models. *Journal of Hydroinformatics*, **5**(2), 75–98, https://doi.org/10.2166/hydro.2003.0007

Dibike Y. B., Solomatine D. and Abbott M. B. (1999). On the encapsulation of numerical-hydraulic models in artificial neural network. *Journal of Hydraulic Research*, **37**(2), 147–161, https://doi.org/10.1080/00221689909498303

Ellul J. (1964). The Technological Society. Vintage Books, New York.

Ewen J., Parkin G. and O'Connell P. E. (2000). SHETRAN: distributed river basin flow and transport modelling system. *Journal of Hydrologic Engineering*, **5**, 250–258, https://doi.org/10.1061/(ASCE)1084-0699(2000)5:3(250)

Feldman D. L. (1991). Water Resources Management: in Search for an Environmental Ethic. The Johns Hopkins University Press, Baltimore, USA.

Heidegger M. (1963). Die Technik, und Die Kehre. Neske, Pfullingen//1977 The Question Concerning Technology and Other Essays (Trans. Lovett, W.). Harper, New York.

Hrachowitz M. and Clark M. P. (2017). HESS Opinions: The complementary merits of competing modelling philosophies in hydrology. *Hydrology and Earth System Sciences*, **21**, 3953–3973, https://doi.org/10.5194/hess-21-3953-2017

Huang R., Ma C., Ma J., Huangfu X. and He Q. (2021). Machine learning in natural and engineered water systems. *Water Research*, **205**, 117666, https://doi.org/10.1016/j.watres.2021.117666

Jameson F. (1991). Postmodernism, or, the Cultural Logic of Late Capitalism. Verso, New York.

Jiang S., Zheng Y. and Solomatine D. P. (2020). Improving AI system awareness of geoscience knowledge: symbiotic integration of physical approaches and deep learning. *Geophysical Research Letters*, **47**(13), e2020GL088229, https://doi.org/10.1029/2020GL088229

Jonoski A. (2002). Hydroinformatics as Sociotechnology: Promoting Individual Stakeholder Participation by Using Network-Distributed Decision Support Systems. Swets and Zeitlinger, Lisse, PhD thesis, UNESCO-IHE Institute for Water Education, Delft/Balkema, Rotterdam.

Jonoski A. and Abbott M. B. (1998). Network distributed decision support systems as multi-agent constructs. In: Hydroinformatics '98, Babovic & Larsen (eds), Balkema, Rotterdam, pp. 1219–1226.

Kapelan Z., Weisbord E. and Babovic V. (2020). Digital Water: Artificial Intelligence Solutions for the Water Sector. International Water Association, London, UK, White paper, © 2020.

Kumar S., Godrej A. N. and Grizzard T. J. (2015). A web-based environmental decision support system for legacy models. *Journal of Hydroinformatics*, **17**(6), 874–890, https://doi.org/10.2166/hydro.2015.007

Latour B. (1993). We Have Never Been Modern. Harvester, London.

Loucks D. P. (1995). Developing and implementing decision support systems: a critique and a challenge. *Journal of the American Water Resources Association*, **31**, 571–582, https://doi.org/10.1111/j.1752-1688.1995.tb03384.x

Loucks D. P. and van Beek E. (2017). Water Resource Systems Planning and Management: An Introduction to Methods, Models, and Applications. Springer in collaboration with Deltares and UNESCO-IHE, Cham, Switzerland.

Lyotard J.-F. (1979). La Condition Postmoderne: Raport sur le Savoir, Minuit, Paris//(1984) The Postmodern Condition: A Report on Knowledge. Manchester University Press, Manchester, UK.

Maier H. R., Kapelan Z., Kasprzyk J., Kollat J., Matott L. S., Cunha V., Dandy V., Gibbs M. S., Keedwell E., Marchi A., Ostfeld A., Savic D., Solomatine D. P., Vrugt J. A., Zecchin A. C., Minsker B. S., Barbour E. J., Kuczera G., Pasha F., Castelletti A., Giuliani M. and Reed P. M. (2014). Evolutionary algorithms and other metaheuristics in water resources: current status, research challenges and future directions. *Environmental Modelling & Software*, **62**, 271–299, https://doi.org/10.1016/j.envsoft.2014.09.013

Marcel G. (1956//1984). The Philosophy of Existentialism, Philosophical Library//(1995). Carol Publishing Group Edition, New York.

Minns A. W. (1998). Artificial Neural Networks as Sub-symbolic Process Descriptors. PhD thesis, Balkema, Rotterdam.

Minns A. W. and Hall M. J. (1996). Artificial neural networks as rainfall-runoff models. *Hydrological Sciences Journal*, **41**, 399–417, http://doi.org/10.1080/02626669609491511

O'Kane J. P. (2015). An enquiry into the place of systems analysis in the politics of water and the environment. *Journal of Hydroinformatics*, **17**(2), 163–175, https://doi.org/10.2166/hydro.2014.200

Oreskes N., Shrader-Frechette K. and Belitz K. (1994). Verification, validation, and confirmation of numerical models in the earth sciences. *Science*, **263**, 641–646, https://doi.org/10.1126/science.263.5147.641

Palmer R. N., Cardwell H. E., Lorie M. A. and Werick W. (2013). Disciplined planning, structured participation, and collaborative modeling – applying shared vision planning to water resources. *Journal of the American Water Resources Association (JAWRA)*, **49**(3), 614–628, https://doi.org/10.1111/jawr.12067

Pedersen A. N., Borup M., Brink-Kjær A., Christiansen L. E. and Mikkelsen P. S. (2021). Living and prototyping digital twins for urban water systems: towards multi-purpose value creation using models and sensors. *Water*, **13**, 592, https://doi.org/10.3390/w13050592

Pianosi F., Beven K., Freer J., Hall J. W., Rougier J., Stephenson D. B. and Wagener T. (2016). Sensitivity analysis of environmental models: a systematic review with practical workflow. *Environmental Modelling & Software*, **79**, 214–232, https://doi.org/10.1016/j.envsoft.2016.02.008

Price R. K. and Vojinović Z. (2011). Urban Hydroinformatics: Data, Models and Decision Support for Integrated Urban Water Management. IWA Publishing, London, UK.

Ramos M. H., van Andel S. J. and Pappenberger F. (2013). Do probabilistic forecasts lead to better decisions?. *Hydrology and Earth System Sciences*, **17**, 2219–2232, https://doi.org/10.5194/hess-17-2219-2013

Reed P. M., Hadka D., Herman J. D., Kasprzyk J. R. and Kollat J. B. (2013). Evolutionary multiobjective optimization in water resources: the past, present, and future. *Advances in Water Resources*, **51**, 438–456, https://doi.org/10.1016/j.advwatres.2012.01.005

Refsgaard J. C. and Storm B. (1995). MIKE SHE. In: Computer Models in Watershed Hydrology, V. J. Singh (ed.), Water Resources Publication, Highlands Ranch, CO, USA, pp. 809–846.

Refsgaard J. C., van der Sluijs J. P., Højberg A. L. and Vanrolleghem P. A. (2007). Uncertainty in the environmental modelling process – a framework and guidance. *Environmental Modelling & Software*, **22**(11), 1543–1556, https://doi.org/10.1016/j.envsoft.2007.02.004

Refsgaard J. C., Storm B. and Clausen T. (2010). Système hydrologique europeén (SHE) – review and perspectives after 30 years development in distributed physically-based hydrological modelling. *Hydrology Research*, **41**(5), 355–377, https://doi.org/10.2166/nh.2010.009

Savenije H. H. G. and Hrachowitz M. (2017). HESS Opinions: 'catchments as meta-organisms – a new blueprint for hydrological modelling'. *Hydrology and Earth System Sciences*, **21**, 1107–1116, https://doi.org/10.5194/hess-21-1107-2017

Sit M., Demiray B. Z., Xiang Z., Ewing G. J., Sermet Y. and Demir I. (2020). A comprehensive review of deep learning applications in hydrology and water resources. *Water Science and Technology*, **82**(12), 2635–2670, https://doi.org/10.2166/wst.2020.369

Sivapalan M., Thompson S. E., Harman C. J., Basu N. B. and Kumar P. (2011). Water cycle dynamics in a changing environment: improving predictability through synthesis. *Water Resources Research*, **47**, W00J01, https://doi.org/10.1029/2011WR011377

Solomatine D. P. and Ostfeld A. (2008). Data-driven modelling: some past experiences and new approaches. *Journal of Hydroinformatics*, **10**(1), 3–22, https://doi.org/10.2166/hydro.2008.015

Solomatine D. P. and Shrestha D. L. (2009). A novel method to estimate model uncertainty using machine learning techniques. *Water Resources Research*, **45**, W00B11, https://doi.org/10.1029/2008WR006839

Solomatine D. P. and Torres L. A. (1996). Neural network approximation of a hydrodynamic model in optimizing reservoir operation. Proceedings of the Second International Conference on Hydroinformatics, September 1996, Zurich. pp. 201–206.

Sušnik J., Masia S., Indriksone D., Brēmere I. and Vamvakeridou-Lydroudia L. (2021). System dynamics modelling to explore the impacts of policies on the water-energy-food-land-climate nexus in Latvia. *Science of the Total Environment*, **775**, 145827, https://doi.org/10.1016/j.scitotenv.2021.145827. Epub 2021 Feb 13. PMID: 33611179.

Thissen W. A. H. and Twaalfhoven P. G. J. (2001). Towards a conceptual structure for evaluating policy analytic activities. *European Journal of Operational Research*, **129**, 627–649, Elsevier, https://doi.org/10.1016/S0377-2217(99)00470-1

Thorkilsen M. and Dynesen C. (2001). An owner's view of hydroinformatics: its role in realising the bridge and tunnel connection between Denmark and Sweden. *Journal of Hydroinformatics*, **3**, 105–135, https://doi.org/10.2166/hydro.2001.0011

Valverde-Pérez B., Johnson B., Wärff C., Lumley D., Torfs E., Nopens I. and Townley L. (2021). Digital Water: Operational Digital Twins in the Urban Water Sector: Case Studies. International Water Association, London, UK, White paper, © 2021.

van Andel S. J., Price R. K., Lobbrecht A. H., van Kruiningen F. and Mureau R. (2008). Ensemble precipitation and water-level forecasts for anticipatory water-system control. *Journal of Hydrometeorology*, **9**(4), 776–788, https://doi.org/10.1175/2008JHM971.1

Voinov A., Kolagani N., McCall M. K., Glynn P. D., Kragt M. E., Ostermann F. O., Pierce S. A. and Ramu P. (2016). Modelling with stakeholders – next generation. *Environmental Modelling & Software*, **77**, 196–220, https://doi.org/10.1016/j.envsoft.2015.11.016

Vojinovic Z. and Abbott M. B. (2012). Flood Risk and Social Justice. IWA Publishing, London, UK.

Wehn U., Ajates R., Fraisl D., Gharesifard M., Gold M., Hager G., Oliver J. L., See L., Shanley L. A., Ferri M., Howitt C., Monego M., Pfeiffer E. and Wood C. (2021). Capturing and communicating impact of citizen science for policy: a storytelling approach. *Journal of Environmental Management*, **295**, 113082, https://doi.org/10.1016/j.jenvman.2021.113082

Yan H., Solomatine D. P., Velickov S. and Abbott M. B. 1999. Distributed environmental impact assessment using internet. *Journal of Hydroinformatics*, **01**(1), 59–70, https://doi.org/10.2166/hydro.1999.0005

Zheng F., Tao R., Maier H. R., See L., Savic D., Zhang T., Chen Q., Assumpção T. H., Yang P., Heidari B., Rieckermann J., Minsker B., Bi W., Cai X., Solomatine D. and Popescu I. (2018). Crowdsourcing methods for data collection in geophysics: state of the art, issues, and future directions. *Reviews of Geophysics*, **56**, 698–740, https://doi.org/10.1029/2018RG000616

doi: 10.2166/9781789062656_0017

Chapter 2

Computational hydraulics: stage for the hydroinformatics act

A. W. Minns[1], A. Verwey[2] and I. R. Warren[3]

[1]Deltares, Marine & Coastal Systems, Delft, the Netherlands
[2]Strategic Advisor Flood Risk Management, World Bank, Rucphen, the Netherlands
[3]Retired, COWI Consulting Engineers, Kgs. Lyngby, Denmark

ABSTRACT

The journey from computational hydraulics to hydroinformatics is described through an overview that starts from the first applications of computer code to civil engineering problems, through the development of the method of characteristics – as a necessary step in the development of numerical methods – to the modern day, wide-spread application of fourth-generation modelling systems that are used to derive solutions to water-related, and particularly environmental problems. Mike Abbott's contributions to the scientific development of numerical methods and computational techniques are described. These contributions led to the publication of his seminal work *Computational Hydraulics – Elements of the Theory of Free Surface Flows* in 1979. At the same time, the authors of this chapter have all worked closely with Mike Abbott to translate his sometimes-esoteric writings into practical and commercially viable solutions. An important milestone in this journey was the establishment of the Computational Hydraulics Centre at the Danish Hydraulics Institute in 1970, which was a result of the recognition of the commercial importance of computational hydraulics and the need for further professionalization of the development of modelling systems. This was witnessed through the development of modelling systems like, for example, System21 (later MIKE 21) and the Système Hydrologique Européen (MIKE SHE). When tracing the development of computational hydraulics, Mike Abbott often referred to a number of generations of modelling methodologies and corresponding models, which has led to the current, fourth-generation modelling systems. Furthermore, as a natural progression of computational hydraulics, he discussed at length the developing social dimension of hydraulics and hydraulic modelling, which was seen as a first and necessary step towards hydroinformatics. This chapter concludes with an assessment of the extent to which the anticipated benefits of fourth-generation modelling may or may not have come to fruition and what this has meant in daily practice. These lessons learned are placed in the context of what this then means for future hydroinformatics applications.

Keywords: Hydroinformatics, computational hydraulics, method of characteristics, numerical methods, hydraulic modelling systems and applications

2.1 INTRODUCTION

It is most likely in the field of computational hydraulics where the vision of Mike Abbott has had the biggest impact on water management practices and, in particular, the way in which we analyse hydrological and hydraulic systems. From his earliest work in which he explored the possibilities opened up in hydraulics by digital machine computation, his vision has been a driving force to establish computational hydraulics as an indispensable discipline for the design and construction of great hydraulic engineering works and water management in the broadest sense. We speak of the 1960s, when Mike Abbott was appointed 'Lector', or Associate Professor as the position would be called these days, at the 'International Courses' in Delft, an institute established to provide training in hydraulics, hydrology and sanitary engineering to professionals from developing countries, now known as IHE Delft Institute of Water Education.

These were the days of the Delta Works in the Netherlands, an enormous investment programme set up to avoid a possible recurrence of the devastating floods of February 1953, which inundated significant parts of the southwestern regions of the Netherlands. Investments were of the order of billions of Euros and their hydraulic engineering designs were substantiated primarily through studies based upon hydraulic scale models built and run at Delft Hydraulics. These studies were commissioned by 'Rijkswaterstaat', part of the Dutch Ministry of Public Works, which also had their own study group, focussed already more on numerical methods and the use of 'alternative' hydraulic simulation methods.

These alternative methods included analogue models, representing the hydraulic conveyance and storage concepts by connecting equivalent electronic devices. Pioneering work was done by Dronkers (1964) who, with his team, built the numerical model of the Dutch Delta which was used to make numerous simulations of conditions that described the operation of the gigantic Eastern Scheldt Storm Surge Barrier.

Such developments and his earlier ground-breaking work on the role of characteristics in hydrodynamics led to Mike Abbott's vision that instead of building such numerical models for each application from scratch, it should be possible to build a modelling system, consisting of a framework of concepts, which would make it much easier to construct new models. In the first instance, Delft Hydraulics was approached for a joint development of such concept. Unfortunately, this did not work out as in those days the focus of Delft Hydraulics was still very much on the application of hydraulic scale models. However, having worked also at the Technical University of Denmark, an open ear was found at the Danish Hydraulic Institute (DHI), in those days a small unit of the Technical University of Denmark. Eventually this led to the development and marketing of the DHI modelling systems. For more detail about these early achievements of Mike Abbott, the reader is referred to Abbott (2002) in which his personal reflections on computational hydraulics and hydroinformatics are recorded.

In the original edition of 'Computational Hydraulics – Elements of the Theory of Free Surface Flows' (Abbott, 1979), computational hydraulics is described as a reformulation of hydraulics to suit digital machine processes. New possibilities for solving mankind's problems were thus centred upon the digital computer. In these cases, the computer is made useful through its ability to obtain quantitative results from mathematical models of the real world. When a mathematical model is described in terms of the processes of numerical analysis, it is commonly called a numerical model. This original edition was directed to a world in which many engineers and scientists built their own numerical–hydraulic models, so that they needed an introduction to the subject that would assist them in these endeavours. The second edition in 1998, together with Tony Minns (Abbott & Minns, 1998), was subsequently directed to a world in which numerical–hydraulic modelling facilities had become widely available in packaged forms. These numerical modelling packages no longer make substantial demands upon the computational–hydraulic knowledge of their users but do require a highly professional understanding of the possibilities and limitations of the hydraulics knowledge that is so encapsulated.

Computational hydraulics was thus transformed from a need for a knowledge of 'how to do things' to a knowledge of 'how things are done'. Present-day modelling system suites provide facilities which far transcend the possibilities of the builder of the individual model, but they make a correspondingly greater demand upon the insight and understanding of their users. The overall tendency is thus towards a more extended range of applications, a greater depth of understanding of relations between physics and numerics and, correspondingly, a sharpening of both analytical and diagnostic abilities.

2.1.1 Historical context

Abbott *et al.* (2001) place the developments of computational hydraulics in an historical context. In the period between the late 15th century and the late 18th century, the works of the great scientists and thinkers of that time like Newton, Euler and Lagrange were based upon the application of a mathematical apparatus in order to describe the outcomes of physical experiments. Prior to these developments, scientific work primarily consisted only of the recording of observations. The modern-scientific approach was principally characterized by two stages: a first one in which a set of observations of the physical system were collected, and a second one in which inductive assertion about the behaviour of the system – a hypothesis – was generated. For the study of water movements, the mathematical formulation of physical phenomena led, along the way, to the theory of ideal fluid motion. Computational fluid dynamics (CFD) is a branch of fluid mechanics that uses numerical analysis and data structures to analyse and solve problems that involve fluid flows. The fundamental basis of almost all CFD problems is the Navier–Stokes equations, which define many single-phase (gas or liquid, but not both) fluid flows. However, during the 19th century it became obvious that there were significant deviations between the behaviours of the observed real-world behaviour of real fluids and the proposed conceptualizations. Mike Abbott once described the Navier–Stokes equations as 'the best description available of an ideal fluid that does not actually exist!' (M. Abbott, pers. comm.). Two-dimensional methods to solve linearized versions of the Navier–Stokes equations for flows around cylinders and air foils were already being developed in the 1930s. Some of the first calculations using finite differences in a physical domain that was divided into cells were done using ENIAC (electronic numerical integrator and computer) in the 1940s. At this time there were many parallel developments going on within industries like the aircraft industry, weather predictions, and mechanical engineering, to develop numerical codes to solve specific problems for that specific industry. Early on, Mike Abbott recognized the potential for the application to water problems. His first works for the most part explored the possibilities opened up in hydraulics by digital machine computation. His first major breakthrough in hydraulics came with the formulation of an implicit scheme with a fast algorithm for its solution applied to the equations of one-dimensional, nearly horizontal, subcritical flow (Abbott & Ionescu, 1967). It is interesting to note that in fact another and different implicit algorithm for the same purpose had been devised by A. Preismann some years before Mike Abbott's own discovery. This issue was reconciled some years later as described in Abbott (2002):

...as he (J. Cunge) most kindly explained, this work had only been published in a place so obscure that I could not possibly have known about it – which was indeed the case.

The need to introduce concepts such as drag, friction and turbulence meant that the study of hydraulics evolved to encompass the use of semi-empirical data about real fluids in real practical situations. The study of computational hydraulics is thus concerned not only with finding mathematical solutions to the governing equations, but also with ensuring that the solutions so obtained are indeed representative of the real-world observations of the very particular fluid we refer to as water. As will be explained in the following sections, advanced modelling systems have now been developed to take care of the former, whereas it is the role of the modelling practitioner to ensure that the latter is sufficiently upheld. It is well known how the results of models of fluid flow can nearly always be fitted to measurements of hydraulic behaviour recorded in a few places by varying the values of certain parameters, and most commonly the roughness coefficients, in the model. This process can however easily hide gross inaccuracies in the description of the modelled domain, such as ill-posed boundaries, errors in topography, the extrapolation of good-looking fits in only a part of the domain to the entire modelling domain, and so on.

2.2 ROLE OF CHARACTERISTICS IN COMPUTATIONAL HYDRAULICS

2.2.1 The meaning of characteristics

In 1966, the book 'An Introduction to the Method of Characteristics' (Abbott, 1966) was published, which took shape during work carried out at the University of Amsterdam (1963–64) and at the Technical University of Denmark (1964–66). For Mike Abbott, the characteristics were fundamental to computational methods in hydraulics and as such deserving frequent discussions with his staff and students. The fundamental issue is that these characteristics represent lines along which information propagates in the space–time domain describing the motion of water, or that of air or fluid in general. This means that anything disturbing the state of a fluid at one point in space would be felt at a later moment at another point, with a time lag defined by the speed of this disturbance, or the characteristic celerity, usually denoted by the character 'c'.

In most cases in fluid dynamics these characteristics travel in two opposite directions from the point of view of a stationary observer, as demonstrated when one throws a stone in a pool of water. The disturbance seen in the water travels with a celerity $c+ = u + \sqrt{(gh)}$ in the downstream direction and $c- = u - \sqrt{(gh)}$ in upstream direction, where: c is the celerity of the disturbance; u is the water velocity; g is the acceleration due to gravity; and h is the water depth. In stagnant water, the celerity of the disturbance is equal in both directions to the left and the right. In the case where the water is moving, the celerity of the disturbance is higher in downstream direction than in upstream direction. It should be noted that for water velocities higher than the value of $\sqrt{(gh)}$ the disturbance is no longer able to travel in the upstream direction. The flow becomes super-critical.

As discussed in the sequel, this understanding of the flow of information is fundamental to the way in which numerical methods perform. It influences the stability behaviour of our numerical methods and must also be taken into account when programming algorithms to implement iterative methods in solutions.

2.2.2 The three- and four-point method of characteristics

Computational hydraulics methods, being used with digital computers, are of necessity discrete methods in which quantities are computed at a finite number of places in distance and time. The places at which numbers are computed are called grid points. The method of characteristics described here needs values in between these grid points in nearly all cases. Between the grid points, nothing is given directly by the computation, but quantities must be determined by a process of interpolation of the quantities given at the grid points. The type of interpolation process used influences not only how quantities are determined from values at the grid points, but also the way in which the computation proceeds, from values of dependent variables at grid points at one time level to their values at the next time level. In principle, the accuracy of the description increases as the accuracy of the interpolation increases, both as concerns the recovery of values between grid-refined values and as concerns the description of the process of change in time. It follows that the choice of grid points, to describe the flow system is intimately related to the methods used to describe the laws determining the behaviour of the flow system.

The following simple example, taken from Abbott and Minns (1998), demonstrates the role of characteristics in the development of numerical solutions. We know that, given values of u and c ($=\sqrt{gh}$) at two points in (x,t), it is possible to calculate the values at a third point situated at the intersection of characteristics through the two given points (Figure 2.1).

The problem here is that the given values of (x,t) pairs usually do not coincide exactly with the user-defined grid pairs of x_j and t_n. The three-point method of characteristics says that if lines AB and BC in Figure 2.1 are characteristic then the problem is reduced to one of determining $(u,c)_b$ and $(u,c)_c$, which is essentially a problem of directly (or *explicitly*) interpolating for these B and C values from values already computed at the previous time level. This algorithm shows immediately one, and only just one, of the weak points of applying the method of characteristics for the solution of real-world problems. Interpolations lead to a loss of information. For example, the most common method for such interpolation is linear, which means that the value collected at time level t_n is extracted from values along a straight line, whereas it should have been extracted from some form of a curved line. Such linear interpolation, therefore, usually leads to a dampening of waves, whatever these waves represent. Higher-order interpolations would reduce this loss of information.

However, such methods are still not perfect and come with the additional problem of handling the interpolations at model boundaries. This method of characteristics, though lacking practical applicability, has provided useful insight into the definition of the so-called explicit numerical schemes. It is well known that, while interpolations lead to some form of dampening of waves, extrapolations would lead to amplification of waves, which show up as uncontrolled amplifications when repeated over many steps in time. This is what is experienced as *unstable* numerical solutions.

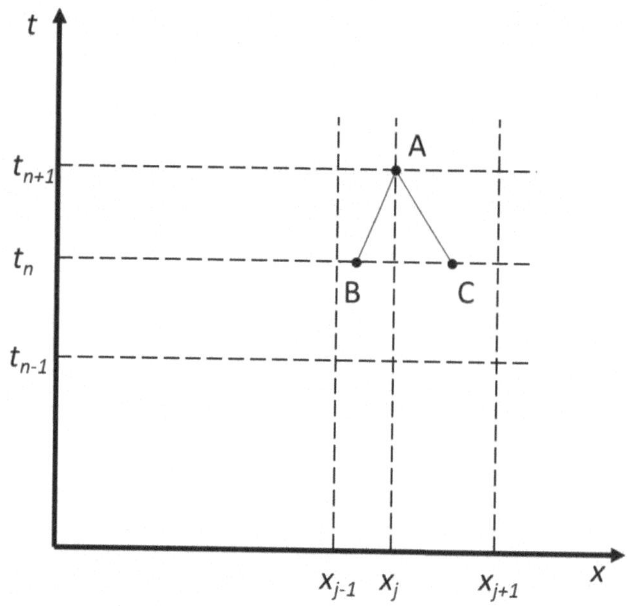

Figure 2.1 Characteristics in a fixed grid system: explicit formulation (From: Computational Hydraulics, 2nd Edition, Abbott and Minns, © 1998, Ashgate Publishing Company. Reproduced by permission of Taylor & Francis Group. Abbott and Minns, 1998).

A variation of the above method is schematized in Figure 2.2. Here the values of $(u,c)_b$ and $(u,c)_c$ are interpolated along the vertical lines of constant x. This will lead to expressions for the unknowns u_j^{n+1} and c_j^{n+1} in terms of the *known* values of (u,c) at time level n as well as values of (u,c) at time level $n + 1$, which are *unknown*. The result is a computational scheme that involves more than one unknown at the forward time level at every grid point, so that the solution of the scheme necessitates the solution of a set of simultaneous equations. This type of scheme is referred to as an *implicit* scheme. This implicit scheme inherits the properties of stability of the explicit approach shown in Figure 2.1. However, here the principle of interpolation applies to any time step, whatever its magnitude beyond the value $\Delta t = \Delta x/c$. In conclusion, this scheme will be *unconditionally stable* for any step in time, though at an immense cost of accuracy.

An alternative method of characteristics, avoiding the problem of interpolations, was based upon a technique where the grid in space and time was not fixed beforehand. The grid would follow the characteristic lines and any new point would be defined with its coordinates in x and t by calculating its position from two already defined points

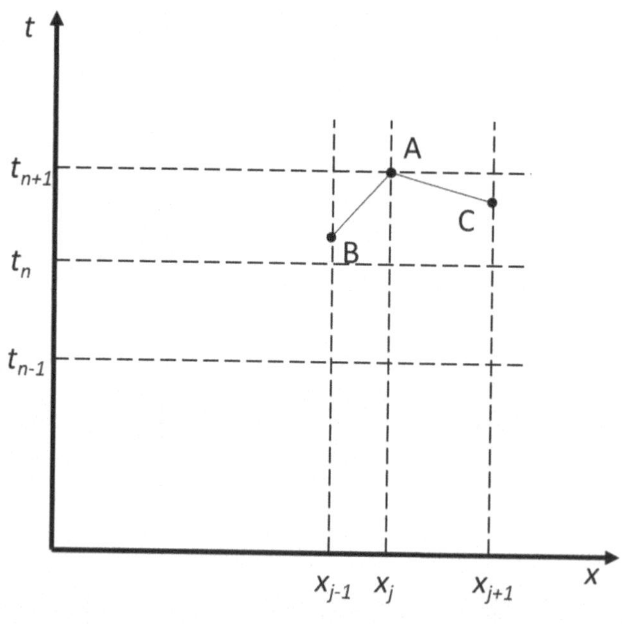

Figure 2.2 Characteristics in a fixed grid system: implicit formulation (From: Computational Hydraulics, 2nd Edition, Abbott and Minns, © 1998, Ashgate Publishing Company. Reproduced by permission of Taylor & Francis Group. Abbott and Minns, 1998).

(Abbott & Verwey, 1970). Although this approach avoided the problem of interpolation and its loss of information, it did not address another important aspect of the modelling of unsteady flow equations: the important role of the gravity and friction terms. In nearly all hydrodynamic calculations these two terms form a decisive role in how the fluid moves. In particular in one-dimensional (1D) hydrodynamics, the inclusion of these terms would increase the inaccuracies already mentioned and would make computational procedures very complex. For these reasons, the method of characteristics never became a numerical tool surviving in the practice of numerical hydrodynamic modelling.

2.2.3 Practical aspects of characteristics

The shortcomings mentioned above should not discourage hydraulic practitioners to turn away from the development of an understanding of the role of characteristics in their practice. This may involve a focus on the development and understanding of numerical techniques or on the understanding of physical processes in engineering practice in general. Characteristics are fundamental to the class of problems characterized in mathematics as 'hyperbolic equations'. This not only applies to the full equations we are dealing with here, but also to parts of these equations, or even their hidden forms. Some examples are discussed below.

Hydrodynamic equations are generally considered to be part of the class of hyperbolic equations, or more particularly, second-order hyperbolic equations. They have two characteristic directions or directions of flow of information. The hydrodynamic equations are composed of a mass or volume conservation principle and a momentum conservation principle. The momentum equation contains a term that is usually described as the *convective momentum term*, describing how momentum is carried by the flow. As part of the second-order hyperbolic equations this term is an underclass of a first-order hyperbolic equation, embedded in the second-order hyperbolic equation, and so has its own peculiar characteristic. Indeed, the momentum is carried by this first-order characteristic with a celerity equal to the flow velocity. In state-of-the-art numerical techniques, the understanding of how this convective momentum terms plays its role and how it should be implemented numerically is of vital importance for the correct representation of flows over wetting and drying fields and the flow over weirs and across channel junctions (Stelling & Verwey, 2005).

Another peculiar role of characteristics is demonstrated by flood waves propagating along rivers. While its flow is generally described by the ingenious de Saint Venant equations, their behaviour may vary according to the location along the river. Arriving at sea, all terms of the de Saint Venant equations are important and the system behaves truly as hyperbolic, with information travelling in both spatial directions. However, further upstream, the nature of the equation changes to advective, due to the dominant roles of gravity and friction. In the momentum equation, these two terms become more or less in balance, overruling the impact of the convective momentum and acceleration terms. This justifies the reduction of the momentum equation to a so-called rating curve (Q–h relation), which simply describes a relation between water level and discharge. Substituted into the volume conservation equation leads to the advective wave form, first-order hyperbolic, with a characteristic celerity given as $c = 1/b_s * dQ/dh$, where b_s is the channel storage width; Q is the discharge and h is the water level.

Any numerical model based upon the full de Saint Venant equations, transform in their behaviour towards the advective wave form of the equations. Consequently, the boundary data requirements transform as well. For this situation, the upstream boundary condition is dominant and of vital importance. However, the downstream boundary becomes a 'weak' condition and may be defined as anything such as a rating curve or a hypothetical water level (Verwey, 1975). The required flow of information is in fact determined by the characteristics of the dominant part of the equations.

2.3 THE GENERATIONS OF MODELLING

2.3.1 First- and second-generation modelling

When computers were first introduced, many of their users employed them only to calculate numerical values from formulae, in much the same way as they had earlier used slide rules and other analogue devices. This use of the digital machines as 'super slide rules' is referred to as 'first-generation modelling'.

Some of these first users subsequently began to use methods that were particularly well suited to the sequential, repetitive, and recursive mode of operation of the digital machine. These particularly 'machine-friendly' methods were especially well adapted to building models that described variations of properties in space and in time. One of the most common was the 'finite difference' method. Over the first decade of modelling, which extended roughly from 1960 to 1970, each model of a particularly physical area was constructed individually, so that the model so constructed applied only to that specific area. This approach, of one-off or customized modelling, is referred to as 'second-generation modelling'.

2.3.2 Third-generation modelling

The development of the third-generation models is described by Abbott and Minns (1998) in the following way:

> ...The needs of engineering practice were such that the second-generation approach proved inadequate: the lead times for model construction and testing were too uncertain and the corresponding costs were too high for most of the applications arising from hydraulic, coastal, and ocean engineering. Accordingly, from about 1970 onwards, a third-generation approach was developed, which consisted of constructing design systems or, as they are nowadays most commonly called, modelling systems. These are software packages that construct and run any model of a given, wide class when the description of the model is provided to the package in a definite, specified format. The modelling system is, in effect, 'a factory for building models'...

In this way, the third-generation modelling tool enabled the user to produce a model in some hours, which previously would have taken several months. Model building became a relatively simple computer coding exercise. The superstructure read in and prepared the data and carried out the post-processing and plotting of the results. It became apparent that a large investment in a third-generation modelling system could now be justified, as these costs would subsequently be amortized over many years through the future creation of countless numbers of individual models. Furthermore, the modellers could turn their attention to the better and more accurate schematizations of their problems, rather than being tied down by specific computer requirements. All of this made numerical models more attractive and accessible to end-users, increasing the user base and making the business model for further investment in the modelling system even more attractive. The feedback from an ever-increasing number of users enabled the developers of the modelling system to learn from practical experience so that improvements could be introduced into the code, new capabilities realized, and the range of applications extended.

The development of third-generation modelling systems thus provided the impetus for a revolution in the way that mathematical modelling studies could be carried out. This was not only in the area of the numerical methods applied, but also in the computer hardware requirements for digitization of input data, including its storage and visualization. This accelerated the movement towards the formation of specialist teams and computer-integrated environments in the centres where such modelling was practised. The changes that this brought about in the practice of computational hydraulics are described in more detail in the following sections.

2.3.2.1 Establishment of the computational hydraulics centre

With these needs in mind, Mike Abbott and Torben Sorensen, director of DHI, applied for, and received a major grant from the Danish State Council for Technical and Scientific Research (STVF) to establish a specialized department, the Computational Hydraulics Centre (CHC) at DHI. This occurred in 1970 and, under Mike Abbott's guidance, the staff of CHC pioneered the development, application and commercialization of the third and subsequent generations of the modelling systems. Foreseeing the need for and establishing CHC was one of Mike Abbott's most far-sighted achievements.

This vision led to the development, more or less in parallel, of two modelling systems (Abbott *et al.*, 1973a):

(1) System 21, Jupiter, enabling the construction of vertically averaged hydrodynamic models in two spatial horizontal directions. This development was taken up by Dr Gaele Rodenhuis, later supported by Ir Ross Warren (Abbott *et al.*, 1973); and
(2) System 11, SIVA, enabling the construction of vertically averaged hydrodynamic models along one spatial horizontal axis. This development was taken up by Ir Adri Verwey (Abbott *et al.*, 1973b; Abbott & Verwey, 1973).

In these developments, Mike Abbott always acted as an inspirational guru, although also quite some challenges had to be resolved on the work floor.

2.3.2.2 Challenges of the development

'On the work floor', so to speak, those who were entrusted with the work of developing the third-generation models (including two of the co-authors of this chapter) were faced with a number of quite 'down to earth challenges' that included, for example, the generic presentation of model topology, application of boundary conditions and the accuracy, stability and speed of computations.

Mainframe computers such as the IBM 360 and 370 series were almost the only available computers in the 1970s. Apart from the big financial businesses, the only computers available for scientific work were at universities, and, due to the enormous cost of the mainframes, users paid for each minute of computer time. For this reason, a high speed of computation was paramount, especially for the two-dimensional models.

At this time the numerical methods for solution of the partial differential equations of conservation of mass and momentum were dominated by finite difference techniques. Accuracy and stability then became a question of the number of terms from the Taylor series expansion of each of the equation terms, model grid resolution and iteration (to be avoided if at all possible) to time centre some of the terms.

The development of the one-dimensional modelling system, System 11 (predecessor of MIKE11), with the objective of building models based upon a one-dimensional and cross-sectional averaged representation of channel flow, started in 1971. It was based upon the Abbott–Ionescu implicit finite difference scheme, defined on a so-called staggered grid, where alternatingly water levels (h) and discharges (Q) were defined (Abbott *et al.*, 1971). For this implicit scheme, the continuity equations were discretized at the water level grid points, whereas the momentum equations were discretized at the discharge grid points. Together with boundary conditions this would define a system of equations which would allow for the computation of a new set of values for h and Q at each time step taken through operations on the coefficient matrix.

For the continuity equation this way of discretizing fitted the grid very well. However, for the momentum equation only the acceleration, water level gradient and friction terms provided a straightforward discretization on this grid. The convective momentum term required special treatment and initially this was handled by defining it explicitly, with a more accurate approximation obtained through a single iteration step. This step was needed anyway, as the dependence on the important hydraulic gradient and friction terms had to be improved through a single iteration step as part of the complete processing of one single time step. The reason for this need was primarily the averaging of model parameters based upon cross-section water level, such as cross-sectional area and conveyance capacity. As discussed in the sequel, this need was less evident in the discretization of two-dimensional (2D) models.

For a single channel, the matrix operations were most efficiently performed through a double sweep algorithm, which eliminates matrix coefficients in the lower of the three diagonals of the resulting matrix in a first sweep, followed by the elimination of the modified coefficients in the upper diagonal. This algorithm was straightforward to program. The team chose to define codes in the PL1 programming language because of its strong pre- and postprocessing facilities already available in the 1960s of the past century. It took at least 10 years, before Fortran, the language subsequently used most for programming technical algorithms, had similar options.

However, nearly all 1D models describe flow in networks of channels. This is particularly true for urban drainage networks, where thousands of drainage open canals and pipes may be connected to drain the runoff of (extreme) storm events precipitating on cities. In the System 11 approach, and later the MIKE Urban approach, connections between the various river channels or drains were defined at water level grid points. At these connection points the water levels were set as equal values for the connecting branches at a junction. It should be noted that this was an approximation of real conditions, which worked well for the connection of river branches with moderate flow velocities. In due time, more refined compatibility conditions at river and channel junctions were developed, which provided a better representation of the real physics of the system, where a more correct description of the passage of convective momentum from outflowing to receiving channels and associated energy losses played an essential role.

Apart from internal connections representing part of the channel boundaries, also the external boundaries form part of the total system of equations to be defined and solved. The principle is straightforward. Each grid point needs the discretization and parameterization of one equation to arrive at the solution of the total set of unknown water levels and discharges at a new level in time. The need for boundary conditions has been described as part of Section 2.2, either defining a water level, a discharge or a relation between discharge and water level, the so-called rating curve.

The implicit numerical schemes led to the parameterization of a system of equations, filling a sparse matrix for their solution. For single channels, the matrix would only fill three main diagonals, for which the required matrix operations could be performed efficiently with the double sweep algorithm described earlier. The solution algorithm required for more complex channel networks became subject to extensive discussions in the scientific community.

In the 1970s, a scientific knowledge exchange platform was formed as part of activities of the IAHR, which discussed developments and needs of computational hydraulics. In addition to Mike Abbott, key players on this platform were Jean Cunge and Alexander Preissmann of Sogreah, Jean-Pierre Benqué of EDF (Electricité de France) and Gerrit Verboom (Delft Hydraulics). A vividly discussed topic was the algorithm for solving the finite difference equations for networks of channels. Distinction was made between simply and multiply connected or looped channels. Simply connected channels were defined as those where only one single route would connect two selected points on the network. The double sweep algorithm could easily be adapted to model such networks.

More complex was the solution for multiply connected networks. Mike Abbott suggested to fall back on the theory of characteristics by creating overlapping branches for parts of loops and take boundary conditions from the previous time step at the ends of these overlaps. In theory, this method should work well. However, in programming this produced very high flows around the loops, with ultimately unstable results. With the knowledge we have these days this was most likely the result of energy created by the compatibility conditions applied at the channel junctions.

For the development of System 11 the ultimate solution adopted was the elimination of equations within each individual branch by a special form of double sweep algorithm, which resulted in a sparse matrix of coefficients defining relations between variables at junctions and boundaries. This remaining system of equations was solved by a standard matrix inversion technique. It was interesting to find several years later that a form of this approach had already been published by Dronkers (1964).

The current approach has evolved much further by entering the coefficients of all equations into one single sparse matrix and applying a minimum connection search algorithm in the matrix elimination process which leads to solutions in a most efficient way. An additional advantage is that no special adaptations are needed in the solution algorithm, when complex additional equations are added to the schematization, such as more complex boundary conditions, for example a surge chamber as an anti-water hammer device. All equations defined this way are simply added as a new row in the matrix. Experience with very large drainage or sewer networks consisting of thousands of pipes shows that the resulting large set of equations can be solved within a fraction of a second.

Currently, many hydrodynamic models are being developed by combining 1D and 2D schematizations in an implicitly coupled way. In such cases, eliminations in the 2D domain can only be partly handled by the minimum search algorithm. This has led to combined algorithms which start with the minimum search method applied, while, due to increasing numbers of connections (growing number of matrix coefficients in a row) resulting from the elimination process, the remaining part of the equations is solved by a conjugate matrix algorithm.

For the two-dimensional modelling system, System 21, once the basic numerical solution was developed, the dominating problem was the reduction of the computer time. The first optimization introduced by Mike Abbott was to develop an implicit numerical solution allowing the model to run with a Courant no. ($Cr = \Delta t/\sqrt{gh}$) greater than 1 (Abbott *et al.*, 1973). A second optimization was to develop a solution which avoided iteration, thereby halving the computer time immediately. The third technique developed was to use the so-called double sweep algorithm to achieve a rapid and stable inversion of the solution matrix. When written like this, it all seems so simple and straightforward, but in reality, it represented major development efforts. Improvement of the model accuracy to enable its application to a wide range of problems continued for many years (Abbott *et al.*, 1981).

2.3.2.3 Stratified flow modelling

The Danish Belts connecting the North Sea with the Baltic are strongly stratified with the low salinity Baltic water overlying the high salinity North Sea water. The study of engineering projects such as harbours and bridge and tunnel connections between the islands and across to Sweden therefore required a model of *stratified flow*. The first proposed tunnel between Elsinor in Denmark and Helsingborg in Sweden was investigated in the early 1970s and Mike Abbott and his staff at the CHC were entrusted with the development of a two-dimensional model of stratified flow, System 22 (Abbott *et al.*, 1974). This was not the first time he had studied stratified flow, and in Abbott (2002) he writes:

> ...Since my first and enduring interest was in hydraulics, however, the first paper that I have selected for this collection concerns the spreading of one fluid over another, describing work that was carried out at the University of Southampton in 1960. Among other things, it represented one of the first, if not the first, application of matrix algebra in this field and was used to investigate a hitherto neglected kind of flow. The methodology followed was one of successive alternations of analytical and geometrical constructions, with no recourse being made to digital machine computation...

The paper here referred to concerns of oil spills on water (Abbott, 1961). This was followed soon after by work on density stratified flows (Abbott & Torbe, 1963). The detailed insights into the phenomena gained through this work were crucial to the development of System 22.

2.3.2.4 Wave modelling

Wave disturbance in harbours and coastal areas had long been a subject of study by coastal engineers, and always in physical models. With the new possibilities for computer computation a range of mathematical models of waves were developed in the 1970s. Apart from simple models of refraction and diffraction, spectral wave energy models and models of short waves in shallow water were added to the third generation. Mike Abbott was closely involved in the latter (Abbott *et al.*, 1978, 1984). The second paper was awarded the ASCE Karl Emil Hilgard Hydraulics Prize in 1986.

2.3.2.5 Applications and commercialization of the third-generation models

The first versions of the third-generation models were concerned with hydraulics only. In this form they found applications to pure flow problems such as flows and flooding in rivers, tidal flow patterns (of interest for example to navigation) and storm surge induced flooding of coastal areas. It quickly became obvious that this was a very limited range of application and there was a much greater range of hydraulic-related problems which could be addressed if the model's hydraulic module (HD) was supplemented with an advection dispersion (AD) module. This involved the solution of the mass conservation equation for a dissolved or suspended substance. (A fully reliable numerical solution with minimized instabilities at point sources took many years to develop, but that is another story.)

The AD module enabled application of the systems to:

- Power station cooling water recirculation studies. The objective of the studies was to determine locations of the cooling water intakes and outfalls to avoid recirculation.
- Sewage outfalls. At this early time, health, and not environment, was the only interest. Therefore, the studies concentrated on computation of *Escherichia coli* concentrations with a simple decay rate for the bacteria.

Sand transport computations using the HD results were also made at this time, but only by hand computation or computerized calculation of formulae.

These developments and applications sparked customer demand for more advanced computations of water quality, eutrophication, waves and sediment transport (ST), thus marking the start of the development of the fourth generation of the modelling systems.

During the late 1980s the so-called mini computers (e.g. IBM 4341 and VAX) became available at prices allowing smaller institutions and consulting engineers to obtain such machines. This sparked the commercialization of the third-generation models. However, the first sale of System 11 and System 21 was to the US Corps of Engineers in 1980 where they were installed on IBM main frames and the famous 'super computer', the Cray. Thereafter followed installation on mini computers at state institutions and consulting engineers around the world. The number was limited, and each installation involved a period of one to three months adaption of the code to the specific computer and training of the users.

2.3.3 Fourth-generation modelling

2.3.3.1 Driving forces for the development

The development of the fourth generation of numerical models occurred during the 1980s. It involved the expansion to a complex of modelling modules to cover the majority of hydraulic-related phenomena and occurred in response to:

- External customer demand, although most did not expect their problems to be investigated with numerical models.

- Internal demand at DHI and its associated Water Quality Institute (WQI). These institutes were already involved in studies of most aquatic phenomena and needed more comprehensive and efficient methods to study entire river systems and large coastal areas.

Another, unrelated development was the appearance of 'microcomputers', first in the form of UNIX work stations and soon after as Personal Computers. This revolutionized and greatly expanded the possibilities for application of the modelling systems and they quickly spread throughout the hydraulic and aquatic environment consulting world.

2.3.3.2 Scope of the development

The fourth generation required the development of user-friendly interfaces, extensive pre- and post-processing facilities and advanced graphical presentations both static and as movies. They also had to be coupled to for example GIS and existing data bases of field measurements and so on.

At DHI it was felt that the resulting complex of modelling modules had outgrown the general name of 'Systems'. In honour of Mike Abbott, the fourth-generation models with which he was connected were given the prefix 'MIKE'. Thus, the names became MIKE 11, MIKE 21 and so on. Figure 2.3 is a convenient illustration of the extent of coverage of hydraulic phenomena of a fourth-generation model (Warren & Bach, 1992).

The further development of all the modules of DHI's fourth-generation modelling systems extended well beyond the input of any single person. The ST module was based on the work of Engelund, Fredsoe and Diegaard of the Danish Technical University and developed by their co-workers during the 1980s (Diegaard *et al.*, 1986; Engelund & Fredsoe, 1976; Engelund & Hansen, 1972; Fredsoe, 1984). Similarly, the water quality (WQ), heavy metal (ME) and eutrophication (EU) modules were developed on the work of Dahl-Madsen (1978) and co-workers at WQI.

This pioneering work of DHI was the frontrunner to the development of fourth-generation modelling systems around the world. In the Netherlands, the pioneering work of Leendertse (1967) was improved by Stelling (1983) in the WAQUA code, which developments were further centralized by the Dutch Ministry of Public Works in order to significantly improve the pre- and postprocessing. Building upon the WAQUA developments in the mid-1980s, Delft Hydraulics started the development of Delft3D-FLOW for three-dimensional modelling of hydrodynamic flows (de Goede, 2020). The fourth-generation modelling systems have enabled the knowledge institutions like DHI (Denmark) and Deltares (the successor to Delft Hydraulics) to become global leaders through the widespread applications of the MIKE Suite and the Delft3D Suite, respectively.

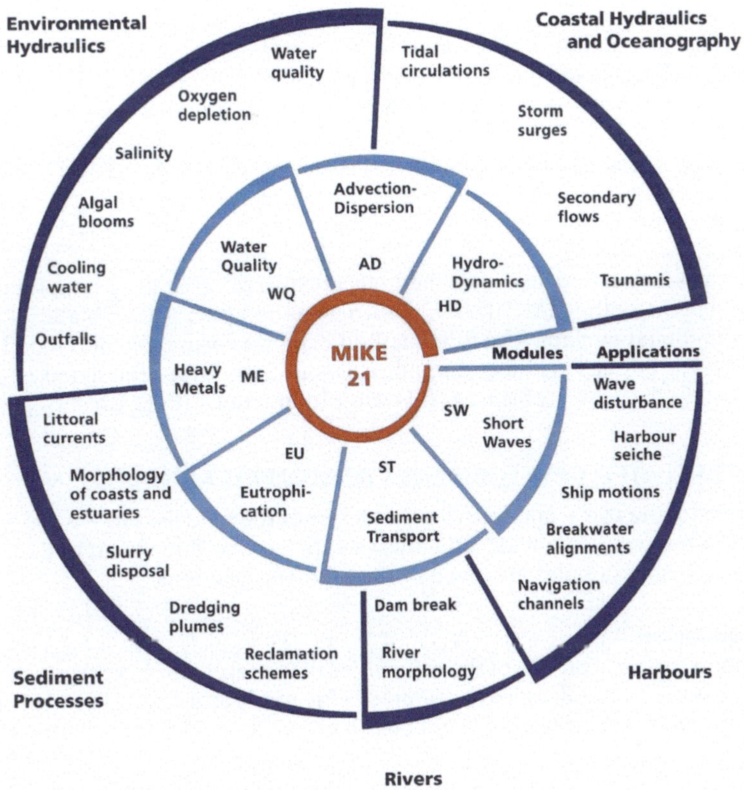

Figure 2.3 DHI's fourth-generation model, MIKE 21 (courtesy of DHI).

2.3.4 New developments in numerical methods

The ongoing developments of computational capacity alongside the changing requirements of users of the modelling systems lead to important new developments in the area of numerical methods.

2.3.4.1 Finite difference methods

Most third-generation, 2D models used finite difference solutions on regular, rectangular grids. Such grids had the disadvantage that they did not realistically fit land or solid boundaries. Furthermore, since fine grids were needed to resolve narrow channels with rapidly changing flows, they also had to be used in large, even depth areas where flows were quite uniform, and a larger grid size would have been sufficient. This led to the use of large area, coarse grid models providing boundary conditions for smaller area, fine grid models in the area of interest: a practical, but not very elegant solution. This also led to the development of curvilinear models which found wide application in rivers. Finite difference techniques dominated the third- and fourth-generation models until around 2000.

2.3.4.2 Unstructured grids

The use of rectangular grids, even in curvilinear models, can lead to poor resolution of land boundaries and narrow channels. A 'staircase' representation of coastlines is sometimes unavoidable in rectangular grids. Furthermore, when using curvilinear grids in meandering rivers, the gridlines may become focussed in the inner bends, leading to unnecessarily small grid cells. The need for a more accurate representation of such features led to the development of unstructured grid methods. Application of the finite-element technique, which was passed over to hydraulics from structural analysis, was developed further for water modelling in the late 1990s. The grid sizes vary depending on the complexity of the bathymetry and thereby provide the required resolution everywhere in the model. Some examples of finite-element modelling systems include ADCIRC (Dietrich *et al.*, 2011), TELEMAC (Hervouet, 2007), UnTRIM (Casulli & Zanolli, 2002), and ELCIRC (Baptista *et al.*, 2005; Zhang *et al.*, 2004).

In practice, many finite-element models are unable to conserve mass (Postma & Hervouet, 2007) and it has become necessary to use the finite volume numerical approaches (Ham *et al.*, 2005; Sleigh *et al.*, 1998), which have been subsequently implemented in MIKE 21 FM and MIKE 3 FM (Sørensen, 2004) and in Delft3D FM (Kernkamp *et al.*, 2011). The numerical stability requirements for these approaches often require quite small time steps, giving larger computational effort for longer simulations. Computational speed can therefore still be one of the limiting factors in modelling using unstructured grids.

2.3.4.3 Particle tracking methods

The finite difference, finite-element and finite volume approaches mentioned above are based on an Eulerian approach to modelling. The Eulerian approach treats the water as a continuum and develops its conservation equations on a control volume basis. Particle tracking, based on a Lagrangian approach, allows processes to be described in a detailed spatial pattern, whereby sub-grid concentration distributions may be resolved. The Lagrangian method considers discrete particles and tracks the pathway of each individual particle (Postma *et al.*, 2013). Particle tracking techniques are often combined in modelling studies using fourth-generation modelling systems for modelling the overall hydrodynamics and water quality processes, and are especially useful when applied to problems of fish larvae, oil spills, and search and rescue operations (Dagestad *et al.*, 2018).

In the field of solid mechanics simulation, meshless methods are also referred to as smoothed particle hydrodynamics (SPH). Similarly, the material point method (MPM) is a meshfree method that utilizes material points and a background mesh to discretize the computational domain. The SPH method itself was originally developed to simulate astrophysics of gas dynamics in the late 1970s but has also found applications in fluid flows (Raymond *et al.*, 2016). More recently, MPM has been applied successfully to the problem of soil–water interactions (Ceccato *et al.*, 2018).

2.4 OTHER MIKE ABBOTT-RELATED DEVELOPMENTS IN NUMERICAL MODELLING

At the time of the development of the third- and fourth generations of the models, Mike Abbott did not restrict his work to nearly horizontal flows. His interests were wide, and always with a social dimension in that they were problems with which our society has to deal. Some examples are given in the following sections.

2.4.1 A dynamic population model

In the early 1970s, Mike Abbott focussed on one of the difficulties of management in agriculture, namely the dependence between an herbivore and its source of food, an easy example being cattle and grass. The abstract of the resulting paper (Abbott & Warren, 1974) is quoted below because it is not possible to write a better description of their brilliant approach to the problem:

> Populations are considered that vary in distance and time and some components of which are capable of motion. The levels and motions of these populations are then described by systems of partial differential equations. The

resulting problems of formulating systems of (energy ed.) conservation laws and behavioural and constitutive laws are discussed, the discontinuous case (of a pest front, or epidemic front) being also considered.

Equations are developed for a system of one herbivore and one vegetation, a system of two competing herbivores and one vegetation, both in fixed habitats, and also for a system of one herbivore and one vegetation in a moving habitat. The local stabilising effects of certain behavioural traits are demonstrated by examples: both a tendency to remain over an existing adequate food supply, so not to be tempted by surrounding opportunities, and a tendency to dispersion improved local stability. The manner in which a more efficient or 'economical' herbivore displaces a less efficient herbivore is also simulated. Finally, a steady state population in a moving stream is described.

2.4.2 European Hydrologic System: Système Hydrologique Européen (SHE)

While employed at DHI, Mike Abbott became convinced that it was possible to develop a model of the entire hydrological cycle, from precipitation impacting on a vegetation canopy to runoff to lakes or seas from streams and groundwater. In 1975–1976 he carried out a market and product research visiting European universities, institutes and consultants. The majority of professional hydrologists agreed that 'it will never work, there will never be enough data'. But Mike Abbott persevered and brought together a group of DHI, Institute of Hydrology (England) and SOGRÉAH (France) who were prepared to take up the challenge. An application to the Commission of European Communities for financial support was prepared. The commission was sufficiently far-sighted to accept the proposal.

The subsequent successful development of the modelling system SHE (see Figure 4.1 in Chapter 4 of this book) is yet another example of Mike Abbott's perseverance and his vision of the possible (Abbott *et al.*, 1979). For more information about the motivation, development and practical applications of SHE (later MIKE SHE), the reader is referred to Chapter 4 of this book.

2.4.3 Laying of marine pipelines

Commercial reserves of oil and gas were discovered in the Danish Sector of the North Sea in 1967. Marine pipelines were required to bring the oil and gas ashore in Denmark. With his background in both structural analysis and hydraulics, Mike Abbott saw the need for a model of the movement and stresses in the pipelines as they were being laid on the seabed (Figure 2.4).

Mike Abbott developed the model within his own company, SOFT (International). He developed the basic equations and under his supervision, these were transformed to fit the physical environment of the pipe positioning domain as part of the development of the modelling system. DHI acquired the system when, in the mid-1970s, it was given responsibility for the engineering of the marine pipelines from the Danish platforms to the coast. Quoting from Danish Hydraulics No. 1 April 1981:

> The resulting modelling system provides simulations of three-dimensional motion and stress characteristics of a sea bed/pipe/lay barge system under any required conditions of bed topography, current field and wave field.
>
> In more technical terms, the system uses a Lagrangian representation of the stresses and motions in a curvilinear coordinate system, expressed as a system of eleven partial-differential equations. Most of the algorithmic and other topographical features of the system have been taken from modern practice in hydraulics. In particular, the solution procedures are modified automatically by the system as the boundary conditions change during certain common operations. The code is subsequently to be used to check-out a number of different laying options, such as flexible stingers, fixed-curved stingers, hinged stingers and tension control devices. The 'thrashing' of the pipe on a bed of any given form and material properties can also be simulated. Non-linear pipe-material properties can also be specified.

The development of the code faced many intriguing challenges (Abbott *et al.*, 1976, 1977). In the first place, there appeared to be a need to transform the equations based upon the pipe angle as dependent variable into a form which recognized the spatial position of the pipe, in particular using the vertical position z as dependent variable. Yet this was not enough. Two particular states presented rapid transitions which needed particular algorithms to keep the motion of the pipe stable.

The first position was the one at the water surface. Within the vertical range of one pipe diameter, buoyancy changed extremely fast. Without capturing the position of the pipe while it passed the surface, it would either shoot up rocket high into the air or be slammed down to even below the sea bed, when it fell back through the water surface. The implicit numerical

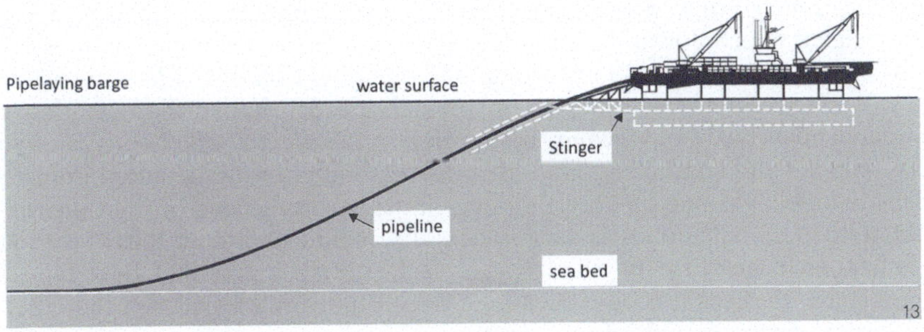

Figure 2.4 Pipelaying barge (courtesy of DHI).

scheme applied did not keep this unstable state under control. A special condition had to be developed to keep the pipe at the water surface, once the solution of the equations would pass it through due to the buoyancy force.

Similar gravity issues were experienced at the sea bed. Touching the bottom while the pipe was sinking, the reaction force described at the sea bed would bounce the pipe back during the next time step in the model, causing it often to shoot through the water surface again. Resolving these problems was achieved through headaches, late night work and stimulating discussions with all involved.

A special version of the code was developed to model the transport of oil and gas pipe lines at sea. The loss of a 2 km long pipe while transporting this at sea led to a special contract with Royal Dutch Shell to study the impacts of currents and waves on the pipe and to define limiting conditions for the safe transport from port to positioning location. This transport took place by fixing it between two tug boats and special boundary conditions had to be developed to complement the equations generated to define conditions along the pipe.

Initially the boundary conditions were formulated as part of the double sweep algorithm to solve the system of equations of this implicit scheme at each time step. It was only much later that it was realized that these boundary conditions for all pipe laying and transporting operations could be defined in a generic way in a boundary matrix, which could be incorporated in the matrix defining the complete system of equations to be solved with a generic minimum search algorithm.

2.5 FROM COMPUTATIONAL HYDRAULICS TO HYDROINFORMATICS

2.5.1 The changing nature of modelling systems

The move towards the fourth-generation modelling systems constituted a clear step in the direction of the 'social dimension' of hydraulics, a first and necessary step towards hydroinformatics.

It soon became clear that what had until that time been seen simply as a model could in fact serve as a device for encapsulating all the hydraulic knowledge and all the application-specific information about any area that was modelled using this knowledge. With this knowledge and information so integrated in one digital electronic form, it became possible to integrate it further with all manner of other devices that similarly encapsulated knowledge and information in a digital-electronic form, such as measuring devices and their transmission and buffer memory devices. Similarly, through the widespread introduction of knowledge-based-systems, it had already become feasible to express the legal, contractual and other kinds of knowledge and information that constrain the construction and operation of works and schemes in what was, in the last analysis, a digital-electronic form. The challenge thus arose of integrating as much as possible the available knowledge and information that would influence decisions (of design, operation, management or whatever) so as to arrive at optimal, or nearly optimal solutions to water resources, and particularly environmental problems.

Up to this point the users of modelling systems have been able to navigate through their professional world by building models of this world. However, the increasing complexity of the problems to be solved and the modelling systems to be applied has led to an increasing need to navigate through a world of models. The problem of navigation in this sense then becomes one of the most pressing problems that arise when applying computational hydraulics within hydroinformatics. In Abbott (1993), reference is made to the 'knowledge revolution' in the water industries. It was postulated that the knowledge industries in the water business would divide into tool-makers and tool-users, which in turn would modify the entire operations of the knowledge institutes themselves. Furthermore, the tool-users (e.g. engineering consultants) would become ever more independent and self-sufficient and able to offer much more sophisticated add-on services to clients. Using the newest tools, it becomes possible to produce results in some hours that would previously have taken several months. New markets appear in the customizing of tools and in tool support, and for a whole range of spin-off activities and special-interest studies. This was elaborated further in Abbott (1994):

> ...we could previously navigate our way, whether as individuals or as a profession, simply by building our models of this world; from now on, however, we must increasingly navigate our way through a world of models. And of course, navigare necesse est....

Abbott and Minns (1998) remark that in hydroinformatics, as in computational hydraulics also, we suppose that every flow of water is associated with a flow of information. For example, in one-dimensional systems, information will flow in two directions when the water flow is subcritical and in one direction only when the water flow is supercritical, while information flow will effectively stop when a pipe or channel runs dry. In hydroinformatics systems, we commonly have two more or less independent ways of viewing the information flows, the one being through actual observations of the system (e.g. measurements and monitoring) and the other through our own domain knowledge. So long as these information flows appear to agree, we may suppose that our model has succeeded in reproducing the properties of the physical domain. But what happens when there is a significant discrepancy between these information flows? The fault may lie on the side of our knowledge of the physical system itself, in that a pipe may have blocked or a pump broken down or a valve becomes stuck. Perhaps it is only because a sensor may have failed. On the other hand, it may be that our knowledge of the physical system is indeed incorrect and that the underlying equations are invalid here. Abbott *et al.* (2001) argue that the way in which hydraulics is viewed and practised is itself now changing as a result of its incorporation into the new paradigm that hydroinformatics provides. A fifth generation of modelling systems is postulated as a precursor to more fully distributed, hydroinformatics systems (Figure 2.5). It is a first rule of hydroinformatics, that no change in application practice is possible without the simultaneous provision of new technical means to support this practice, and it is the role of hydroinformatics to develop and provide these new techniques to the modelling community. Some of these new techniques are explored more in the following chapters.

The move from the fourth-generation modelling systems to the fifth generation and finally towards full hydroinformatics services brings with it a 'paradigm shift' in the way that this technology is applied by the user to solve their 'technical' problems. As discussed already in Chapter 1, there are no longer just technical problems, but all problems are in fact sociotechnical problems. However, the social application of hydraulic and related knowledge cannot simply be realized without the introduction of new technical means. Insofar as non-expert members of the general public wish to make use of such knowledge in order to assess the impact of a particular development upon their quality of life, so they need new classes of tools to assist them. They need tools that will enable them to relate their own personal and group interests, concerns and intentions to any proposed intervention in their environment if they are to participate at all effectively in the stakeholder interactions that govern such interventions. Hydroinformatics tools and services provide new opportunities to electronically encapsulate not only knowledge about the physical system, but also all manner of social, legal, personal and anecdotal information. A hydroinformatics system facilitates a synergy between the individual practitioner and his or her store of electronically encapsulated knowledge. In this relation, it is the encapsulated knowledge that adds the most immediate benefit, as it allows the practitioner to mobilize unprecedented knowledge resources as and when these are required at costs that are only a tiny part of what would be required were this knowledge to be made available in an unencapsulated form. Hydroinformatics also provides new tools and services for data and knowledge acquisition that proceed with great rapidity and often without direct human intervention. The practitioner now becomes less a personal carrier of all manner of detailed knowledge and information and more one who is adept at organizing and integrating electronically encapsulated knowledge and information. Such was the impact of this paradigm shift upon the traditional hydraulics research community that, in 1994, the *IAHR Journal of Hydraulic Research* devoted an entire special edition to just this challenge (see Abbott, 1994 and all of the other contributions in *Journal of Hydraulic Research*, Vol. 32, 1994, Extra Issue). Many of the authors of the contributions to this special edition are also co-authors of several chapters of this current book.

2.5.2 Challenges to model users

To date, the full promise of a fifth generation of modelling systems (as described in Figure 2.5) is yet to be completely realized. In the description of fourth-generation models, Abbott and Minns (1998) write:

> ...At the same time, a considerable effort had also to be put into determining default values of parameters that allow these systems to continue functioning in inexperienced hands while still providing results of adequate accuracy...

Not surprisingly, the expectation that the models could be successfully applied by 'inexperienced hands' has not proved to be universally possible. The authors of this chapter have often been asked to review model applications by other individuals or

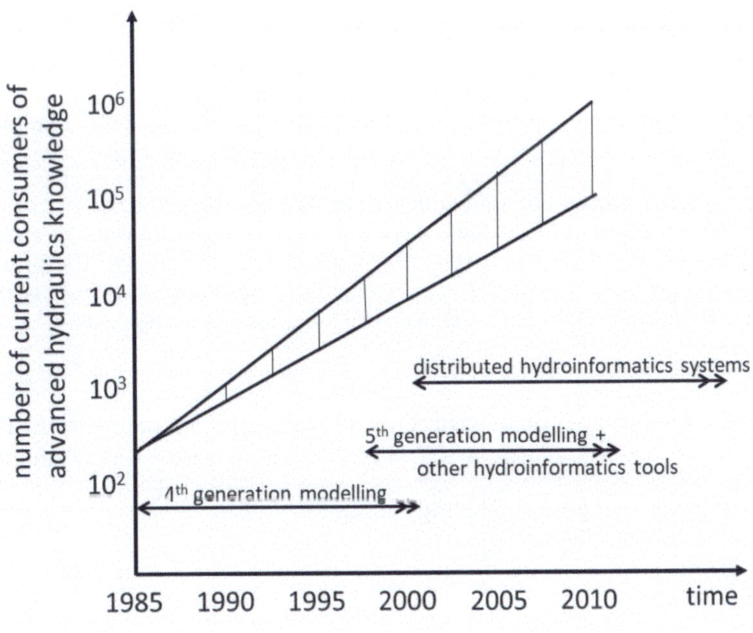

Figure 2.5 Historical and predicted growths in the numbers of consumers of high-level knowledge in hydraulics, hydrology and water resources (from Abbott *et al.*, 2001).

groups. Too often, they have discovered serious errors in the applications that have been carried out by 'inexperienced hands'. The errors can be due to:

- Poor and incorrect input data. For example, the users do not have the specialist training to enable them to scrutinize and analyse boundary data even for obvious errors. It could also be a poor model bathymetry or inappropriate time steps. Some eutrophication models require the user to choose a set of equations to be solved depending on the specific biological/chemical variables of the case in question. Only a specialist in aquatic biology can make such a choice. The situation is similar with ST modelling where knowledge of the sediment properties, wave and current conditions are required for correct simulations.
- Blind acceptance of default parameter values. It requires specialized knowledge and experience to assign the correct parameter values for a specific application.

With such errors in the applications, the resulting model output will be incorrect. Warren has on numerous occasions ironically asked the user, 'Do you really believe that water can do that?' The 'inexperienced user' is often not capable of seeing that the model results are at least incorrect, sometime quite impossible.

The need to use simple common sense in urban drainage modelling has been described by Verwey (2009). Furthermore, Verwey and Doan (2010) undertook a test to investigate peak runoff values produced by a variety of urban rainfall–runoff model concepts available in a range of frequently used modelling systems by using their built-in default values for the simulation. Knowledge of peak discharge values is essential in the design of hydraulic structures, as this parameter determines their dimensions and associated construction costs. A correct prediction of such values is most important in urbanized areas. General experience is that in this environment, discharge data are rarely available for model calibration.

The test compared results produced by simulations with MIKE Urban, SOBEK, InfoWorks and SWMM producing runoff from a 1 ha terrain, applying a standard design storm of 2 h duration and a total rain depth of 199 mm, representing a 1-in-100-year storm event in Hong Kong. Although the authors expected differences in resulting peak values, they were even amazed to see a factor of 2 between the highest and lowest peak values produced. Used as a basis for the design of a hydraulic structure, the outcome would also imply a difference of a factor of 2 in the investment cost. Most likely one of the models would lead to an overdesigned and another one to an under designed structure. Either representing a partly wasted budget or a risk of failure of the structure at a higher frequency event than used in the design. Given the fact that many cities in the world are facing the need for huge investments to reduce their flood hazards and risks, both cases are highly undesirable.

This places a burden on the shoulders of modellers as part of design teams. It is their responsibility to get the order of magnitude of the peak flows right. Installing discharge measuring stations is often no option, as the design usually has to be completed within a single year. The costs associated with water level and discharge monitoring is in most cases prohibitive so that measured rating curves are unavailable. For this reason, prescribing the use of several rainfall–runoff modelling concepts in the Terms of Reference for the associated study is a creative way to reduce uncertainties about peak flows to be used for the designs. The principal lesson learned in this study has been the conclusion that default parameters incorporated in the modelling system are not replacing the need to bring in well-trained and experienced modelling staff and the need for thorough reviews of the outcome of model applications.

The basis for good modelling practices is provided by good education, a need always emphasized by Mike Abbott, promoted in his role within the IAHR and passed on to his staff. Verwey (2009) states:

> …Above examples demonstrate the need for good education in the development and use of mathematical models. This issue has been a primary concern of the former IAHR Section on Computational Hydraulics (chaired by Mike Abbott for many years), raised at various meetings as far back as the seventies of the previous century. It even becomes more of a concern nowadays, as modelling systems have developed into robust tools that easily mask errors in the model schematization and do not crash as easily anymore as in the past…

And further down:

> …Models of urban drainage systems are just an abstraction of reality and not reality itself. The art of modelling is to arrive at a model which represents reality as closely as possible. Such achievement is limited by various factors, such as:
>
> - limitations of the storm water routing modelling concepts;
> - level of detailing of the model schematisation;
> - quality of the underlying data;
> - skill of the modelling team…

The role of good training and education for the ongoing successful application and development of computational hydraulics and hydroinformatics is addressed in Chapter 7 of this book.

2.5.3 The way ahead

As we move from 'traditional' computational hydraulics towards distributed hydroinformatics services we often state that we are now living in the 'hydroinformatics era'. However, Abbott *et al.* (2001) point out that hydraulics in the hydroinformatics era

> ...cannot be regarded merely as a new way of applying an otherwise unchanged hydraulics within society. Instead, hydroinformatics, while building upon existing hydraulics and taking existing hydraulics for the most part as its central core, introduces its own paradigm into hydraulics itself.
>
> ...Hydroinformatics takes up many activities of hydraulics research and practice that were hitherto considered as rather separated and brings them together in new and innovative ways using the new information and communication technologies. By these means it increases the value of the individual activities, often markedly. Thus, for example, the bringing together and integration of field measurements, numerical models and remotely sensed data provide new and highly valuable new facilities that increase greatly the value of the component activities....

This chapter has explored the early developments of computational hydraulics and described how these developments were indeed central to the initial developments of hydroinformatics. However, sociotechnical developments promise ultimately to lead to the most profound changes in the practice of hydraulics itself. The following chapters in this book describe how hydroinformatics has made it possible for traditional methods and sciences to evolve (e.g. Chapter 4) through the use of new communication technologies and machine-learning techniques (e.g. Chapter 3). This in turn is opening up new opportunities in business (e.g. Chapters 5 and 6) and in education and training (Chapter 7). This can be summarized quite succinctly once more by Mike Abbott himself (Abbott *et al.*, 2001):

> ...From being one who personally 'thinks about water', the hydraulician is drawn (or even dragged!) by degrees to become one who has also to 'think about how other persons think about water.

REFERENCES

Abbott M. B. (1961). On the spreading of one fluid over another. La Houille Blanche, **16**(5/6), pp. 622–635/827–846. - October 1961 and No. 6 - December 1961.

Abbott M. B. (1966). An Introduction to the Method of Characteristics. Thames & Hudson, London, and American Elsevier, New York.

Abbott M. B. (1979). Computational Hydraulics. Elements of the Theory of Free Surface Flows. Pitman Publishing Ltd, London, UK.

Abbott M. B. (1993). The impact of the knowledge revolution in the water industries. Proc. Watercomp 1993. Aust. Inst of Engineers, Melbourne, pp. 123–131.

Abbott M. B. (1994). Hydroinformatics: a Copernican revolution in hydraulics. *Journal of Hydraulic Research*, **32**(S1), 3–13. https://doi.org/10.1080/00221689409498800

Abbott M. B. (2002). Personal Reflections on Computational Hydraulics and Hydroinformatics. On Behalf of Honoris Causa Doctorate of Professor Michael Barry Abbott by the Civil Engineering Department of the Aristotle University of Thessaloniki, Thursday, 10 October 2002.

Abbott M. B. and Ionescu F. (1967). On the numerical computation of nearly-horizontal flows. *Journal of Hydraulic Research*, **5**(2), 97–117. https://doi.org/10.1080/00221686709500195

Abbott M. B. and Minns A. W. (1998). Computational Hydraulics. 2nd edn. Ashgate Publishing Company, UK.

Abbott M. B. and Torbe I. (1963). On flows and fronts in a stratified fluid. *Proceedings of the Royal Society A*, **273**, 12–40.

Abbott M. B. and Verwey A. (1970). Four-point method of characteristics. *Proceedings of the American Society of Civil Engineers*, **96**(HY 12), paper 7763, pp. 2549–2564.

Abbott M. B. and Verwey A. (1973). Some aspects of the design system SIVA. Proceedings of the 15th Congress of the IAHR, Istanbul, vol. **5**, paper 4.3, pp. 127–133.

Abbott M. B. and Warren I. R. (1974). A dynamic population model. *Journal of Environmental Management*, **2**, 281–297.

Abbott M. B., Rodenhuis G. S. and Verwey A. (1971). Further development of the implicit difference scheme for flood wave calculation. Proceedings of the 14th Congress of the IAHR, Paris, vol. **5**, paper 310, pp. 183–186.

Abbott M. B., Damsgaard A. and Rodenhuis G. S. (1973). System 21, JUPITER, a design system for two-dimensional nearly horizontal flows. *Journal of Hydraulics Research*, **11**(1), 1. https://doi.org/10.1080/00221687309499788

Abbott M. B., Rodenhuis G. S. and Verwey A. (1973a). Mathematical modelling of wave propagation using design systems. Proceedings of the 15th Congress of the IAHR, Istanbul, vol. **5**, paper 4.2, pp. 123–126.

Abbott M. B., Rodenhuis G. S. and Verwey A. (1973b). System 11, SIVA, A design system for rivers and estuaries. Proceedings of the International IAHR Symposium on River Mechanics, Bangkok, Thailand, vol. **III**, paper C-18, pp. 207–215.

Abbott M. B., Bertelsen J. A. and Warren I. R. (1974). The System 22/23 'NEPTUNE': application to the mathematical modelling of the Danish Belts. ICES Special Meeting on Water Circulation in the Baltic, Paper No. 7.

Abbott M. B., Larsen I. and Verwey A. (1976). Computer forecast of laying procedure. Proceedings of the IFAC/IFIP Symposium on Automation in Offshore Oil Field Operation, Bergen, Norway.

Abbott M. B., Larsen I. and Verwey A. (1977). Computer control of pipe laying procedure. Proceedings of the 17th Congress of the IAHR, Baden-Baden, Germany.

Abbott M. B., Petersen H. M. and Skovgaard O. (1978). Numerical modelling of short waves in shallow water. *Journal of Hydraulic Research*, **16**, 173–204. https://doi.org/10.1080/00221687809499616

Abbott M. B., O'Connell P. E. and Preissmann A. (1979). The European hydrologic system – an advanced hydrologic mathematical modelling system. Proceedings of the XVIII IAHR Congress, Cagliari, Italy.

Abbott M. B., McCowan A. and Warren I. R. (1981). Numerical modelling of free surface flows that are two-dimensional in plan. In: Transport Models for Inland and Coastal Waters, H. B. Fisher (ed.), Academic Press, New York, USA, pp. 222–283.

Abbott M. B., McCowan A. D. and Warren I. R. (1984). Accuracy of short wave numerical models. *Journal of Hydraulic Engineering*, **110**(10), pp. 1287–1301. October 1984 ASCE. (Awarded the ASCE Karl Emil Hilgard Hydraulics Prize, 1986).

Abbott M. B., Babovic V. and Cunge J. (2001). Towards the hydraulics of the hydroinformatics era. *Journal of Hydraulic Research*, **39**(4), 339–349. https://doi.org/10.1080/00221680109499839

Baptista A. M., Zhang Y.-L., Chawla A., Zulauf M. A., Seaton C., Myers E. P., Kindle J., Wilkin M., Burla M. and Turner P. J. (2005). A cross-scale model for 3D baroclinic circulation in estuary-plume-shelf systems: II. Application to the Columbia river. *Continental Shelf Research*, **25**, 935–972. https://doi.org/10.1016/j.csr.2004.12.003

Casulli V. and Zanolli P. (2002). Semi-implicit numerical modeling of nonhydrostatic free-surface flows for environmental problems. *Mathematical and Computer Modelling*, **36**, 1131–1149. https://doi.org/10.1016/S0895-7177(02)00264-9

Ceccato F., Yerro A. and Martinelli M. (2018). Modelling soil-water interaction with the Material Point Method: evaluation of single-point and double-point formulations, Numerical Methods in Geotechnical Engineering IX: Proceedings of the 9th European Conference on Numerical Methods in Geotechnical Engineering – NUMGE 2018, Porto, Portugal, pp. 351–358.

Dagestad K.-F., Röhrs J., Breivik Ø and Ådlandsvik B. (2018). OpenDrift v1.0: a generic framework for trajectory modelling. *Geoscientific Model Development*, **11**, 1405–1420. https://doi.org/10.5194/gmd-11-1405-2018

Dahl-Madsen K. I. (1978). Mathematical modelling of eutrophied coastal areas. *Progress in Water Technology*, **10**(5/6), 217–236.

de Goede E. D. (2020). Historical overview of 2D and 3D hydrodynamic modelling of shallow water flows in the Netherlands. *Ocean Dynamics*, **70**, 521–539. https://doi.org/10.1007/s10236-019-01336-5

Diegaard R., Fredsoe J. and Broker Hedegaard I. (1986). Suspended sediment in the surf zone. *Journal of Waterway, Port, Coastal and Ocean Engineering Div.*, ASCE, **112**(1), 115–128. https://doi.org/10.1061/(ASCE)0733-950X(1986)112:1(115)

Dietrich J. C., Zijlema M., Westerink J. J., Holthuijsen L. T., Dawson C., Luettich R. A., Jensen R. E., Smith J. M., Stelling G. S. and Stone G. W. (2011). Modelling hurricane waves and storm surge using integrally-coupled, scalable computations. *Coastal Engineering*, **58**, 45–65. https://doi.org/10.1016/j.coastaleng.2010.08.001

Dronkers J. J. (1964). Tidal Computations in Rivers and Coastal Waters. North Holland Publ. Cy, Amsterdam, 518 pp.

Engelund E. and Fredsoe J. (1976). A sediment transport model for straight alluvial channels. *Nordic Hydrology*, **7**, 293–306. https://doi.org/10.2166/nh.1976.0019

Engelund F. and Hansen E. (1972). A Monograph on Sediment Transport in Alluvial Streams. Teknisk Forlag, Copenhagen.

Fredsoe J. (1984). Turbulent boundary layer in wave and current motion. *Journal of Hydraulic Engineering*, ASCE, **110**(8), 1103–1120. https://doi.org/10.1061/(ASCE)0733-9429(1984)110:8(1103)

Ham D. A., Pietrzak J. and Stelling G. S. (2005). A scalable unstructured grid 3-dimensional finite volume model for the shallow water equations. *Ocean Modelling*, **10**(1–2), 153–169. https://doi.org/10.1016/j.ocemod.2004.08.004

Hervouet J. M. (2007). Hydrodynamics of Free Surface Flows: Modelling with the Finite Element Method. John Wiley & Sons, Ltd, Chichester, UK.

Kernkamp H. W. J., van Dam A., Stelling G. S. and de Goede E. D. (2011). Efficient scheme for the shallow water equations on unstructured grids with application to the Continental Shelf. *Ocean Dynamics*, **61**, 1175–1188. https://doi.org/10.1007/s10236-011-0423-6

Leendertse J. J. (1967). Aspects of a Computational Model for Long-Period Water-Wave Propagation. PhD thesis, RM-5294-RR, Rand Corporation, Santa Monica.

Postma L. and Hervouet J.-M. (2007). Compatibility between finite volumes and finite elements using solutions of shallow water equations for substance transport. *International Journal for Numerical Methods in Fluids*, **53**, 1495–1507. https://doi.org/10.1002/fld.1373

Postma L., van Beek J. K. L., van den Boogaard H. F. P. and Stelling G. S. (2013). Consistent and efficient particle tracking on curvilinear grids for environmental problems. *Numerical Methods in Fluids*, **71**(10), 1226–1237. https://doi.org/10.1002/fld.3705

Raymond S. J., Jones B. and Williams J. R. (2016). A strategy to couple the material point method (MPM) and smoothed particle hydrodynamics (SPH) computational techniques. *Computational Particle Mechanics*, **5**(1), 49–58. https://doi.org/10.1007/s40571-016-0149-9

Sleigh P. A., Gaskell P. H., Bersins M. and Wright N. G. (1998). An unstructured finite-volume algorithm for predicting flow in rivers and estuaries. *Computers & Fluids*, **27**(4), 479–508. https://doi.org/10.1016/S0045-7930(97)00071-6

Sørensen O. (2004). *Floodplain modelling using unstructured finite volume technique.* DHI Technical Note, January 2004.

Stelling G. S. (1983). On the Construction of Computational Methods for Shallow Water Problems. PhD thesis, Delft University of Technology, Delft, The Netherlands.

Stelling G. S. and Verwey A. (2005). Numerical flood simulation. In: Encyclopaedia of Hydrological Sciences, G. M. Anderson and J. J. McDonnell (eds), John Wiley & Sons Ltd., UK, 257–270.

Verwey A. (1975). On external and internal boundary conditions and their influence on net flows and mean water levels in systems of channels. Proceedings of the 16th Congress of the IAHR, Sao Paulo.

Verwey A. (2009). The use of common sense in urban drainage modelling. Proceedings of the 33rd IAHR Conference, 9–14 August 2009, Vancouver, Canada.

Verwey A. and Doan C. D. (2010). Sobre Parâmetros Chuva-Vazâo em Modelos de Cheias Urbanas – On rainfall-runoff parameters in urban flooding. 14th Latin American IAHR Conference, 21–25 November 2010, Punta del Este, Uruguay.

Warren I. R. and Bach H. K. (1992). MIKE 21: A modelling system for estuaries, coastal waters and seas. *Environmental Software*, **7**, 229–240. https://doi.org/10.1016/0266-9838(92)90006-P

Zhang Y.-L., Baptista A. M. and Myers E. P. (2004). A cross-scale model for 3D baroclinic circulation in estuary-plume-shelf systems: I. Formulation and skill assessment. *Continental Shelf Research*, **24**, 2187–2214. https://doi.org/10.1016/j.csr.2004.07.021

doi: 10.2166/9781789062656_0033

Chapter 3

Hydroinformatics opening new horizons: union of computational hydraulics and artificial intelligence

V. Babovic[1] and A. W. Minns[2]

[1]National University of Singapore, Department of Civil and Environmental Engineering, Singapore
[2]Deltares, Marine & Coastal Systems, Delft, the Netherlands

ABSTRACT

The strengthening dominance of electronics over past several decades has resulted in an almost total dominance of digital representation of the hydro-environment. Recognizing these developments, Mike Abbott introduced the notion of electronic encapsulation of information and knowledge. The act of encapsulating information and knowledge changes the very nature of the information and knowledge involved. Suffice to say that electronic encapsulation must also change the way in which an engineer accesses and uses the available information and knowledge. There is a shift in paradigm away from a model-based approach to a more data-based approach. Adopting such foundational considerations as its core, hydroinformatics opened research to the latest IT developments in the fields of artificial intelligence (including machine learning, evolutionary algorithms and artificial neural networks), artificial life, cellular or finite-state automata and other, previously unrelated sciences and technologies. Through studying and exploiting elements of these, at first sight unrelated, sciences, hydroinformatics produced new and innovative solutions to hydraulic and hydrological problems, as represented by real-time control and diagnosis, real-time forecasting, calibration of numerical models, data analysis and parameter estimation. In particular, the new approaches are able to generate important components of physically based, modelling systems by inducing models or sub-models of individual physical processes based only upon measured data. These (sub)models may then replace whole systems of complex, non-linear, differential equations that would otherwise require great skills from the modeller to calibrate, and powerful computing devices to solve. This chapter captures evolution of data science and AI within the field of hydroinformatics and provides an outline of the present state-of-affairs together with some ideas for future directions. The chapter outlines tangible solutions that have been applied by the hydroinformatics community to address specific challenges in hydro-environmental systems.

Keywords: Hydroinformatics, data-driven modelling, deterministic modelling, artificial intelligence, machine learning, theory-guided data science, computer vision

3.1 INTRODUCTION

Traditionally, the fields of hydraulics, hydrology and hydro-environment in general are concerned with natural phenomena that are described through deterministic equations derived from our best understanding of conservation laws and the underlying physics, chemistry and biology.

Artificial intelligence (AI) is an area of computer science that focuses on creating intelligent machines that can perceive their environment and make decisions to optimise towards a given goal. Two disciplines worth highlighting in the context of AI are machine learning (ML) and computer vision. ML aims to provide computers with the ability to learn iteratively, improve predictive models and find insights from data without being explicitly programmed. Computer vision relies on a set of AI methods to train computers to interpret digital images and videos.

Initially, there was only a limited interest from scientists and from communities of practice in the hydraulics and hydrology circles. In the early 1990s, there was even a degree of scepticism about AI. At that time, AI was going through its own formative phase (Goldberg, 1989; Koza, 1992; Rumelhart & McClelland, 1986).

With its origins in computational hydraulics, hydroinformatics was defined by Mike Abbott as the study of the flows of knowledge and data related to the flow of water and all that it transports, together with interactions with both natural and man-made, or artificial, environments (Abbott, 1991). As described in Chapter 2, the development of hydroinformatics was seen as the next logical progression beyond the fourth-generation modelling systems, which had started to enter regular service in the mid-1980s. Initially, the field of hydroinformatics was broadly defined as the application of information and communication technologies to solve water-related problems. With time, hydroinformatics came to recognise the inherently social nature of the problems of water management and of decision-making processes, and to understand the social processes by which technologies are brought into use. Hydroinformatics provided a logical early platform for (the then younger) researchers to introduce AI topics in the context of hydraulics and hydrology.

Mike Abbott's vision and intellectual leadership in the mid-1990s was ahead of its time. The strengthening dominance of electronics from the early 1990s resulted in an almost total dominance of digital representation. Not only hydraulics, but also other fields were quite naturally affected by these developments. Recognising these developments, Abbott (1993) wrote of an *electronic encapsulation of information and knowledge.* The act of encapsulating information and knowledge was changing the very nature of the information and knowledge involved. As such, he outlined a vision for hydroinformatics that drew upon the integration of hydraulics, hydrology, environmental engineering, and many other disciplines. According to Mike Abbott's vision, hydroinformatics played an important application at all points in the water cycle from atmosphere to ocean, and in artificial interventions in that cycle such as urban drainage and water supply systems. Suffice to say, the electronic encapsulation then also changed the way in which an engineer accesses and uses the available information and knowledge. Abbott (1994) describes the many ways in which knowledge can be encapsulated electronically in hydraulics and its related fields. One of the earliest forms in historical terms, which is indeed relevant to hydroinformatics, is that of measuring and remote sensing instruments. The information gathered by these sensors is encapsulated in the form of databases that can then be interrogated by the user. Another form of encapsulation of knowledge is the encapsulation of our beliefs concerning the physics, chemistry, biology, economics and other forms of scientific knowledge in our numerical models.

Vojinovic and Abbott (2017) stressed that the primary role of hydroinformatics nowadays is in the development and installation of sociotechnical arrangements that can truly enable the right balance between quantities and qualities and apply them meaningfully in our research and practice. This was exemplified in Figure 3.1.

Today, we witness confirmation of Mike Abbott's leadership and the contributions of his Hydroinformatics Group – particularly in relation to AI. On the heel of successes of AI over the last two decades, there has been a growing interest and accelerating adoption by the water community of state-of-the-art AI techniques that originated in computer science. In a recent editorial of Nature Reviews | Physics (2021), it is pointed out that the use of machine learning is not at all new to physicists. The use of machine learning during the 2011–2012 analysis of the Large Hadron Collider data underlying the discovery of the Higgs boson, led to an increase in sensitivity equivalent to collecting 50% more data. Furthermore, the authors of the editorial indicate that the number of physics papers that refer to the use of machine learning, which have been submitted to the American Physical Society, has grown from less than 0.5% in 2016 up to almost 3.5% in 2021.

Hydroinformatics has also utilised research on the latest IT developments in the fields of artificial intelligence (including machine learning, evolutionary algorithms and artificial neural networks), artificial life, cellular or finite-state automata and others, previously unrelated sciences and technologies. Through studying and exploiting elements of these, at first sight unrelated, sciences, hydroinformatics has produced new and innovative solutions to hydraulic and hydrological problems, examples of which include real-time control and diagnosis, real-time forecasting, calibration of numerical models, data analysis and parameter estimation. At the International Workshop on Advances in Hydroinformatics, which took place in June 2007 in Niagara Falls, Canada, about 50 invited speakers and keynote speakers met for four days to assess the state of the art in hydroinformatics, and to discuss new directions for future research in hydroinformatics. This led to a special issue of the *Journal of Hydroinformatics* entitled *'Advances in Hydroinformatics'*, which included papers selected from the workshop contributions, providing a state-of-the-art overview as well as emergent directions for future developments in hydroinformatics (Coulibaly *et al.*, 2009).

3.2 EARLIEST AI EFFORTS IN HYDRAULICS, HYDROLOGY AND HYDROINFORMATICS

The debate about the implications of technology on work and jobs is as old as the industrial era. In early 19th century, English textile workers known as the Luddites protested against the introduction of spinning frames and power looms, fearing that the machines would leave them without jobs. Since then, new technological advancements are also associated with another wave of concern about a possible displacement of labour. Seen from such a historical perspective, it is then not too surprising that the hydraulics and hydrology communities were initially quite sceptical about hydroinformatics, and in particular about AI.

Irrespective of the early distrust, the opportunities that AI brings to the hydro-environment community remain very significant. Current applications of AI are the result of just the first 30 years of a paradigm shift that is helping humans to advance discoveries and to leverage scientific knowledge accumulated during the past century through improved computing power and enhanced datasets. As a result, we are now just witnessing the early successes of practical applications, and only beginning to grasp the potential of AI technologies to enhance and/or match human intelligence.

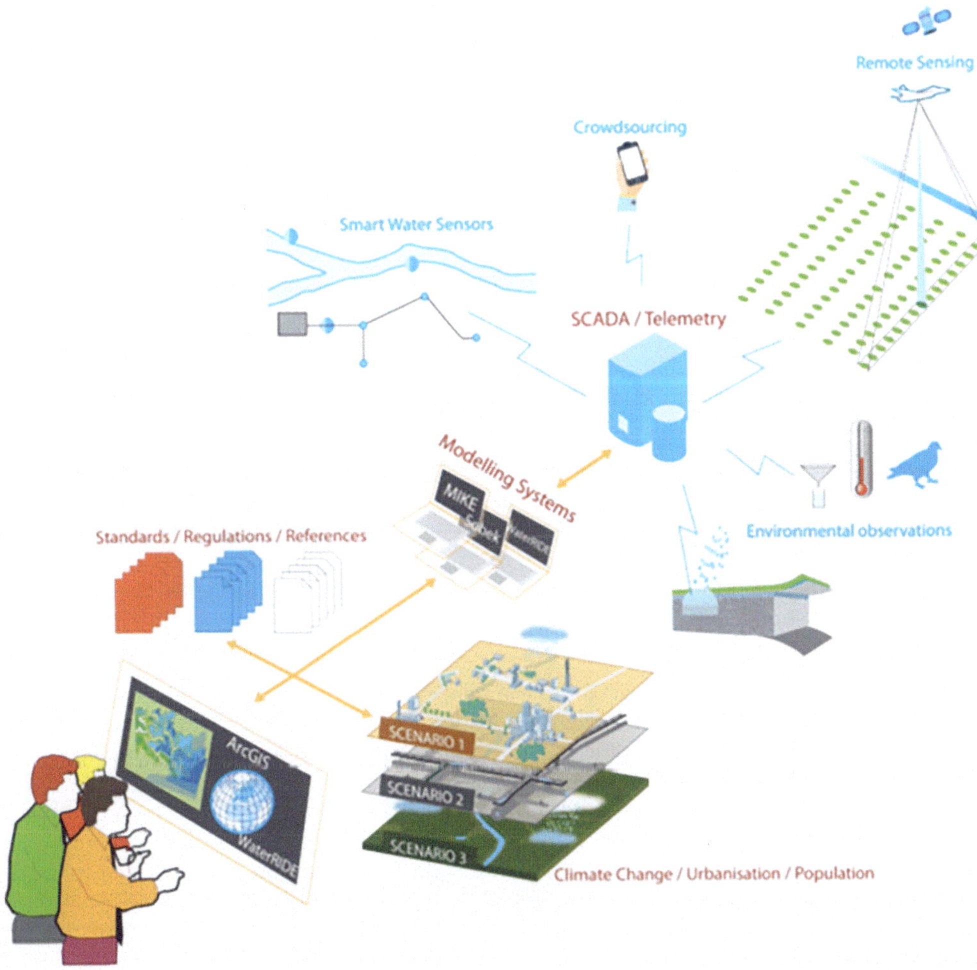

Figure 3.1 Illustration of hydroinformatics environments which have the primary aim of enabling stakeholders to co-develop their water-related solutions in a holistic way (from Vojinovic and Abbott, 2017).

It is now time to fully embrace these opportunities, which not only includes the enhancement of existing capabilities, but also the opening up of completely new research and application avenues.

The First International Conference on Hydroinformatics held in Delft in 1994 (Verwey *et al.*, 1994) offered early glimpses of applications of neural networks (Babovic & Minns, 1994) and evolutionary computation (Babovic *et al.*, 1994). In fact, at that stage, the field was not ready to refer to the work as being AI at all. Instead, the presented research was referred to as 'adaptive computing methods' (Minns & Babovic, 1996).

It took several more years before more substantial work started to appear in peer-reviewed literature. A first influential paper on the application of neural networks in hydrology appeared in 1996 (Minns & Hall, 1996), followed a by two papers on evolutionary computation (Babovic & Abbott, 1997a, 1997b) that appeared in the *IAHR Journal of Hydraulic Research*.

3.3 PICKING-UP MOMENTUM

The 3rd International Conference on Hydroinformatics, held in Copenhagen in 1998, was a super-spreader event for a wider adoption of AI within IAHR circles. One reason, in particular, was that one of the keynote speakers at the conference was David Goldberg, a global authority of Genetic Algorithms. Goldberg, himself a hydraulic engineer with a PhD degree on pipeline operations (Goldberg, 1983) delivered a passionate and engaging address, which eloquently articulated a need for use of more intelligent algorithms in hydraulics. His own research was largely shaped under mentorship of his PhD supervisors: hydraulician Benjamin Wylie (Wylie & Streeter, 1993) and computer scientist John Holland (1975). During the conference, Goldberg also offered a short course on evolutionary computing.

At its meeting held during this conference, the IAHR Hydroinformatics Section decided to broaden its base by becoming a joint section together with International Water Association (IWA) and International Association for Hydrological Sciences (IAHS). By doing so, hydroinformatics and IAHR were able to attract even more researchers and to create the momentum and the critical mass needed to advance hydroinformatics and AI applications. In 1999, Mike Abbott and Roger Falconer subsequently established the *Journal of Hydroinformatics*, published by the international Water Association (IWA) Press, and sponsored by IAHR, IAHS and IWA. This publication provided, and still provides, the platform for the publication and communication about less traditional research and applications, including AI-related work.

Conferences in Iowa in 2000, Cardiff in 2002 and Singapore in 2004 followed. The opening address of Her Royal Highness The Princess Royal, Princess Anne, made the Cardiff Conference a landmark event in another sense again. In effect these events catalysed the formation of a hydroinformatics community with a wide range of highly variegated but still connected interests.

One of the main obstacles for a more widespread acceptance of AI in the early years was the initial scepticism with which it was met by the traditional modelling world. The proposed techniques were completely new, and to a large extent performed as 'black boxes'. A commonly held misconception was that the purely data-driven models would replace physically based models, which are developed based on our best understanding of the physics of the processes involved. Although the earlier descriptions of data-driven modelling and data mining described the new approaches as complementary and an 'enrichment' to traditional techniques (Minns, 2000; Minns & Babovic, 1996), the scepticism remained for many years (if not decades). Minns and Hall (2004) provided some very specific warnings about the use of these black-box models where they stated:

> ...the potential user [of a black box model] is not absolved from devoting some thought to the mode of presentation of data to the network. The principal question to be posed is: what exactly constitutes the pattern of inputs that produces the pattern of outputs? Moreover, do the selected input and output patterns contain additional information that is not strictly relevant to the purpose of the exercise? These questions are inevitably problem-dependent, but in all cases the selection of inputs, whether data as recorded or variables derived from operations on recorded data, requires the application of hydrological insight as much as any conventional physical/conceptual rainfall-runoff model...

The lack of easily accessible, large 'raw' data sets and the relatively poor computational capacity generally available also hindered the further development of the science. In addition, much of the early applications in AI involved the use of expert systems. However, Minns and Babovic (1996) pointed out that in addition to a 'symbol grounding, many expert systems suffer the so-called 'completeness issue'. If a particular situation is not represented in the knowledge tree, this will result in the failure of the inference engine to find an appropriate instance upon which to draw corresponding conclusions. Due to this problem, expert systems are said to be brittle systems, in the sense that as long as every question has an explicitly coded answer they will perform well, but as soon as a situation is not explicitly represented in the knowledge base they fail quite suddenly, in a brittle fashion.

These problems and the associated general distrust of the new and unknown meant that it took many years before a common ground was found.

3.4 THE FORMATIVE YEARS

As the momentum began to pick up, a larger group of authors started to apply data-driven techniques to a growing range of problems. Typical examples include runoff prediction (Elshorbagy *et al.*, 2000) and downscaling of climate models (Coulibaly *et al.*, 2005). Also, interest in the algorithmic aspects of AI broadened. The authors started exploring and reporting on the performance of a wider range of machine-learning algorithms: Support vector machines (Dibike *et al.*, 2001), fuzzy logic (Abebe *et al.*, 2000), model tree induction (Bhattacharya & Solomatine, 2005; Solomatine & Khada, 2003; Solomatine & Xue, 2004), chaos theory (Babovic & Keijzer, 2000b; Elshorbagy *et al.*, 2002) are but few examples.

One of the greatest strengths of these hydroinformatics approaches is their ability to identify relationships and to induce models based upon measured data without requiring a detailed preconceived knowledge of physical system characteristics a priori. One of the reasons for this is that many of these approaches manipulate the data at the level of the computer representation of the numbers. That is, the data are represented in our digital computers as bits and the operations upon this data then take place upon these individual bits. The modeller in this case has no direct influence upon the bits that convey the basic knowledge. After translating the underlying data that characterises our natural world into bit strings, the original symbols are then further irrelevant for the subsequent manipulations of the bits. As such, algorithms operate at the level of the bits, they in fact operate at the so-called *sub-symbolic* level (Minns, 1998, 2000). This represented a profound shift in paradigm away from a physical-based, deterministic, model-based approach to a more data-based approach. In particular, these innovative approaches may be used to generate important components of physically based modelling systems by inducing models or sub-models of individual physical processes based only upon measured data. These (sub)models may then replace whole systems of complex, non-linear, differential equations that would otherwise require great skills from the modeller to calibrate, and powerful computing devices to solve.

Abbott (1993) carried out a rather thorough analysis of the effect that this paradigm shift would have on hydraulic engineering, water resources management and similar areas of application. A study of historical parallels indicated that, rather than just increasing the number of traditional, deterministic modelling systems, this development would inevitably justify the introduction of a wide variety of quite other and quite new products. It was clear that what had until that time been seen simply as a 'model' could now be seen as a 'knowledge encapsulator'. It was now possible to encapsulate all the hydraulic knowledge and all the application-specific information about any area that was being modelled with this very knowledge. It is the encapsulated knowledge which adds the most immediate benefit, as it allows the practitioner to mobilise unprecedented knowledge resources as and when these are required at costs that are only a tiny part of what would be required were this knowledge to be made available in an unencapsulated form (Abbott *et al.*, 2001).

Abbott *et al.* (2001) warn, however, that the very great success of modelling and simulation tools working within the new paradigm of *knowledge consumption* in hydroinformatics also brings many problems with it. Although not strictly speaking 'new', these problems have become exacerbated by inadequacies in the knowledge frames of many of the users of such systems. Some of the issues raised in this article, related to the calibration of hydraulic models, illustrate the sort of problems that users could encounter if they do not address these inadequacies appropriately. It is, for example, well known how the results of hydraulic models can nearly always be fitted to measurements of hydraulic behaviour recorded in a few places by varying the values of certain model parameters. The most common calibration parameter is the roughness coefficient in a hydraulic model. This process can however easily hide gross inaccuracies in the description of the modelled domain, such as in topography, local head losses and/or dimensions of structures. In these cases, automatic calibration procedures may lead to unrealistic values of roughness coefficients in an attempt to compensate for unknown information or, worse, for physical phenomena not represented by the model laws, as expressed in the equations used in the model. In the latter case, a model with excellent calibration results may in fact have no predictive capabilities whatsoever. Such issues are not directly apparent when data-driven models are built, as they do not have physics-related parameters. However, as already discussed in the previous section, the need for similar caution does not disappear, and physical realism is still required when selecting data for building data-driven models, analysing their outputs, and for their testing before potential use.

Regardless of the knowledge encapsulation approach, this has opened doors for hydroinformatics to provide systems for real-time control and other services that proceed with great rapidity without direct human intervention, which would otherwise be quite inconceivable. The large, technological service organisations have become major suppliers of tools and other specialised products and associated services, while the hydraulics practitioner generally has become a user of tools on a rapidly increasing scale (Cunge, 1991). In this way, the practitioner now becomes less a personal carrier of all manner of detailed knowledge and information and more one who is adept at organising and integrating electronically encapsulated knowledge and information.

3.5 OPENING THE BLACK BOX

Perhaps, the key obstacle to earlier and broader acceptance of AI within the larger hydraulics community was the opposition of practitioners to the use of models perceived as being 'black boxes'. It was often claimed that such models could not add to scientific knowledge or improve understanding of the field of hydro-environment. Early applications of AI were primarily focused on enhancing forecasting abilities and non-linear approximation of input–output relationships. Minns and Hall (1997) even refer to the use of artificial neural networks as the '*ultimate black box*' as the machine-learning process does not depend upon any assumptions relating to the form of the input–output transfer function, the number of (active) parameters or their possible physical interpretation.

However, the years that followed brought an increasing trend towards opening up the black boxes and trying to understand how these models work and, more importantly, how can they indeed enhance our understanding of the underlying processes. In an attempt to extract actual 'domain knowledge' from the results of a machine-learning algorithm, Minns (1998) demonstrated that it was indeed possible to recreate the underlying physical equations from the weights and connections that had been learned by a neural network. Dibike *et al.* (1999a, 1999b) developed these concepts further and explored the entire encapsulation of numerical-hydraulic models in neural networks. Babovic and Keijzer (2000a) dedicated considerable attention to exploration of methods for incorporation of domain knowledge into machine learning. Giustolisi (2004), Giustolisi and Savic (2006) reported on a range of machine-learning techniques generating interpretable equations thus having the potential to contribute to knowledge discovery in different hydro-environment disciplines.

In his review of the so-called human-competitive results produced by genetic programming Koza (2010) highlights the work of Babovic (2000) and Baptist *et al.* (2007) as examples of machine-learning outcomes that are matching, or are better than the results produced by human experts. One might even argue that these efforts were pre-cursors to the so-called theory-guided data science (TGDS) (Karpatne *et al.*, 2017) as an emerging paradigm that aims to leverage the wealth of scientific knowledge for improving the effectiveness of data science models in enabling scientific discovery. The overarching vision of TGDS is to introduce scientific consistency as an essential component for learning generalisable models. By producing scientifically interpretable models, TGDS aims to advance our scientific understanding by discovering novel domain insights. Indeed, the paradigm of TGDS has started to gain prominence in a number of scientific disciplines. This includes not only hydroinformatics, but also turbulence modelling, material discovery, quantum chemistry, bio-medical science, bio-marker discovery, climate science, and hydrology. TGDS is in fact closely related to the field of Explainable AI (see e.g., Arrieta *et al.*, 2020).

3.6 GROWING VOLUMES OF DATA AND ACCELERATING COMPUTING POWER

The capabilities of digital devices continue to increase, and Internet of Things (IoT) sensors provide even greater amounts of information than ever before, and at lower costs and with greater reliability than previously possible (Mao *et al.*, 2018). The confluence of these two trends only increases the relevance of AI to the hydro-environment community.

Abbott *et al.* (2001) highlighted the changing nature of hydraulics research in the hydroinformatics era, in which information technology can be employed to assist the human analyst in the process of hypothesis generation. This computer-assisted analysis, usually of large, multi-dimensional data sets, is sometimes referred to as data mining for knowledge discovery. This new approach is coupled with the development of tools to facilitate the conversion of data into a number of forms that provide a better understanding of the physical, biological and other processes that generated or produced these data. These new models, when combined with the already available understanding of the physical processes result in an improved understanding and novel formulations of physical and other laws, and so provide an improved predictive capability.

Furthermore, we are witnessing an increase in the so-called *opportunistic sensing* facilitated, for example, by crowdsourcing, social media and citizen science in which the general public provide additional observations of local conditions (Mazzoleni *et al.*, 2017; Seibert *et al.*, 2019; Wagener *et al.*, 2021). In addition, earth observation (EO) technologies now gather information about the planet's physical, chemical and biological systems. When these two sources of information are combined, we get the perfect ingredients to assess the status of, and changes in, the natural and man-made environment. In recent years, EO has become more and more sophisticated with the advancement of remote-sensing satellites and increasingly high-tech 'in-situ' instruments. Today's EO instruments include floating buoys for monitoring ocean currents, temperature and salinity; rainfall trends and similar.

3.7 RECENT EXAMPLES OF AI-ENABLED SYSTEMS

Some recent AI-enabled systems based on a very large EO data sets coupled with crowd sensed data are described below. This is not an exhaustive overview, but simply serves to provide the reader with some recent, innovative applications related to water and the hydro-environment.

3.7.1 Planetary-scale surface water detection from space

Donchyts (2018) has studied automated methods of surface water detection from satellite imagery in order to monitor changes in surface water worldwide to help to understand how our planet is changing as a result of human activities or climate change. A new method for accurate surface water detection was developed that performed better than existing methods to discriminate surface water from noisy satellite images. A probabilistic method was developed to reconstruct surface water from satellite images where surface water is only partially observed (due to limited swath, atmospheric noise or snow/ice cover). This approach was implemented in a machine-learning algorithm to estimate global-scale surface water changes directly from medium-resolution multitemporal and multispectral satellite data. The method was applied without the need for direct human intervention to process more than a petabyte of Landsat data to identify surface water changes globally. This is an excellent example of the use of truly big data, in which a planetary-scale analysis of three decades of satellite images has been performed, quantifying Earth's surface water changes at the 30m spatial resolution and occurring during the last three decades. The results of the study were made freely available with the help of the parallel satellite data processing platform Google Earth Engine (Donchyts *et al.*, 2016).

3.7.2 Water quality sensing

EO data have been utilised to retrieve water quality information since 1970s (Holyer, 1978; Ritchie *et al.*, 1976). Many of the remote-sensing-based water quality monitoring methods focus on optically active parameters such as turbidity, chlorophyll-a, suspended particulate matter (SPM), coloured dissolved organic matter (CDOM), and so on (Hou *et al.*, 2017; Shahzad *et al.*, 2018). However, based on the fact that non-optically active parameters may be highly correlated with optically active parameters, Guo *et al.* (2021) have compared 255 possible Sentinel-2 imagery band compositions to identify the most appropriate ones for total phosphorous (TP), total nitrogen (TN), and chemical oxygen demand (COD) retrieval. Three machine-learning models, namely Random Forest (RF), Support Vector Regression (SVR) and Neural Networks (NN), were compared to seek the most robust ones for retrieving the above non-optically active parameters.

Yang *et al.* (2022) provide an overview of several techniques for water quality parameter retrieval from remote sensing data (including multispectral and hyperspectral data). They discuss retrieval algorithms for several specific water quality variables, including total suspended matter (TSM), chlorophyll-a (Chl–a), coloured dissolved organic matter (CDOM), chemical oxygen demand (COD), total nitrogen (TN) and total phosphorus (TP). Retrieval algorithms include the use of empirical, analytical, and/or semi-empirical techniques, and also AI techniques (including artificial neural networks and support vector machines). Furthermore, the authors discuss issues surrounding atmospheric correction of EO data and the associated challenges for inland water environments compared to ocean waters.

3.7.3 Computer vision for opportunistic rainfall monitoring

The quantity and quality of precipitation data are crucial in meteorological and water resource management applications. Using rain gauges is the classic approach to measuring rainfall. However, as we enter the age of the IoT in which 'anything may become data' the so-called opportunistic sensing using unconventional data sources offers the promise to enhance the spatiotemporal representation of existing observation networks (Kapelan *et al.*, 2020). One particular area attracting attention is the estimation of quantitative and analytical rainfall intensity from video feeds acquired by smart phones or CCTV surveillance cameras. Technological advances in image processing and computer vision enable extraction of diverse features, including identification of rain streaks enabling the estimation of the instantaneous rainfall intensity (Allamano *et al.*, 2015). Recent AI and machine-learning approaches rely on the use of autoencoders, deep learning and convolutional neural networks to address the problems. Companies such as WaterView (Italy), the Hydroinformatics Institute (Singapore), as well as universities (Southern University of Science and Technology China, Shenzhen) have proposed and implemented practical approaches to weather hazards in energy, automotive and smart cities application domains (Jiang *et al.*, 2019).

3.7.4 Hydrologically informed machine learning for rainfall–runoff modelling

Minns and Hall (1996) provide a number of examples of quite accurate lumped, rainfall–runoff models using artificial neural networks on data from gauged catchments. However, the extraction of semantic content from the ANN models is by no means obvious, or even possible. Minns and Babovic (1996) had some early success in deriving meaningful expressions from the same data used by Minns and Hall by applying genetic programming techniques. The potential for extending these techniques beyond individual catchments was mentioned but not explored further. The meaningfulness and reliability of hydrological inferences gained from lumped models may tend to deteriorate within large catchments where the spatial heterogeneity of forcing variables and watershed properties is significant.

Inspired by the theory-guided data science (TGDS) and physics-informed machine-learning approaches, a new rainfall–runoff model induction framework has been developed using genetic programming (GP). The framework is referred to as Machine Learning Rainfall–Runoff Model Induction (ML-RR-MI) (Chadalawada *et al.*, 2020). This research is part of a broader effort aiming to develop a methodology for automated model induction built from data and using existing background knowledge. ML-RR-MI relies on elements of well-established hydrological modelling frameworks (SUPERFLEX and FUSE) and uses these as the building blocks constituting the background hydrological knowledge. ML-RR-MI has been shown to be capable of developing fully fledged lumped conceptual rainfall–runoff models. The framework was subsequently extended towards semi-distributed rainfall–runoff models. This approach for distributed rainfall–runoff modelling using hydrology-informed machine learning is referred to as Machine Induction Knowledge Augmented – System Hydrologique Asiatique (MIKA-SHA) (Herath *et al.*, 2021).

Simplistic application of data-driven black-box machine-learning techniques may lead to development of accurate yet meaningless models, which do not satisfy basic hydrological insights and may have severe difficulties with interpretation. Yet, it is the value associated with interpretability of resulting models that utilise elements of background knowledge that distinguishes this new approach from other methods. The approach makes it possible to rely on the resulting models with more than just statistical confidence. It is only in this way that we can take full advantage of machine-aided knowledge discovery and advance our understanding of physical processes.

3.8 OPPORTUNITIES FOR FUTURE DEVELOPMENTS

It is quite obvious from this brief and incomplete review of AI-related developments related to hydroinformatics that significant progress has been made over the past 25–30 years. The hydraulics research community has embraced and significantly benefitted from data science. The following sections provide an overview of some of the most important research and application directions that are driving the ongoing innovations in hydroinformatics.

3.8.1 Machine learning

Depending on the problem to be solved, there are several machine-learning techniques that can be used (Figure 3.2). Take, for example, the situation in which the user wishes to predict the value of a variable. If that variable is continuous (i.e. water level; discharge; subsidence rate), regression could be used. If the variable is categorical (i.e. type of subsoil; levels of poverty) classification should be used. Both regression and classification algorithms belong to supervised learning. On the other hand, if the user is not interested in prediction of a given variable but is interested in looking for similarities in the data, clustering should be used. Dimensionality reduction techniques then need to be applied in order to analyse correlations among variables. Both clustering and dimensionality reduction are methods belonging to unsupervised learning. In reinforcement learning, the user will apply a method of rewarding desired behaviours and punishing negative behaviours. The result is to maximise long-term rewards and to achieve an optimal solution.

3.8.2 Enabling technologies

Knowledge institutions like Deltares (The Netherlands), DHI (Denmark) and the Hydroinformatics Institute (Singapore) have recognised the potential for new technologies to be game changers in the delta technology sector given their potential to make ground-breaking innovations. Their use in other sectors illustrates the disruptive power of artificial intelligence, big data, cloud computing, and of advanced materials and new monitoring techniques. For example, in 2018, the focus on enabling technologies became a new element in the Deltares research strategy. The objective is to explore these technologies and to make them applicable to a wider range of social issues inasmuch that these technologies also encourage the involvement of the general public and stakeholders in citizen science projects and crowdsourcing activities (Deltares, 2018).

Staeb *et al.* (2021) introduce the concept of a digital twin of the North Sea, in which a ready-to-use platform is being created that brings together new technologies such as open source data and mapping services, cloud computing, gaming and virtual reality. It is intended to bring a better understanding of all of the stakeholder pressures and to subsequently lead to better decision making. A virtual reality (VR) tool allows the user to be immersed in the world of the sea. Citizens and politicians only rarely go out on the sea and have difficulty in understanding the challenges. VR helps these people to see the outcome of

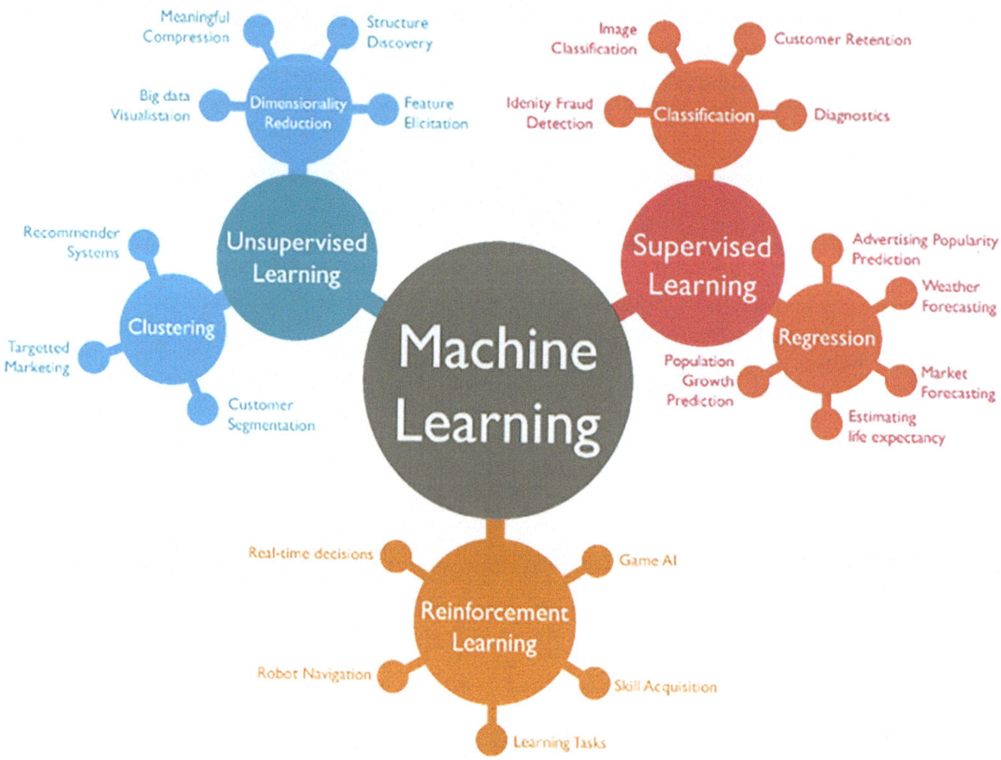

Figure 3.2 Overview of machine-learning methods and application areas (from Mayo, 2019).

decisions that are made. With 3D glasses on your head you can be the captain of a big oil tanker or even a sea bird trying to find its way in the turmoil of shipping lanes and windfarms.

3.8.3 Freedom to the data!

In a recent White Paper, Valero *et al.* (2021) stress that the democratisation of observations and data is an area that needs significant attention. Issues associated with data privacy on one hand, and the need for openness and data exchange on the other are essential – particularly within the hydro-environment community. The authors point out that IAHR was founded by 66 hydraulic laboratories, making the science of measuring and data acquisition entrenched in the origins of the association. Observations and measurements are fundamental for scientific progress in general and remain essential for advancing our insights and knowledge. Induction of relationships and processing of data to improve the understanding of the processes that generated or produced these data has always been at the very heart of hydraulics research, as it is also in many other domains.

These developments open avenues for taking advantage of big data, an area that is gaining attention in the hydroinformatics community (Chen & Han, 2016). Big data research is driven by the need to process extremely large datasets that cannot be processed using traditional data-processing methods within a reasonable time frame. Big data technology and analytical tools enable researchers to handle increasingly larger datasets from which creative ideas and new insights can be mined.

Sharing the data as a community will result in not only more affordable access to quality-assured data, but also would accelerate scientific and technological advances within the hydro-environment community at large.

3.8.4 Deep learning

With increasing data availability, deep learning is gaining popularity. Deep learning is a class of machine-learning algorithms that uses multiple layers to progressively extract higher-level features from raw inputs. For example, in image processing, lower layers may identify edges, while higher layers may identify the concepts relevant to a human such as digits, or letters, or faces. It has been used to replace the conventional hydrodynamic models, which are very slow to run due to complex numerical computations, in simulating flood inundations and vulnerabilities (Bui *et al.*, 2020).

3.8.5 Hydroinformatics-informed machine learning?

As we enter the true digital information era, one of the greatest challenges facing organisations and individuals is related to turning the rapidly expanding data stores into accessible, and actionable knowledge. Without such developments, we risk missing most of what the data have to offer. The traditional approach of a human analyst, intimately familiar with a data set, serving as a conduit between raw data and synthesised knowledge by producing useful analyses and reports, is breaking down.

What is to be done with all these data? Ignoring whatever we cannot analyse would be wasteful and unwise. This is particularly pronounced in scientific endeavours, where data represent carefully collected observations about particular phenomena that are under study. Data science models, although successful in a number of commercial domains, have had limited applicability in scientific problems involving complex physical phenomena.

Hydroinformatics must deal with the relationship between models and data. A common saying is that '*...all models are wrong, but some models are useful...*'. Although this judgement is often applied to hydraulic models, it is remarkable that the same partiality is not applied to the quality of the measurements. Skogen *et al.* (2021) concede that models may not be perfect, and that there are indeed examples where models have been used improperly to provide misleading answers with great confidence. However, the authors question to what extent an observation may be said to actually represent the truth. The precision of the observational gear may be high, but what about representativeness? The interpretation of observations is simply another model, but this time not coded in a computer language but rather formed by the individual observer. It is proffered that it would be more productive to initiate a process where the norm is that models and observations are joined to strengthen both.

As already observed in this chapter, theory-guided data science (Karpatne *et al.*, 2017) is an emerging paradigm that aims to leverage the wealth of scientific knowledge for improving the effectiveness of data science models in enabling scientific discovery.

Theory-driven machine learning (TDML) recognises that applying the AI alone is not the entire story. At least not in scientific domains such as hydro-environment! Scientific theories encourage the acquisition of new data and these data in turn leads to the generation of new theories. Traditionally, the emphasis is on a theory that demands that appropriate data be obtained through observation or experiment. In such an approach, the process is what we call *theory-driven*. Especially when a theory is expressed in mathematical form, theory-driven discovery may make extensive use of strong methods associated with mathematics or with the subject matter of the theory itself. The converse view takes a body of data as its starting point and searches for a set of generalisations, or a theory, to describe the data parsimoniously or even to explain it. Usually such a theory takes the form of a concise mathematical statement of the relations existing among the data. This would be the *AI-driven* discovery process, or more particularly the *ML-driven* process. The new, data-driven ML-discovered models, combined with the understanding of the physical processes – the theory – can result in an improved understanding and novel formulations of physical laws and an improved predictive capability.

It is quite apparent that early activities along these lines are starting to take shape (Babovic, 2009; Jiang *et al.*, 2020), but there is still a long road ahead to realise the full potential of combining the theory-driven, understanding-rich approach with the state-of-the-art AI-algorithms to accelerate knowledge discovery in the hydro-environment.

The capabilities of our digital devices continue to increase, and IoT sensors provide ever-increasing amounts of information, at lower cost and with greater reliability than previously possible. In addition to this, we witness an increase in the so-called opportunistic sensing, lately facilitated by crowdsourcing, social media and citizen science that enables the general public to observe local environmental conditions. As mentioned earlier, when these are coupled with EO developments, unprecedented opportunities may be presented for assessing the status of, and changes in, the natural and man-made environment.

Some exciting developments are certainly beginning to emerge as the tech giants like Microsoft and Google make their global computing capacity available to monitor, model, and manage Earth's natural systems. See, for example, Microsoft's AI for Earth initiative (https://www.microsoft.com/en-us/ai/ai-for-earth) and Google's Crisis Response initiative (https://crisisresponse.google/). As part of the latter initiative, Google has launched a flood prediction service using machine learning to identify areas of land prone to flooding. In 2018, the system was used to alert users in India's Patna region before the waters arrived with some success (see https://blog.google/technology/ai/flood-forecasts-india-bangladesh/). Realising the limits of machine-learning tools, Google has built a forum to connect computer data scientists with hydrologists/hydraulicians in order to merge ML and process knowledge to achieve the best results. In 2019, they hosted a workshop in Tel Aviv entitled Google Flood Forecasting Meets Machine Learning (see: https://ai.googleblog.com/2019/03/a-summary-of-google-flood-forecasting.html).

3.9 THE WAY FORWARD

The current chapter has provided a less than exhaustive overview of some innovative applications of AI in the area of hydroinformatics and some future challenges related to the modelling of water systems. It is obvious that future efforts related to the development and use of hydro-environment modelling systems will be increasingly directed towards coupling of different kinds of models and encapsulated knowledge via web-based applications as well as better data processing activities (including quantification of uncertainties) that can enable more accurate model results.

Gathering and processing large quantities of domain data with new technologies and tools is yet another important ongoing activity in hydroinformatics. Vojinovic and Abbott (2017) also point out that the future of hydroinformatics will be directed at supporting and indeed enabling holistic analysis, design, installation and development of on-line and real-time hydro-environment modelling systems that will be highly adaptive to changing conditions, such as those that may occur slowly over years (e.g., climate change effects) and over a few hours (e.g., flood conditions), or in extreme cases even over some minutes (e.g., evacuation of people in advance of disastrous events). In many cases, new developments in AI and machine-learning techniques will continue to change and be refined by the hydroinformatics community depending on the application area and on the available technologies.

REFERENCES

Abbott M. B. (1991). Hydroinformatics: Information Technology and the Aquatic Environment. Brookfield, Aldershot, UK. USA: Ashgate.

Abbott M. B. (1993). The electronic encapsulation of knowledge in hydraulics, hydrology and water resources. *Advances in Water Resources*, **16**, 21–39, https://doi.org/10.1016/0309-1708(93)90027-D

Abbott M. B. (1994). Hydroinformatics: a Copernican revolution in hydraulics. *Journal of Hydraulic Research*, **32**(S1), 3–13, https://doi.org/10.1080/00221689409498800

Abbott M. B., Babovic V. and Cunge J. (2001). Towards the hydraulics of the hydroinformatics era. *Journal of Hydraulic Research*, **39**(4), pp. 339–349, https://doi.org/10.1080/00221680109499839

Abebe A. J., Solomatine D. P. and Venneker R. G. W. (2000). Application of adaptive fuzy rule-based models for reconstruction of missing precipitation events. *Hydrological Sciences Journal*, **45**(3), 425–436, https://doi.org/10.1080/02626660009492339

Allamano P., Croci A. and Laio F. (2015). Toward the camera rain gauge. *Water Resources Research*, **51**, 1744–1757, https://doi.org/10.1002/2014WR016298

Arrieta A. B., Díaz-Rodríguez N., Del Ser J., Bennetot A., Tabik S., Barbado A., Garcia S., Gil-Lopez S., Molina D., Benjamins R., Chatila R. and Herrera F. (2020). Explainable artificial intelligence (XAI): concepts, taxonomies, opportunities and challenges toward responsible AI. *Information Fusion*, **58**, 82–115, https://doi.org/10.1016/j.inffus.2019.12.012

Babovic V. (2000). Data mining and knowledge discovery in sediment transport. *Computer-Aided Civil and Infrastructure Engineering*, **15**(5), 383–389, https://doi.org/10.1111/0885-9507.00202

Babovic V. (2009). Introducing knowledge into learning based on genetic programming. *Journal of Hydroinformatics*, **11**(3–4), 181–193, https://doi.org/10.2166/hydro.2009.041

Babovic V. and Abbott M. B. (1997a). The evolution of equations from hydraulic data part I: theory. *Journal of Hydraulic Research*, **35**(3), 397–410, https://doi.org/10.1080/00221689709498420

Babovic V. and Abbott M. B. (1997b). The evolution of equations from hydraulic data part II: applications. *Journal of Hydraulic Research*, **35**(3), 411–430, https://doi.org/10.1080/00221689709498421

Babovic V. and Keijzer M. (2000a). Genetic programming as a model induction engine. *Journal of Hydroinformatics*, **2**(1), 35–60, https://doi.org/10.2166/hydro.2000.0004

Babovic V. and Keijzer M. (2000b). Forecasting of river discharges in the presence of chaos and noise. In: *Flood issues in contemporary water management*, NATO Science Series 2 Environmental Security 71, J. Marsalek, W.E. Watt, E. Zeman and F. Sieker (eds.), Springer Science+Business Media, Dordrecht, the Netherlands, pp 405–420.

Babovic V. and Minns A. (1994). Use of computational adaptive methodologies in hydroinformatics. In: Proceedings of the First International Conference on Hydroinformatics, A. Verwey, A. W. Minns, V. Babovic and C. Maksimovic (eds.), Balkema Publishers, Delft, Netherlands, Rotterdam, pp. 201–210.

Babovic V., Wu Z. Y. and Larsen L. C. (1994). Calibrating hydrodynamic models by means of simulated evolutions. In: Proceedings F' International Conference on Hydroinformatics, A. Verwey, A. W. Minns, V. Babovic and C. Maksimovic (eds.), Balkema, Rotterdam, pp. 193–200.

Baptist M. J., Babovic V., Rodríguez Uthurburu J., Keijzer M., Uittenbogaard R. E., Mynett A. and Verwey A. (2007). On inducing equations for vegetation resistance. *Journal of Hydraulic Research*, **45**(4), 435–450, https://doi.org/10.1080/00221686.2007.9521778

Bhattacharya B. and Solomatine D. P. (2005). Neural networks and M5 model trees in modelling water level–discharge relationship. *Neurocomputing*, **63**, 381–396, https://doi.org/10.1016/j.neucom.2004.04.016

Bui D. T., Hoang N. D., Martínez-Álvarez F., Ngo P. T. T., Hoa P. V., Pham T. D., Samui P. and Costache R. (2020). A novel deep learning neural network approach for predicting flash flood susceptibility: A case study at a high frequency tropical storm area. *Science of the Total Environment*, **701**, 134413, https://doi.org/10.1016/j.scitotenv.2019.134413

Chadalawada J., Herath H. and Babovic V. (2020). Hydrologically informed machine learning for rainfall-runoff modeling: a genetic programming-based toolkit for automatic model induction. *Water Resources Research*, **56**(4), e2019WR026933, https://doi.org/10.1029/2019WR026933

Chen Y. and Han D. (2016). Big data and hydroinformatics. *Journal of Hydroinformatics*, **18**, 599–614, https://doi.org/10.2166/hydro.2016.180

Coulibaly P., Dibike Y. B. and Anctil F. (2005). Downscaling precipitation and temperature with temporal neural networks. *Journal of Hydrometeorology*, **6**(4), 483–496, https://doi.org/10.1175/JHM409.1

Coulibaly P., Babovic V., Cluckie I., Mynett A. and Ball J. (2009). 2009 Special issue of the journal of hydroinformatics on advances in hydroinformatics. *Journal of Hydroinformatics*, **11**(3–4), 165–165, https://doi.org/10.2166/hydro.2009.000

Cunge J. A. (1991). Informatique, hydraulique numerique, hydroinformatique: que devrait savoir un hydraulicien a la fin des études en 1995? *La Houille Blanche*, **77**(3/4), 253–256, https://doi.org/10.1051/lhb/1991023

Deltares (2018). R&D Highlights, 2018 edn, Deltares, Delft.

Dibike Y. B., Minns A. and Abbott M. B. (1999a). Applications of artificial neural networks to the generation of wave equations from hydraulic data. *Journal of Hydraulic Research*, **37**(1), 81–97, https://doi.org/10.1080/00221689909498533

Dibike Y. B., Solomatine D. P. and Abbott M. B. (1999b). On the encapsulation of numerical-hydraulic models in artificial neural network. *Journal of Hydraulic Research*, **37**(2), 147–161, https://doi.org/10.1080/00221689909498303

Dibike Y. B., Velickov S., Solomatine D. P. and Abbott M. B. (2001). Model induction with support vector machines: introduction and applications. *Journal of Computing in Civil Engineering*, **15**(3), 208–216, https://doi.org/10.1061/(ASCE)0887-3801(2001)15:3(208)

Donchyts G. (2018). Planetary-Scale Surface Water Detection from Space. PhD thesis, TU Delft, http://doi.org/10.4233/uuid:510bd39f-407d-4bb6-958e-dea363c5e2a8

Donchyts G., Baart F., Winsemius H. C., Gorelick N., Kwadijk J. C. J. and van de Giesen N. (2016). Earth's surface water change over the past 30 years. *Nature Climate Change*, **6**(9), 810–813, http://doi.org/10.1038/nclimate3111

Elshorbagy A., Simonovic S. P. and Panu U. S. (2000). Performance evaluation of artificial neural networks for runoff prediction. *Journal of Hydrologic Engineering*, **5**(4), 424–427, https://doi.org/10.1061/(ASCE)1084-0699(2000)5:4(424)

Elshorbagy A., Simonovic S. P. and Panu U. S. (2002). Estimation of missing streamflow data using principles of chaos theory. *Journal of Hydrology*, **255**(1–4), 123–133, https://doi.org/10.1016/S0022-1694(01)00513-3

Giustolisi O. (2004). Using genetic programming to determine Chezy resistance coefficient in corrugated channels. *Journal of Hydroinformatics*, **6**(3), 157–173, https://doi.org/10.2166/hydro.2004.0013

Giustolisi O. and Savic D. A. (2006). A symbolic data-driven technique based on evolutionary polynomial regression. *Journal of Hydroinformatics*, **8**(3), 207–222, https://doi.org/10.2166/hydro.2006.020b

Goldberg D. (1983). Computer-Aided Gas Pipeline Operation Using Genetic Algorithms and Rule Learning. PhD thesis, University of Michigan, Ann Arbor, MI.

Goldberg D. (1989). Genetic Algorithms in Search, Optimization and Machine Learning. Addison-Wesley, Reading, MA, USA.

Guo H., Huang J., Chen B., Guo X. and Singh V. (2021). A machine learning-based strategy for estimating non-optically active water quality parameters using Sentinel-2 imagery. *International Journal of Remote Sensing*, **42**, 1841–1866, https://doi.org/10.1080/01431161.2020.1846222

Herath H. M. V. V., Chadalawada J. and Babovic V. (2021). Hydrologically informed machine learning for rainfall-runoff modelling: towards distributed modelling. *Hydrology and Earth System Sciences*, **25,** 4373–4401, https://doi.org/10.5194/hess-25-4373-2021

Holland J. H. (1975). Adaptation in Natural and Artificial Systems. University of Michigan Press, Ann Arbor, MI, USA.

Holyer R. J. (1978). Toward universal multispectral suspended sediment algorithms. *Remote Sensing of Environment*, **7**(4), 323–338, https://doi.org/10.1016/0034-4257(78)90023-8

Hou X., Feng L., Duan H., Chen X., Sun D. and Shi K. (2017). Fifteen-year monitoring of the turbidity dynamics in large lakes and reservoirs in the middle and lower basin of the Yangtze river, China. *Remote Sensing of Environment*, **190**(March), 107–121, https://doi.org/10.1016/j.rse.2016.12.006

Jiang S., Babovic V., Zheng Y. and Xiong J. (2019). Advancing opportunistic sensing in hydrology: A novel approach to measuring rainfall with ordinary surveillance cameras. *Water Resources Research*, **55**(4), 3004–3027, https://doi.org/10.1029/2018WR024480

Jiang S., Zheng Y. and Solomatine D. P. (2020). Improving AI system awareness of geoscience knowledge: symbiotic integration of physical approaches and deep learning. *Geophysical Research Letters*, **46**(13), e2020GL088229, https://doi.org/10.1029/2020GL088229

Kapelan Z., Weisbord E. and Babovic V. (2020). Digital Water: Artificial Intelligence Solutions for the Water Sector. White paper, © 2020 International Water Association, London, UK.

Karpatne A., Atluri G., Faghmous J. H., Steinbach M., Banerjee A., Ganguly A., Shekhar S., Samatova N. and Kumar V. (2017). Theory-guided data science: A new paradigm for scientific discovery from data. *IEEE Transactions on Knowledge and Data Engineering*, **29**, 2318–2331, https://doi.org/10.1109/TKDE.2017.2720168

Koza J. R. (1992). Genetic Programming: On the Programming of Computers by Means of Natural Selection. MIT Press, Cambridge, MA, USA.

Koza J. R. (2010). Human-competitive results produced by genetic programming. *Genetic Programming and Evolvable Machines*, **11**, 251–284, https://doi.org/10.1007/s10710-010-9112-3

Mao F., Clark J., Buytaert W., Krause S. and Hannah D. M. (2018). Water sensor network applications: time to move beyond the technical? *Hydrological Processes*, **32**, 2612–2615, https://doi.org/10.1002/hyp.13179

Mayo M. (2019). Which Data Science/Machine Learning methods and algorithms did you use in 2018/2019 for a real-world application?, KDnuggets.com, https://www.kdnuggets.com/2019/04/poll-data-science-machine-learning-methods-algorithms-use-2018-2019.html (accessed 15 July 2022).

Mazzoleni M., Verlaan M., Alfonso L., Monego M., Norbiato D., Ferri M. and Solomatine D. P. (2017). Can assimilation of crowd sourced data in hydrological modelling improve flood prediction? *Hydrology and Earth System Sciences*, **21**(2), 839–861, https://doi.org/10.5194/hess-21-839-2017

Minns A. W. (1998). Artificial Neural Networks as Sub-symbolic Process Descriptors. PhD thesis, Balkema, Rotterdam.

Minns A. W. (2000). Subsymbolic methods for data mining in hydraulic engineering. *Journal of Hydroinformatics*, **2**(1), 3–13, https://doi.org/10.2166/hydro.2000.0002

Minns A. W. and Babovic V. (1996). Hydrological modelling in a hydroinformatics context. In: Distributed Hydrological Modelling, M. B. Abbott and J. C. Refsgaard (eds.), Kluwer Academic Publishers, Dordrecht, the Netherlands, pp. 297–312.

Minns A. W. and Hall M. J. (1996). Artificial neural networks as rainfall-runoff models. *Hydrological Sciences Journal*, **41**(3), 399–417, https://doi.org/10.1080/02626669609491511

Minns A. W. and Hall M. J. (1997). Living with the ultimate black box: more on artificial neural networks. Proc. National Hydrology Symposium, British Hydrological Society, Salford.

Minns A. W. and Hall M. J. (2004). Rainfall-runoff modelling. In: Neural Networks for Hydrological Modelling, Chapter 9, R. J. Abrahart, P. E. Kneale and L. See (eds.), Balkema, Leiden, pp. 157–175.

Nature Reviews Physics (2021). The rise of data-driven modelling. *Nature Reviews Physics*, **3**, 383–383, https://doi.org/10.1038/s42254-021-00336-z

Ritchie J. C., Schiebe F. R. and McHenry J. R. (1976). Remote sensing of suspended sediments in surface waters. *Photogrammetric Engineering and Remote Sensing*, **42**(12), 1539–1545.

Rumelhart D. E. and McClelland J. L. (1986) Parallel Distributed Processing. The MIT Press, Vols **1** and **2**, Cambridge, MA, USA.

Seibert J., Strobl B., Etter S., Hummer P. and van Meerveld H. J. (2019). Virtual staff gauges for crowd-based stream level observations. *Frontiers in Earth Science*, **7**, 1–, https://doi.org/10.3389/feart.2019.00070

Shahzad M. I., Meraj M., Nazeer M., Zia I., Imam A., Mahmood K. and Zafar H. (2018). Empirical estimation of suspended solids concentration in the Indus delta region using landsat-7 ETM + imagery. *Journal of Environmental Management*, **209**, 254–261, https://doi.org/10.1016/j.jenvman.2017.12.070

Skogen M. D., Ji R., Akimova A., Daewel U., Hansen C., Hjøllo S. S., van Leeuwen S. M., Maar M., Macias D., Mousing E. A., Almroth-Rosell E., Sailley S. F., Spence M. A., Troost T. A. and van de Wolfshaar K. (2021). Disclosing the truth: Are models better than observations? Mar Ecol Prog Ser, **680**, 7–13, https://doi.org/10.3354/meps13574

Solomatine D. P. and Khada N. D. (2003). Model trees as an alternative to neural networks in rainfall–runoff modelling. *Hydrological Sciences Journal*, **48**(3), 399–411, https://doi.org/10.1623/hysj.48.3.399.45291

Solomatine D. P. and Xue Y. (2004). M5 model trees and neural networks: application to flood forecasting in the upper reach of the Huai river in China. *Journal of Hydrologic Engineering*, **9**(6), 491–501, https://doi.org/10.1061/(ASCE)1084-0699(2004)9:6(491)

Staeb J. A., Mayer I., Koldenhof Y., Baart F., Jeurissen P. and Thijsse P. (2021) Digital twin of the North Sea: a visual geo tool with computer modelling to support stakeholder engagement, to aid decision makers and support citizen involvement. Proc. IMDIS 2021 – International Conference on Marine Data and Information Systems, 12–14 April 2021, pp. 224–222.

Valero D., Schalko I., Friedrich H., Abad J. D., Bung D., Donchyts G., Felder S., Ferreira R. M. L., Hohermuth B., Kramer M., Li D., Mendes L., Moreno-Rodenas A., Nones M., Paron P., Ruiz-Villanueva V., Wang R. Q. and Franca M. J. (2021) Pathways Towards Democratization of Hydro-Environment Observations and Data, IAHR, Madrid, Spain and Beijing, China, White Paper 2021-1.

Verwey A., Minns A. W., Babovic V. and Maksimovic C. (eds.) (1994). Hydroinformatics '94. In: Proc. 1st International Conference on Hydroinformatics, IHE-Delft, Balkema, Rotterdam.

Vojinovic Z. and Abbott M. B. (2017). Twenty-five years of hydroinformatics. *Water*, **9**(1), 59, https://doi.org/10.3390/w9010059

Wagener T., Savic D., Butler D., Ahmadian R., Arnot T., Dawes J., Djordjevic S., Falconer R., Farmani R., Ford D., Hofman J., Kapelan Z., Pan S. and Woods R. (2021). Hydroinformatics education – the water informatics in science and engineering (WISE) Centre for Doctoral Training. *Hydrology and Earth System Sciences*, **25**, 2721–2738, https://doi.org/10.5194/hess-25-2721-2021

Wylie E. B. and Streeter V. L. (1993). Fluid Transients in Systems. Prentice-Hall, New York, p. 463.

Yang H., Kong J., Hu H., Du Y., Gao M. and Chen F. (2022). A review of remote sensing for water quality retrieval: progress and challenges. *Remote Sensing*, **14**, 1770, https://doi.org/10.3390/rs14081770

doi: 10.2166/9781789062656_0045

Chapter 4

Hydroinformatics impact on hydrological modelling

J. C. Refsgaard[1], M. Erlich[2] and J. Bathurst[3]

[1]Professor Emeritus, Geological Survey of Denmark and Greenland (GEUS), Copenhagen, Denmark
[2]Freelance Consultant in Flood Risk, Former Project Director at ARTELIA, Grenoble, France
[3]Retired, Newcastle University, School of Engineering, Newcastle upon Tyne, UK

ABSTRACT

Mike Abbott was an outsider to hydrological science, who nevertheless fundamentally advanced hydrological modelling by introducing knowledge from computational hydraulics and later hydroinformatics. His main contribution was the development of the concept of physically based distributed modelling and the European Hydrological System – Système Hydrologique Européen 'SHE', in a programme, which he initiated and led during the period 1975–1986, by forging a strong collaboration between public and private European research institutes and companies. The development of the SHE was a pioneering effort resulting in a quantum leap in hydrological modelling at its time of development. Technologically, the SHE remained for the next 1–2 decades the most advanced hydrological modelling system and it is still in use today, around 40 years after its birth. The SHE modelling philosophy was both challenged and imitated during these years. This chapter describes the obstacles to and the achievements of the SHE development, along with the first application studies and the debates generated by the scientific challenges to the SHE concept. The chapter also describes how Mike Abbott's ideas on encapsulation of knowledge in the so-called fourth-generation software systems inspired the development of user-friendly software packages, enabling professionals with domain understanding but without computer knowledge to use models. Further, the chapter describes Mike Abbott's vision of using intelligent hydroinformatics software systems to empower stakeholders and in this way democratise the decision-making processes and ensure social justice in the water sector, and briefly discusses why this vision, in contrast to the very successful ideas related to the SHE and the fourth-generation systems, had limited impact so far. Finally, the chapter discusses the strengths and weaknesses of Mike Abbott's contributions to hydrological modelling seen from today's state of the art and the impacts his contributions continue to have on hydrological modelling today.

Keywords: Integrated hydrological cycle, distributed physically based model, computational hydraulics, hydroinformatics, SHE, software systems, water management, modelling philosophy, scientific debate

4.1 INTRODUCTION

As described in Chapter 2 of this book, Michael B. Abbott (in the following referred to as Mike Abbott) was a hydraulic engineer with a strong mathematical theoretical basis and a wish to combine research, teaching and practical applications. During the 1960s he established himself as one of the leading experts in computational hydraulics. This led to the establishment of the Computational Hydraulics Centre at the Danish Hydraulic Institute (DHI) in 1970, where he pioneered the development of advanced modelling systems describing free surface hydraulic flows such as one-dimensional (1D), one-layer rivers/channels (System11, later MIKE11) and two-dimensional (2D), one-layer estuaries/oceans (System21, later MIKE21). These modelling systems turned out to become technological and commercial success stories.

At the same time, mathematical modelling in the neighbouring discipline, hydrology, was dominated by lumped conceptual modelling systems like SSARR (Rockwood, 1964), Stanford Watershed IV (Crawford & Linsley, 1966), Tank (Sugawara, 1967), NAM (Nielsen & Hansen, 1973), Sacramento (Burnash *et al.*, 1973) and HBV (Bergström & Forsman, 1973). While Mike Abbott's hydraulic modelling systems were based on complex numerical solutions to the partial differential equations describing the hydraulic flow processes (e.g., St. Venant equations in 1D and 2D), typically using small time steps and fine spatial discretisations, the hydrological models were much simpler with lumping of process descriptions to catchment units and explicit book-keeping calculations to keep track of water flows typically on a daily basis. At that time, spatially distributed, process-based hydrological models existed for only one of the hydrological sub-domains, namely groundwater (Freeze & Witherspoon, 1968; Tyson & Weber, 1964). The first attempt in the hydrological scientific community to outline the potentials for developing process based, spatially distributed catchment models was made by Freeze and Harlan (1969).

Mike Abbott considered hydrology as an immature science that was only a means of interface between rather separate and distinct, but nevertheless well-established, sciences such as soil physics, plant physiology, hydrogeology and hydraulics (Abbott, 1992). In Mike Abbott's view, hydrology lacked a solid foundation of its own, instead of relying on separate theories and taxonomies from these other, independent sciences. In a philosophical discussion, he argues that developments like the Système Hydrologique Européen (SHE) could help create a unifying hydrological science (Abbott, 1992).

It was based on this view of the hydrological science, and on the scientific, technological and commercial successes of System11 and System21, that Mike Abbott felt that the same modelling paradigms and integrated methods could contribute to advancing the technological status of catchment hydrological models. This resulted in the development of the European Hydrological System – SHE (Abbott *et al.*, 1986a, 1986b), which turned out to be a major step forward in hydrological modelling science and which, during the following decades, contributed to enhancing engineering practice.

Mike Abbott never became a hydrologist and never worked with many of the key challenges that make hydrology different from free surface flow in a homogeneous media (water), such as spatial heterogeneity of the soil and geological domains and the interaction with heterogeneously developed vegetation. Like the case of Vice Admiral Nelson, who in the Battle of Copenhagen in 1801 put the telescope to his blind eye to avoid receiving an order to retreat and ultimately won the battle, Mike Abbott's lack of attention to major hydrological challenges was perhaps an advantage for at all daring to enter into the very ambitious and risky SHE development, where he brought new thinking and new knowledge from computational hydraulics into hydrological modelling. As discussed in Section 4.3, some of the key challenges and recognised uncertainties that were not addressed in the SHE development subsequently caused the SHE approach to be controversial and subject to heavy dispute (Beven, 1989).

In addition to introducing computational methodologies in hydrological modelling Mike Abbott also advocated for encapsulation of knowledge in the so-called fourth-generation software systems which inspired the development of user-friendly software packages, enabling professionals with understanding of hydrology or hydraulics but without computer knowledge, to use models (Abbott, 1993; Abbott *et al.*, 1991). Further, Mike Abbott had a vision of using intelligent hydroinformatics software systems (Fifth Generation Software systems) to empower stakeholders and in this way democratise the decision-making processes (Abbott, 2001; Abbott & Jonoski, 2001) and ensure social justice in the water sector (Abbott & Vojinovic, 2010a, 2010b).

The objectives of this chapter are (1) to describe Mike Abbott's key contributions to hydrological modelling, and (2) to discuss Mike Abbott's achievements as seen in the context of the state of the art at that time and the state of the art today.

4.2 TERMINOLOGY AND MODEL CLASSIFICATIONS

4.2.1 Classification according to hydrological process description

Fundamentally, we distinguish between a *model code* (also called modelling system) as a generic software package and a model as a site and purpose specific setup of a model code for a particular catchment. This implies that we for instance talk about the SHE model code, but about a specific SHE model, when the code is populated with catchment-specific data and parameter values for a catchment application. This distinction is important so that we can talk about the reliability of a particular site-specific model without providing universal claims about the model code (Refsgaard & Henriksen, 2004).

Numerous hydrological model codes have been developed and a large number of classification schemes have been proposed such as Woolhiser (1973), Todini (2007), Hrachowitz and Clark (2017). In this chapter we use a classification scheme based on Refsgaard *et al.* (2022).

- *Black-box models (BB)* are empirical with mathematical equations and coefficients that are developed and assessed purely from analyses of concurrent input and output time series without considering hydrological process understanding.
- *Lumped models (LU)* include some concepts reflecting hydrological process understanding such as root zone capacity, overland flow and baseflow that are upscaled to represent hydrological response at catchment scale. These models have often been denoted conceptual models, but we do not prefer this term here, because the term conceptual is generally used to describe the understanding of the system (Hrachowitz & Clark, 2017; Refsgaard & Henriksen, 2004) and other hydrological model types also include conceptual understanding.
- *Process-based models (PB)* apply point scale (or very small scale) equations of hydrological processes and a spatially distributed representation of the catchment, enabling hydrological inputs as well as simulated outputs also to be spatially distributed. These models have often been denoted physically based.

4.2.2 Classification according to technological level

Model code software packages can be categorised according to their technological level in five generations of modelling systems (Abbott *et al.*, 1991; Chapter 2 of this book).

First Generation – Computerised Formulae. The first generation dates back to the early 1950s. These computer codes mainly aimed at making the traditional hand calculation methods easier and quicker through application of very simple numerical techniques.

Second Generation – One-Off Numerical Models. The second-generation systems appeared in the late 1950s. Computer codes were developed and applied, mostly by universities or research institutes, for solving one specific problem, that is a model for a specific catchment, aquifer or estuary. The codes were not generally applicable for use by other persons or to similar problems for example for other catchments under other external conditions. Second-generation models generally required comprehensive user experience in computer science (hardware and software) and the related numerical techniques.

Third Generation – Generalised Numerical Modelling Systems. A third-generation system was generalised so that many different problems could be solved by one and the same computer code. Third-generation modelling systems were introduced in the early 1960s. For example, 1D and 2D systems of this category were developed in 1961–1963 by the French consulting company SOGREAH, Électricité de France EDF and the Dutch research and specialist consultancy institute Delft Hydraulics and applied to various rivers and estuaries. These systems were dependent on the computers, programming languages and developer teams. They could be transferred to other computer facilities, but only with considerable efforts. These systems were developed and applied mainly by hydraulics–hydrology engineers who had knowledge of applied mathematics and experience in computer programming.

Fourth Generation – Industrial User-friendly Software Products. These are user-friendly software products, which can be applied by engineering and scientific professionals that have domain-, but not necessarily computer-, experience. Often a fourth-generation system is based on a third-generation modelling system supplemented by a graphical user interface, thorough technical documentation and built-in help facilities. Fourth-generation systems, which emerged in the 1980s, today constitute the most commonly used professional modelling software.

Fifth Generation – 'Intelligent' Modelling Systems. Here a fourth-generation modelling system is built into a generalised decision support system that can enable a technically skilled person without extensive modelling experience to calibrate and execute model runs, analyse simulation results and combine these with models or database information from other domains such as economics and ecology. Fifth generation systems are aimed at being used by multiple groups of users such as water managers, stakeholders with professional domain knowledge and interested citizens. As advocated by Abbott (1991) machine-learning techniques provide opportunities to support such systems.

4.3 THE SHE VENTURE

The presentation in this section is based on already published material (Abbott *et al.*, 1986a, 1986b; Beven *et al.*, 2015; Refsgaard *et al.*, 2010), supplemented through dialogues with two of Mike Abbott's senior partners in the development of SHE, Enda O'Connell (formerly at the then Institute of Hydrology (IH), UK, latterly at Newcastle University) and Jean Cunge (SOGREAH) as well as our own memories and unpublished notes.

4.3.1 State of the art in hydrological modelling in the early 1970s

In the early 1970s lumped models (LU) such as Stanford Watershed IV (Crawford & Linsley, 1966) and Sacramento (Burnash *et al.*, 1973) constituted the state of the art in hydrological modelling. Spatially distributed process-based models had been developed for the groundwater domain (Freeze & Witherspoon, 1968; Tyson & Weber, 1964), but not for catchment modelling. While LU models had proven capabilities in operational rainfall–runoff simulations, for example for extensions of streamflow time series and for real-time flow forecasting (WMO, 1975), there was an increasing need for assessing problems arising from the adverse impacts of human activities on the hydrological cycle, such as impacts of land-use change and simulating sediment and water quality processes at field and catchment scale, accounting for both surface water and groundwater processes. As LU process descriptions reflect not local, but catchment scale, conditions, and as LU parameter values cannot be linked directly to local field conditions, LU models were not well suited to predicting impacts of human-induced changes in specific subareas within a catchment.

Hence, there was a societal need and a science gap motivating the development of spatially distributed, process-based models using numerical techniques similar to those that had been successfully applied in computational hydraulics. The first attempt to outline the potentials and feasibility for such a model type was made by Freeze and Harlan (1969), but no operational model reflecting these ideas had yet been developed.

4.3.2 Motivation of the SHE development and creation of the SHE partnership

As described by Beven *et al.* (2015) the genesis of SHE can be traced to 1975, when the World Meteorological Organisation (WMO) invited tenders for the development of a mathematical model of the Upper Nile basin. The DHI was one of the bidders, and Enda O'Connell at the UK's then IH (now the UK Centre for Ecology and Hydrology) was invited by Mike Abbott to advise on DHI's hydrological catchment modelling. He took the view that a relatively simple systems modelling approach would be appropriate, given the limited data that would be available. However, the contract was awarded to the Australian consulting company Snowy Mountains Corporation, which proposed the use of the Sacramento model for predicting the impacts of land-use changes.

This outcome led to discussions about the need to develop a general physically based modelling system that would supersede the generation of existing lumped models, such as the Sacramento model, that had evolved from the Stanford Watershed Model. The vision was to provide a sounder physical basis for predicting the impacts of increasing human activities on the flow, water quality and sediment regimes of catchments. IH had already embarked on the development of the Institute of Hydrology Distributed Model (IHDM), while O'Connell had been influenced by his contact with Allan Freeze and his blueprint for a general physically based modelling system (Freeze & Harlan, 1969). Mike Abbott was of the view that a competitive future position for Europe depended on a major initiative that would mobilise the complementary strengths of a consortium of European partners with the relevant expertise and build a modelling system that could exploit associated technological developments in computing power, remote sensing/data capture, and so on (Beven *et al.*, 2015). The French private consulting engineering company SOGREAH, also a leader in computational hydraulic modelling, was invited to join the consortium, drawing in leading figures such as Alexander Preissmann and Jean Cunge, and the title SHE (European Hydrological System) was conceived to give an identity to the ambitious modelling system that was to be developed.

The major funding needed to develop SHE was obtained through a loan from one of the European Commission's IT programmes. The EC loan was supplemented with some co-funding from national funding sources. The EC loan implied that SHE would have to be exploited commercially to generate the necessary revenue streams to repay the loan. An association was formed in 1976 (ASHE, Association pour le SHE) to create the necessary legal entity to manage the development of the SHE and to satisfy the requirements of the EC. In 1985 the British responsibility for the SHE was transferred from the Institute of Hydrology to the Water Resources Systems Research Unit at Newcastle University. In 1990 the French responsibility was transferred from SOGREAH to the Laboratoire d'Hydraulique de France (LHF), a joint venture subsidiary of SOGREAH and the Institut National Polytechnique de Grenoble (INPG). In both cases the key scientists were also transferred.

4.3.3 The initial SHE development (1976–1986)

A series of meetings of the ASHE partners were held in 1976, during which the physical and logical structure and numerical architecture of the SHE was conceived. A major challenge at the time was the need to reconcile the structure, and numerical solution schemes chosen with the available computing power. Because the mainframe computers used by the ASHE partners offered operational memory up to about only 1 MB, a fully 3D approach was not feasible. Therefore, it was decided to simplify the water cycle processes by linking governing equations of the 2D representations of horizontal surface flow and of saturated groundwater flow with a 1D vertical representation of flow in the unsaturated zone, with the solutions of the governing physically based equations obtained on a rectangular finite difference grid (Figure 4.1). It is interesting to note that the requirements of cooperation sometimes led to solutions that were not ideal. As an example, SOGREAH offered its 2D surface flow simulation system CARIMA to be integrated into SHE. CARIMA was a new advanced system based on an irregular finite difference grid allowing for faithful representation of topographic

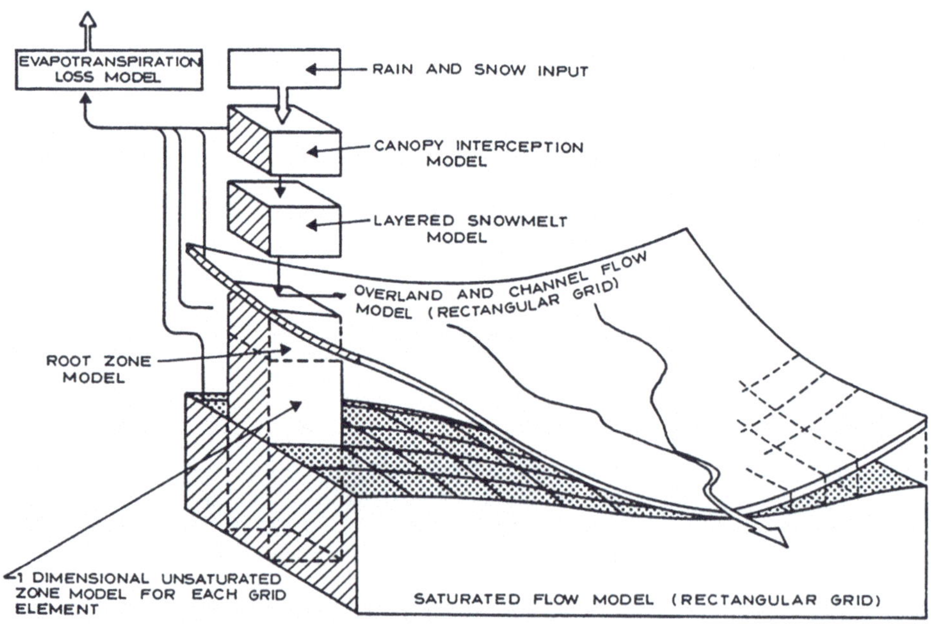

Figure 4.1 The SHE structure illustrated during the initial stage. From Abbott *et al.* (1986b), adapted from Beven *et al.* (1980).

characteristics of inundated areas. This offer was refused by the other partners, who wanted to use rectangular grids, as with the other SHE components. Thus, the SHE missed the possibility of representing overland flow realistically other than by using a very fine grid resolution, which in practice often turned out to be computationally impossible.

The first stage of the SHE development has been described by Abbott *et al.* (1986a) and Beven *et al.* (2015) and only a brief summary is given here. The SHE was structured in components with individual numerical solutions to partial differential equations for each of the principal processes for saturated zone, unsaturated zone, evapotranspiration, snow melt, overland flow and river flow. The detailed design and coding of the individual components were allocated to partner organisations according to their respective expertise. These components were then coupled through a central control element, the FRAME, steering exchange of data between components. While the numerical solutions represented state of the art for the respective components, the couplings soon turned out to pose the main challenges and probably were responsible for half of the coding efforts. In particular, the interface between the 2D saturated zone and the 1D unsaturated zone required major efforts (Storm, 1991).

In those pre-fax and pre-Internet days, communication was maintained by letter, telephone and telex. Versions of programmes and data sets were carried to meetings via magnetic tape and computer printout. Computing was carried out on main-frame computers and model runs typically occupied so much computer memory that they had to be carried out during night-time. Compared with today's working process this was not very efficient. Therefore, the development of the concept and the first prototype ready for testing took approximately five years until 1981. During the following couple of years, the code was tested and improved, before it was able to produce its first real hydrograph in 1984.

The first scientific SHE paper (Beven *et al.*, 1980) appeared after four years, but it took 10 years until the development efforts and the first model results were published in a peer-reviewed scientific journal (Abbott *et al.*, 1986a, 1986b; Bathurst, 1986a, 1986b). The simulations reported by Bathurst (1986a, 1986b) showed results for five rainfall events with durations of 3–5 days for the 10.55 km^2 River Wye research catchment in Wales. The model had a spatial grid cell resolution of 250 m × 250 m (169 grid cells). The small number of grid cells and the short simulation periods were dictated by the limited capacity and limited computational speed of the available IT infrastructure in those days. Similar experiences were gained by some of the initial researchers outside the ASHE organisation who tested the SHE code, for example the Hydrology Centre in New Zealand (Ibbitt, 1985) and the University of Braunschweig, Germany (Rohdenburg *et al.*, 1986).

4.3.4 From research code to practical applications (1986–1990s)

The four journal papers in 1986, which have been exceptionally well cited (a total of 1797 ISI Web of Science citations by 10th January 2022), marked the end of the initial SHE development phase. During the subsequent years, the code was further developed and applied to a variety of research and commercial projects, either by the individual partner organisations (Bathurst *et al.*, 1996; Erlich, 1988; Refsgaard & Storm, 1995) or in some cases jointly by all three partners (Refsgaard *et al.*, 1992).

The single most important project during this period was a collaborative effort between the three ASHE partners and the National Institute of Hydrology in India (Jain *et al.*, 1992; Lohani *et al.*, 1993; Refsgaard *et al.*, 1992). By the end of this project the SHE had matured to a well proven third-generation modelling system with a solid technical documentation and comprehensive utility programs for pre-processing of data and post-processing of simulation results. Owing to faster computers and fine-tuning of the model code it was now possible to perform simulations for catchments of several thousand km^2 for periods of several years rather than the event simulations for a small catchment performed by Bathurst (1986a). In France, SHE was used as a simulation platform to study the impact of spatial variability of the infiltration processes through crusted soils on the water resources availability (Aboujaudé, 1991).

After 1992 the SHE code development followed two different pathways. DHI developed a graphical user interface and commercialised the code into MIKE SHE, which was sold worldwide to research organisations, consulting companies and water managers. In the mid-1990s, LHF applied MIKE SHE in its flagship EUREKA research project ISMAP dealing with the problem of contamination of drinking water catchments in aquifers by phytosanitary products and the resultant need to modify the agricultural practices (Guinot, 1995). Today, MIKE SHE is an integrated part of DHI's software system (https://www.mikepoweredbydhi.com/products/mike-she) and a well-proven high-end fourth-generation software product used by numerous professionals worldwide. Meanwhile, Newcastle University further developed the SHE towards SHESED (Wicks & Bathurst, 1996) and into SHETRAN (Ewen *et al.*, 2000) with an eye towards scientific research applications, incorporating a revised subsurface hydrology component and transport components for sediment, nutrients and contaminants. This transformation was facilitated in the late 1980s and the 1990s by a major contract with UK NIREX Ltd (formerly the Nuclear Industry Radioactive Waste Executive), to simulate the potential flux of radionuclides towards the biosphere from deep underground disposal sites for radioactive waste, and by a further technology transfer project with the Chilean National Forestry Corporation (Corporación Nacional Forestal – CONAF). SHETRAN was subsequently applied in a range of research studies concerned with land use and climate change impacts and, supported by documentation and free availability (http://research.ncl.ac.uk/shetran), it has been taken up by research institutions around the world.

Nevertheless, in spite of differences in some functionalities the daughter products fundamentally reflect the original SHE structure. The ASHE organisation was officially dissolved in 1995, when the loan to the EC had been repaid.

4.4 EVALUATION OF THE SHE ACHIEVEMENTS

4.4.1 SHE: a contentious quantum leap

The SHE was the first model code to follow fully the Freeze and Harlan blueprint in integrating surface and subsurface water components into an operational, generalised modelling system of the hydrological cycle. SHE went beyond both the previous generation of LU models and the few existing PB models covering either surface runoff (Beven, 1977), variably saturated/unsaturated flow (Stephenson & Freeze, 1974) or groundwater flow (Freeze & Witherspoon, 1968). In this way, the SHE represented a quantum leap in hydrological modelling techniques. This is reflected by the fact that the two original SHE papers (Abbott *et al.*, 1986a, 1986b) with 867 and 779 ISI Web of Science citations (10th January 2022) are both among the top 25 most cited papers in *Journal of Hydrology*.

The SHE, through its daughter products MIKE SHE and SHETRAN, maintained its role as the most advanced, among well-proven hydrological model codes, for two decades. Many codes existed that were superior to SHE on single domains. However, it was not until the emergence of a number of new model codes during the first decade of this millennium that we saw codes going beyond the SHE with respect to concepts, versatility, flexible grids and numerical engines. Examples of such codes, which were based on the fully 3D Richards equation for the combined saturated/unsaturated zone, were INHM (VanderKwark & Logue, 2001), CATHY (Bixio *et al.*, 2002; Camporese *et al.*, 2010), GEOtop (Rigon *et al.*, 2006), ParFlow (Kollet & Maxwell, 2006) and HydroGeoSphere (Therrien *et al.*, 2010).

The SHE development and the underlying philosophy was, however, at the same time challenged by several prominent scientists, who considered the concept as being naïve and an expression of overselling. The following summary of this debate has been adapted from Refsgaard *et al.* (2010).

Right from the beginning the SHE concept was controversial in the scientific community. Initially, a key criticism was that the concept was too unrealistic ever to be implemented. If the SHE process equations that were based upon the conservation of mass and momentum laws had been fully correct, the parameters used in the model should have physical meaning. Based on the results of the first applications this turned out not to be the case and classical split-sample calibration–validation approaches revealed systematic errors indicating structural error due to inadequate schematisation (Todini, 1982).

In addition, scarcity of spatially distributed data at a catchment scale was considered a major barrier for SHE operational applications by the engineering companies. The 10-year long initial development stage and the severe limitations dictated by the available computer facilities indicated that the SHE was not, as originally anticipated, realistic for practical applications in the 1970s or early 1980s. Eventually, towards the late 1980s, with the progress of computational capacity and GIS technology, when it became possible to produce simulations for ordinary catchments for periods of several years through overnight computer runs, this debate ceased.

A few years after the first SHE publications Beven (1989) raised a more fundamental critique against the whole concept and the way physically based models had been promoted by for example Abbott *et al.* (1986a). He did not agree with the expectations about the capabilities and potential achievements of physically based models described in these early SHE papers. Beven pointed amongst others to the following key problems: (1) the process equations were simplifications leading to model structure uncertainty; (2) spatial heterogeneity at subgrid scale was not included in the physically based models – the current generation of distributed physically based models were in reality lumped conceptual models at a more sophisticated level; and (3) there was a great danger of over parameterisation, if it was attempted to simulate all the hydrological processes thought to be relevant while fitting the related parameters against observed discharge data only. As a conclusion, Beven argued that for future applications, attempts should be made to obtain realistic estimates of the uncertainty associated with their predictions, particularly in the case of evaluating future scenarios of the effects of management strategies.

Several other authors have contributed to the debate on the validity and usefulness of the SHE type of approach. As an example, Grayson *et al.* (1992), discussing the limitations of the process descriptions and the often quite poor accuracy of simulations internally in a catchment by physically based models, noted that 'there is a certain arrogance associated with physically-based models regarding their superiority over lumped parameter or empirical models' and that 'the real issue is whether, in practice, there is any difference between such models except for the vastly increased time required to calibrate the numerous parameters associated with physically-based models'. As another example, Seyfried and Wilcox (1995) discussed the critical importance of spatial variability in relation to process descriptions, grid scales and measurement scales. They noted the irony that while description of spatial variability is presumed to be a strength of physically based models, it is possible to include spatial variability only if it can be explicitly described. They concluded that it is necessary to include stochastic approaches to account for spatial variability of key physical parameters and state variables when using physically based models.

These criticisms were rebutted and generated new scientific discussions during the following years (Beven, 1996a, 1996b; Ewen & Parkin, 1996; Refsgaard *et al.*, 1992, 1996) and eventually led to new scientific knowledge and discourses (Bathurst *et al.*, 2004; Lukey *et al.*, 2000; Refsgaard & Henriksen, 2004; Refsgaard *et al.*, 2010, 2022) that incorporated important arguments from both sides of the original dispute.

As a conclusion, there is no doubt that the SHE development was a pioneering achievement that generated important scientific debates/disputes and in this way contributed to enhancing the hydrological modelling science.

4.4.2 Evaluation of the impact of SHE on hydrological modelling today

The SHE development has generated impacts on hydrological modelling that can be seen today, both directly in terms of model codes and indirectly through contributions to the enhancement of hydrological modelling practices.

Table 4.1 Statistics from Web of Science on journal papers citations when searching with code name as keywords.

Model Code	Search Words	ISI Web of Science Search 30th June 2009 (Refsgaard *et al.*, 2010)		ISI Web of Science Search 10th January 2022	
		Number of ISI Papers	Number of ISI Citations to Papers	Number of ISI Papers	Number of ISI Citations to Papers
SHE	'European Hydrological System' or SHE for 1982–1993	20	1206	26	3023
MIKE SHE	'MIKE SHE' or 'MIKESHE' or 'MIKE-SHE'	108	1079	432	11 237
SHETRAN	'SHETRAN' or 'SHESED'	41	437	91	2365
HydroGeoSphere	'HydroGeoSphere'	7	12	153	3210
GSFLOW	'GSFLOW'	0	–	21	525
SWAT	'SWAT' after 1990	808	5534	6700	138 193
TOPMODEL	'TOPMODEL'	370	7122	664	31 787
HBV	'HBV' and {'hydrology' or 'water'}	136	1574	594	15 932
MIKE11	'MIKE 11' or 'MIKE11' or 'MIKE–11'	76	353	457	5957
MODFLOW	'MODFLOW'	536	2110	2213	26 153
FEFLOW	'FEFLOW'	44	167	361	3911
ParFlow	'ParFlow'	Not included		119	3125

SHE model codes – scientific impact. The SHE daughter products (MIKE SHE and SHETRAN) are still widely used today. Table 4.1 shows the citations to peer-reviewed journal papers for a few selected model codes. The citation numbers are based on two searches on ISI Web of Science made in 2009 and in 2021, respectively. From the table it is seen that the SHE family of models in 2021 was represented by more than 500 journal papers that had received more than 15 000 citations. The large increase in number of SHE papers from 2009 to 2021 documents the still active use of the model codes in the hydrological science today. It is also noted that the number of SHE papers and citations during the last decade are higher than those for the more recent, and in some respects more advanced, model codes such as HydroGeoSphere and ParFlow. The longevity of MIKE SHE and SHETRAN arises from the two dedicated teams at DHI and University of Newcastle, who further developed and maintained the codes and supported their application by disseminating them to the industry and the research modelling communities.

SHE model codes – used by practitioners. The SHE daughter products, and in particular MIKE SHE, are also widely used by practitioners. An indication of the economic/commercial impact of MIKE SHE is its use in commercial projects. On the basis of the sales numbers and the size of a typical project, DHI evaluates that MIKE SHE supports a project base of about 40–50 M€/year. This is only the value of the projects to establish the MIKE SHE models and deliver the results, while the value for the end projects, where the results are being used, would be much larger (Douglas Graham, 2022). One example of an organisation that uses MIKE SHE as its standard tool is the Geological Survey of Denmark and Greenland, which during the past 25 years, has developed and maintained a national hydrological model based on MIKE SHE software – this model is the key model tool for the Danish government in relation to the Water Framework Directive implementation (Højberg *et al.*, 2013; Stisen *et al.*, 2012).

Hydrological modelling practices – simulation of discharge. An underlying key hypothesis motivating the SHE development was that more detailed process description with improved hydrological knowledge and use of more catchment data would lead to more accurate model simulations compared with what could be achieved by lumped conceptual models. This discourse was advocated by many of us when SHE was promoted at conferences up to 1990. In scientific publications such as Abbott *et al.* (1986a), the phrasing was qualified by only stating that SHE types of model were expected to be superior only for situations where lack of data made calibration impossible or confined calibration to very short periods. This hypothesis was questioned, for example by Beven (1989). Since then, several studies have documented the inability to confirm the hypothesis for cases where the simulation target is confined to river discharge (Perrin *et al.*, 2001; Reed *et al.*, 2004; Refsgaard & Knudsen, 1996).

Hydrological modelling practices – simulation of multiple variables. While the SHE approach could not document improved simulations of discharge it has been very successful in simulating and being calibrated against multiple variables. As an example, Birkinshaw (2008) in an application of SHETRAN to the 0.94 km^2 Slapton Wood catchment in southwest England successfully adopted a three-stage (multi scale) calibration strategy involving use of field data on soil water potential, phreatic surface levels as well as discharges. At a larger spatial scale, Stisen *et al.* (2012) calibrated a model for entire Denmark (43 000 km^2) using observation data from 191 discharge stations and 9198 groundwater wells. In another study, Stisen *et al.* (2018) calibrated and evaluated a model using observed data on discharge, soil moisture, evapotranspiration, land surface temperature and groundwater heads concluding that introduction of additional types of calibration data reduced the uncertainty on the calibrated parameter values and hence resulted in a more robust model.

Hydrological modelling practices – model evaluation. While the focus during the initial SHE development was on performing deterministic model simulations without consideration of simulation uncertainties, SHE models are today often used as tools in connection with comprehensive model evaluations and uncertainty assessments (Bathurst, 2011; Stisen *et al.*, 2018; Vazquez *et al.*, 2009). This reflects a general trend in modelling practice among process-based models like the SHE (Refsgaard *et al.*, 2022).

Development strategies. Different research groups are following different strategies to improve hydrological modelling (Refsgaard *et al.*, 2022). One strategy (brute force) is the one outlined by Freeze and Harlan (1969) and followed by Abbott *et al.* (1986a), namely, to use the ever-increasing computational power to add process complexity and apply finer and finer spatial grids in deterministic simulations. Another strategy (comprehensive evaluation) is to consider explicitly the uncertainty in data and models and use advanced model evaluation tools to assess the reliability of parameter values and model simulations. The two approaches are competing and complementary. Both can provide valuable scientific insights. Although the brute force strategy may not result in more reliable models as such, it is even today pushing the borders for model applications with more sophisticated and detailed process descriptions, larger numbers of small-scale computational grids and couplings with other domains (e.g. Kollet *et al.*, 2018). The brute force strategy underlying the SHE development (Abbott *et al.*, 1986a) was for a couple of decades the dominating strategy adopted by a large part of the hydrological modelling community, and it is still followed today by some research groups.

4.5 OTHER HYDROINFORMATICS CONTRIBUTIONS

4.5.1 Encapsulation of knowledge in digital modelling systems

Mike Abbott's overall aim was to bring hydraulic and hydrological science to practical use by professionals and ultimately ordinary citizens. He was therefore continuously striving towards developing technologies and software systems that were generically applicable for a variety of purposes and contexts. The key approach in this endeavour was to encapsulate knowledge in software, so that the users did not need to be experts in everything (Abbott, 1993; Abbott *et al.*, 1991).

Inspired by experience from computational hydraulics, the SHE was designed and became a third-generation modelling system. Practical use of the original version of the SHE code required considerable skills in computer and numerical techniques and was therefore constrained to relatively few, mainly academic, organisations.

The next step towards meeting the ambition of making the modelling technology available for a larger audience was to develop fourth-generation software systems. Inspired by the great technological and commercial successes of transferring some of DHI's hydraulic modelling systems to software packages, such as MOUSE (urban modelling) and MIKE 11 (river modelling), DHI decided in the mid-1990s to further develop SHE into a fourth-generation system denoted MIKE SHE. MIKE SHE was initially running on Unix-based computers but was soon converted to a Microsoft platform. The incomes from selling MIKE SHE to organisations worldwide generated the necessary income to maintain and further develop the MIKE SHE software. Without these sales incomes DHI would not have been able to keep its version of SHE alive (Refsgaard *et al.*, 2010). MIKE SHE never became the same commercial success as for example MOUSE and MIKE 11 (Refsgaard *et al.*, 2010) and relative to for example MIKE 11 it had a larger share of its users in research institutions, which is reflected by the scientific contributions shown in Table 4.1. Nevertheless, insofar as there is an industry standard for a physically based catchment modelling system, MIKE SHE may be considered as the best example.

4.5.2 Intelligent software systems to support stakeholders in water management

The next step in making the modelling systems useful for society was development of fifth-generation modelling systems. In this respect, Mike Abbott had a vision of using intelligent hydroinformatics software systems to empower stakeholders and in this way democratise the decision-making processes (Abbott, 2001; Abbott & Jonoski, 2001) and ensure social justice in the water sector (Abbott & Vojinovic, 2010a, 2010b). The functionality aimed at making this possible was the integration of databases, models and other decision support elements including what we today would classify as machine-learning tools – everything coupled with tools to ease dialogues with a variety of stakeholders.

The underlying philosophy behind Mike Abbott's vision was anchored in European philosophical literature (Kierkegaard, Heidegger, etc.) and as such theoretically well founded. It is, however, remarkable that the ideas and discussion in his papers (Abbott, 2001; Abbott & Jonoski, 2001; Abbott & Vojinovic, 2010a, 2010b) were completely detached from the discussion and discourse that during the same years emerged in the international water management community leading to the much broader concept of integrated water resources management (ICWE, 1992; Jønch-Clausen, 2004). A possible explanation for this detachment may be the different point of departures. IWRM was driven by social and political science and the main barrier for improved water management practices was not considered to be a lack of technology. Hydroinformatics and Mike Abbott's thinking, on the other hand, had a focus on technological developments. Another explanation could be that Mike Abbott's aim of empowering citizens and stakeholders represented a more radical change in power structure than IWRM, which operated within existing government structures.

The development of such systems has proven more difficult than the third- and fourth-generation systems, maybe because they operate in a complex world of often conflicting use of available water resources, beyond technical logic, where political power aspects often trump sophisticated technology. In any case, it is clear that, while third- and fourth-generation systems became mainstream soon after their introduction, this was not the case for fifth-generation systems.

Recent developments with increasing use of water-related serious gaming, in which often collaborative modelling is part of the supporting technology and decision-making process (Aubert *et al.*, 2018), may be seen as an indication that fifth-generation systems after several decades may achieve an important role related to citizen science – in line with Mike Abbott's original vision. However, it is probably fair to state that Mike Abbott did not live long enough to see his fifth-generation vision fully unfold and materialise.

4.6 CONCLUSIONS

Mike Abbott's ideas on hydrology reflected his background in computational hydraulics as well as the thinking in major parts of the hydrological modelling community in the 1970s. This implied that Mike Abbott never addressed the many uncertainties in hydrological modelling caused by limited data availability, parameter uncertainty and model structure uncertainty that are at the top of the scientific agenda today. At the same time, Mike Abbott was very optimistic about technology and the capabilities of model codes like SHE and, as many other model developers at that time, presented the ideas in a way that in hindsight would be called 'overselling'.

Mike Abbott was of the opinion that a single deterministic model could provide the right answers. One of the authors remembers a discussion he had with Mike Abbott around 1990. On his remark that approximate solutions of Navier–Stokes equations led to construction of aircraft while distributed modelling with SHE (at that time) did not bring reliable, generic and transferable solutions, Mike Abbott concluded that hydrological processes are much more complicated, and their limited observability has not allowed the expected progress.

Mike Abbott made two significant novel contributions to hydrological modelling. Most importantly, the development of SHE (Abbott *et al.*, 1986a, 1986b) was a pioneering achievement creating a quantum leap in hydrological modelling science. The two SHE daughter codes, still widely used today in both commercial and research applications, provided a new and practical means of investigating land-use and climate change impacts for future altered catchment states. This stimulated both discussion and further development in the hydrological modelling community that is recorded in some of the most widely referenced journal papers in the subject. In addition, he promoted the development of MIKE SHE as the first hydrological fourth-generation modelling system of its kind, which resulted in a widespread dissemination of SHE technology worldwide and contributed to enhancing modelling practice among professionals.

As an engineer, Mike Abbott was a counter example of George Pompidou's assertion that, of the several ways to ruin, while the fastest and the pleasantest could be found amongst the vices, the surest is with technicians! Although contested, Mike Abbott constantly stressed the importance for the hydrological community of leaving its comfort zone, pushing the technical and technological boundaries and thereby having to face numerous challenges and take innovative risks.

In addition to his knowledge on computational hydraulics, Mike Abbott's key contribution to the SHE development was that he formed a strong international consortium that was capable of developing this ground-breaking catchment modelling code. Mike Abbott had a charismatic personality and the scientific credibility required to persuade the top management in three European organisations to engage and commit to this unknown, highly ambitious and risky adventure that turned out to take 10 years until the first scientific benefits emerged and even longer for the commercial benefits to materialise. In part this was achieved through a philosophical approach that enabled him to bypass the self-imposed constraints of narrower viewpoints and to maintain a vision of the higher destination of hydroinformatics.

ACKNOWLEDGEMENTS

We appreciate the valuable review comments and suggestions from Andreja Jonoski and Schalk Jan van Andel.

REFERENCES

Abbott M. B. (1991). Hydroinformatics: Information Technology and the Aquatic Environment. Avebury Technical, Aldershot, Brookfield, USA.

Abbott M. B. (1992). The theory of hydrologic model, or: the struggle for the soul of hydrology. In: Advances in Theoretical Hydrology, a Tribute to James Dooge, J. P. O'Kane (ed.), Elsevier, Amsterdam, pp. 237–254.

Abbott M. B. (1993). The electronic encapsulation of knowledge in hydraulics, hydrology and water resources. *Advances in Water Resources*, **16**, 21–39, https://doi.org/10.1016/0309-1708(93)90027-D

Abbott M. B. (2001). The democratization of decision making processes in the water sector I. *Journal of Hydroinformatics*, **3**(1), 23–34, https://doi.org/10.2166/hydro.2001.0004

Abbott M. B. and Jonoski A. (2001). The democratization of decision making processes in the water sector II. *Journal of Hydroinformatics*, **3**(1), 35–48, https://doi.org/10.2166/hydro.2001.0005

Abbott M. B. and Vojinovic Z. (2010a). Realising social justice in the water sector: 1. *Journal of Hydroinformatics*, **12**(1), 97–117, https://doi.org/10.2166/hydro.2010.065

Abbott M. B. and Vojinovic Z. (2010b). Realising social justice in the water sector: 2. *Journal of Hydroinformatics*, **12**(2), 225–239, https://doi.org/10.2166/hydro.2009.065

Abbott M. B., Bathurst J. C., Cunge J. A., O'Connell P. E. and Rasmussen J. (1986a). An introduction to the European Hydrological System – Système Hydrologique Européen 'SHE', 1: history and philosophy of a physically-based distributed modelling system. *Journal of Hydrology*, **87**, 45–59, https://doi.org/10.1016/0022-1694(86)90114-9

Abbott M. B., Bathurst J. C., Cunge J. A., O'Connell P. E. and Rasmussen J. (1986b). An introduction to the European Hydrological System – Système Hydrologique Européen 'SHE', 2: structure of a physically-based distributed modelling system. *Journal of Hydrology*, **87**, 61–77, https://doi.org/10.1016/0022-1694(86)90115-0

Abbott M. B., Havno K. and Lindberg S. (1991). The fourth generation of modelling in hydraulics. *Journal of Hydraulic Research*, **29**(5), 581–600, https://doi.org/10.1080/00221689109498978

Aboujaudé A. (1991). Modélisation hydrologique de l'infiltration dans les sols encroûtés à différentes échelles spatiales (Hydrological Modelling of Infiltration Through Crusted Soils at Different Spatial Scales). PhD thesis, Grenoble University.

Aubert A. H., Bauer R. and Lienert J. (2018). A review of water-related serious games to specify use in environmental multi-criteria decision analysis. *Environmental Modelling & Software*, **105**, 64–78, https://doi.org/10.1016/j.envsoft.2018.03.023

Bathurst J. C. (1986a). Physically-based distributed modelling of an upland catchment using the Système Hydrologique Européen. *Journal of Hydrology*, **87**, 79–102, https://doi.org/10.1016/0022-1694(86)90116-2

Bathurst J. C. (1986b). Sensitivity analysis of the Système Hydrologique Européen for an upland catchment. *Journal of Hydrology*, **87**, 103–123, https://doi.org/10.1016/0022-1694(86)90117-4

Bathurst J. C. (2011). Predicting impacts of land use and climate change on erosion and sediment yield in river basins using SHETRAN. In: Handbook of Erosion Modelling, R. P. C. Morgan and M. A. Nearing (eds), Wiley-Blackwell Publishing Ltd, New Jersey, pp. 263–288.

Bathurst J. C., Kilsby C. and White S. (1996). Modelling the impacts of climate and land-use change on basin hydrology and soil erosion in Mediterranean Europe. In: Mediterranean Desertification and Land Use, C. J. Brandt and J. B. Thornes (eds), John Wiley and Sons Ltd., Chichester, UK, pp. 355–387.

Bathurst J. C., Ewen J., Parkin G., O'Connell P. E. and Cooper J. D. (2004). Validation of catchment models for predicting land-use and climate change impacts. 3. Blind validation for internal and outlet responses. *Journal of Hydrology*, **287**, 74–94, https://doi.org/10.1016/j.jhydrol.2003.09.021

Bergström S. and Forsman A. (1973). Development of a conceptual deterministic rainfall-runoff model. *Nordic Hydrology*, **4**, 147–170, https://doi.org/10.2166/nh.1973.0012

Beven K. J. (1977). Hillslope hydrographs by the finite element method. *Earth Surface Processes and Landforms*, **2**, 13–28, https://doi.org/10.1002/esp.3290020103

Beven K. (1989). Changing ideas in hydrology – the case of physically based models. *Journal of Hydrology*, **105**, 157–172, https://doi.org/10.1016/0022-1694(89)90101-7

Beven K. J. (1996a). A discussion of distributed hydrological modelling. In: Distributed Hydrological Modelling, M. B. Abbott and J. C. Refsgaard (eds), Kluwer Academics Publishers, Dordrecht, the Netherlands, pp. 255–278.

Beven K. J. (1996b). Response to comments on 'A discussion of distributed hydrological modelling' by J.C. Refsgaard *et al.* In: Distributed Hydrological Modelling, M. B. Abbott and J. C. Refsgaard (eds), Kluwer Academics Publishers, Dordrecht, the Netherlands, pp. 289–295.

Beven K., Warren R. and Zaoui J. (1980). SHE: towards a methodology for physically-based distributed forecasting in hydrology. *IAHS Publication*, **129**, 133–137.

Beven K., Bathurst J., O'Connell E., Littlewood I., Blackie J. and Robinson M. (2015). Hydrological modelling. Chapter 7. In: Progress in Modern Hydrology: Past, Present and Future, J. C. Rodda and M. Robinson (eds), Wiley & Sons Ltd, Chichester, UK, pp. 216-239.

Birkinshaw S. J. (2008). Physically-based modelling of double-peak discharge responses at Slapton Wood catchment. *Hydrological Processes*, **22**, 1419–1430, https://doi.org/10.1002/hyp.6694

Bixio A. C., Gambolati G., Paniconi C., Putti M., Shestopalov V. M., Bublias V. N., Bohuslavsky A. S., Kasteltseva N. B. and Rudenko Y. F. (2002). Modeling groundwater-surface water interactions including effects of morphogenetic depressions in the Chernobyl exclusion zone. *Environmental Geology*, **42**(2–3), 162–177, https://doi.org/10.1007/s00254-001-0486-7

Burnash R. F. J., Ferral R. L. and McGuire R. A. (1973). A General Streamflow Simulation System – Conceptual Modelling for Digital Computers. US Department of Commerce, National Weather Service and State of California, Department of Water Resources, Sacramento.

Camporese M., Paniconi C., Putti M. and Orlandini S. (2010). Surface-subsurface flow modeling with path-based runoff routing, boundary condition-based coupling, and assimilation of multisource observation data. *Water Resources Research*, **46**, W02512, https://doi.org/10.1029/2008WR007536

Crawford N. H. and Linsley R. K. (1966). Digital simulation in hydrology, Stanford Watershed Model IV. Department of Civil Engineering, Stanford University, Technical Report 39.

Erlich M. (1988). On the modelling of the infiltration process in arid zones for irrigation project purposes with the aid of the 'Système Hydrologique Européen' (SHE). *Agricultural Water Management*, **13**, 195–210, https://doi.org/10.1016/0378-3774(88)90154-0

Ewen J. and Parkin G. (1996). Validation of catchment models for predicting land-use and climate change impacts. 1. Method. *Journal of Hydrology*, **175**, 583–594, https://doi.org/10.1016/S0022-1694(96)80026-6

Ewen J., Parkin G. and O'Connell P. E. (2000). SHETRAN: distributed river basin flow and transport modelling system. Proceedings of American Society of Civil Engineers. *Journal of Hydrologic Engineering*, **5**, 250–258, https://doi.org/10.1061/(ASCE)1084-0699(2000)5:3(250)

Freeze R. A. and Harlan R. L. (1969). Blueprint for a physically-based digitally-simulated hydrologic response model. *Journal of Hydrology*, **9**, 237–258, https://doi.org/10.1016/0022-1694(69)90020-1

Freeze R. A. and Witherspoon P. A. (1968). Theoretical analysis of regional groundwater flow: 3. Quantitative interpretations. *Water Resources Research*, **4**, 581–590, https://doi.org/10.1029/WR004i003p00581

Graham D. (2022). Personal communication. DHI, Hørsholm, Denmark.

Grayson R. B., Moore I. D. and McHahon T. A. (1992). Physically based hydrologic modelling, 2. Is the concept realistic?. *Water Resources Research*, **26**(10), 2659–2666, https://doi.org/10.1029/92WR01259

Guinot V. (1995). Modélisation mécaniste du devenir des produits phytosanitaires dans l'environnement souterrain: application à la protection des captages en aquifère. PhD thesis, Université Joseph-Fourier – Grenoble I.

Hrachowitz M. and Clark M. P. (2017). HESS Opinions: the complementary merits of competing modelling philosophies in hydrology. *Hydrology and Earth System Sciences*, **21**, 3953–3873, https://doi.org/10.5194/hess-21-3953-2017

Højberg A. L., Troldborg L., Stisen S., Christensen B. B. S. and Henriksen H. J. (2013). Stakeholder driven update and improvement of a national water resources model. *Environmental Modelling & Software*, **40**, 202–213, https://doi.org/10.1016/j.envsoft.2012.09.010

Ibbitt R. (1985). SHE – Model of the '80s. STREAMLAND 39, Ministry of Works and Developments, New Zealand.

ICWE. (1992). The Dublin statement and report of the conference. International Conference on Water and the Environment: Development Issues for the 21st Century, 26–31 January 1992, Dublin, Ireland.

Jain S. K., Storm B., Bathurst J. C., Refsgaard J. C. and Singh R. D. (1992). Application of the SHE to catchments in India – part 2: field experiments and simulation studies with the SHE on the Kolar subcatchment of the Narmada River. *Journal of Hydrology*, **140**, 25–47, https://doi.org/10.1016/0022-1694(92)90233-L

Jønch-Clausen T. (2004). Integrated Water Resources Management (IWRM) and Water Efficiency Plans by 2005. Global Water Partnership. TEC Background Papers, No 10. Stockholm.

Kollet S. J. and Maxwell R. M. (2006). Integrated surface-groundwater flow modeling: A free-surface overland flow boundary condition in a parallel groundwater flow model. *Advances in Water Resources*, **29**(7), 945–958, https://doi.org/10.1016/j.advwatres.2005.08.006

Kollet S. J., Gasper F., Brdar S., Goergen K., Hendricks-Franssen H. J., Keune J., Kurtz W., Küll V., Pappenberger F., Poll S., Trömel S., Shrestha P., Simmer C. and Sulis M. (2018). Introduction of an experimental terrestrial forecasting/monitoring system at regional to continental scales bases on the terrestrial system modeling platform (v1.1.0). *Water*, **10**, 1697, https://doi.org/10.3390/w10111697

Lohani V. K., Refsgaard J. C., Clausen T., Erlich M. and Storm B. (1993). Application of the SHE for irrigation command area studies in India. *Journal of Irrigation and Drainage Engineering*, **119**(1), 34–49, https://doi.org/10.1061/(ASCE)0733-9437(1993)119:1(34)

Lukey B. T., Sheffield J., Bathurst J. C., Hiley R. A. and Mathys N. (2000). Test of the SHETRAN technology for modelling the impact of reforestation on badlands runoff and sediment yield at Draix, France. *Journal of Hydrology*, **235**, 44–62, https://doi.org/10.1016/S0022-1694(00)00260-2

Nielsen S. A. and Hansen E. (1973). Numerical simulation of the rainfall-runoff process on a daily basis. *Nordic Hydrology*, **4**, 171–190, https://doi.org/10.2166/nh.1973.0013

Perrin C., Michel C. and Andréassian V. (2001). Does a large number of parameters enhance model performance? Comparative assessment of common catchment model structures on 429 catchments. *Journal of Hydrology*, **242**, 275–301, https://doi.org/10.1016/S0022-1694(00)00393-0

Reed S., Koren V., Smith M., Zhang Z., Moreda F. and Seo D. J. (2004). Overall distributed model intercomparison project results. *Journal of Hydrology*, **298**, 27–60, https://doi.org/10.1016/j.jhydrol.2004.03.031

Refsgaard J. C. and Henriksen H. J. (2004). Modelling guidelines – terminology and guiding principles. *Advances in Water Resources*, **27**(1), 71–82, https://doi.org/10.1016/j.advwatres.2003.08.006

Refsgaard J. C. and Knudsen J. (1996). Operational validation and intercomparison of different types of hydrological models. *Water Resources Research*, **32**(7), 2189–2202, https://doi.org/10.1029/96WR00896

Refsgaard J. C. and Storm B. (1995). MIKE SHE. In: Computer Models in Watershed Hydrology, V. J. Singh (ed.), Water Resources Publication, Highlands Ranch, CO, USA, pp. 809–846.

Refsgaard J. C., Seth S. M., Bathurst J. C., Erlich M., Storm B., Jørgensen G. H. and Chandra S. (1992). Application of the SHE to catchments in India – part 1: general results. *Journal of Hydrology*, **140**, 1–23, https://doi.org/10.1016/0022-1694(92)90232-K

Refsgaard J. C., Storm B. and Abbott M. B. (1996). Comment on 'A discussion of distributed hydrological modelling' by K. Beven. In: Distributed Hydrological Modelling, M. B. Abbott and J. C. Refsgaard (eds), Kluwer Academics Publishers, Dordrecht, the Netherlands, pp. 279–287.

Refsgaard J. C., Storm B. and Clausen T. (2010). Système Hydrologique Europeén (SHE) – review and perspectives after 30 years development in distributed physically-based hydrological modelling. *Hydrology Research*, **41**(5), 355–377, https://doi.org/10.2166/nh.2010.009

Refsgaard J. C., Stisen S. and Koch J. (2022). Hydrological process knowledge in catchment modelling – lessons and perspectives from 60 years development. *Hydrological Processes*, **36**, e14463, https://doi.org/10.1002/hyp.14463

Rigon R., Bertoldi G. and Over T. M. (2006). GEOTOp: a distributed hydrological model with coupled water and energy budgets. *Journal of Hydrometeorology*, **7**, 371–388, https://doi.org/10.1175/JHM497.1

Rockwood D. M. (1964). Streamflow Synthesis and Reservoir Regulation. US Army Engineer Division, North Pacific, Portland, Oregon. Engineering Studies Project 171, Technical Bulletin No. 22.

Rohdenburg H. B., Diekkrüger B. and Bork H. R. (1986). Deterministic hydrological site and catchment models for the analysis of agro-ecosystems. *Catena*, **13**, 119–137, https://doi.org/10.1016/S0341-8162(86)80008-X

Seyfried M. S. and Wilcox B. P. (1995). Scale and the nature of spatial variability: field examples having implications for hydrologic modelling. *Water Resources Research*, **31**(1), 173–184, https://doi.org/10.1029/94WR02025

Stephenson G. R. and Freeze R. A. (1974). Mathematical simulation of subsurface flow contributions to snowmelt runoff, Reynolds Creek watershed, Idaho. *Water Resources Research*, **10**(2), 284–294, https://doi.org/10.1029/WR010i002p00284

Stisen S., Højberg A. L., Troldborg L., Refsgaard J. C., Christensen B. S. B., Olsen M. and Henriksen H. J. (2012). On the importance of appropriate precipitation gauge catch correction for hydrological modelling at mid to high latitudes. *Hydrology and Earth System Sciences*, **16**, 4157–4176, https://doi.org/10.5194/hess-16-4157-2012

Stisen S., Koch J., Sonnenborg T. O., Refsgaard J. C., Bircher S., Ringgaard R. and Jensen K. H. (2018). Moving beyond runoff calibration – multi-variable optimization of a surface-subsurface-atmosphere model. *Hydrological Processes*, **32**(17), 2654–2668, https://doi.org/10.1002/hyp.13177

Storm B. (1991). Modelling of saturated flow and the coupling of the surface and subsurface flow. In: Recent Advances in the Modelling of Hydrologic Systems, D. S. Bowles and P. E. O'Connell (eds), Kluwer, Dordrecht, pp. 185–203.

Sugawara M. (1967). The Flood Forecasting by a Series Storage Type Model. IAHS Publication, Wallingford, UK, **85**, pp. 1–6.

Therrien R., McLaren R. G., Sudicky E. A. and Park Y. J. (2010). HydroGeoSphere: A Three-Dimensional Numerical Model Describing Fully-Integrated Subsurface and Surface Flow and Solute Transport. Aquanty Inc., Waterloo, Ontario, Canada.

Todini E. (1982). *Sensitivity analysis of SHE* (private communication).

Todini E. (2007). Hydrological catchment modelling: past, present and future. *Hydrological Earth System Sciences*, **11**(1), 468–482, https://doi.org/10.5194/hess-11-468-2007

Tyson H. N. and Weber E. M. (1964). Computer simulation of groundwater basins. *Journal of Hydraulics Division*, **90**(HY4), 59–77, Proceedings of the ASCE, https://doi.org/10.1061/JYCEAJ.0001093

VanderKwark J. E. and Logue K. (2001). Hydrologic-response simulations for the R-5 catchment with a comprehensive physics-based model. *Water Resources Research*, **37**(4), 999–1013, https://doi.org/10.1029/2000WR900272

Vazquez R. F., Beven K. and Feyen J. (2009). GLUE Based assessment on the overall predictions of a MIKE SHE application. *Water Resources Management*, **23**, 1325–1349, https://doi.org/10.1007/s11269-008-9329-6

Wicks J. M. and Bathurst J. C. (1996). SHESED: A physically-based, distributed erosion and sediment yield component for the SHE hydrological modelling system. *Journal of Hydrology*, **175**(1–4), 213–238, https://doi.org/10.1016/S0022-1694(96)80012-6

WMO. (1975). Intercomparison of conceptual models used in operational hydrological forecasting. WMO Operational Hydrology Report No 7, WMO No 429. World Meteorological Organization, Geneva.

Woolhiser D. A. (1973). Hydrologic and watershed modeling – state of the art. *Transactions of the ASAE*, **16**, 533–559, https://doi.org/10.13031/2013.37562

doi: 10.2166/9781789062656_0057

Chapter 5

Hydroinformatics as a 'game changer' in the water business

E. Zeman[1], *T. Metelka*[2], *P. Ingeduld*[3] *and S. Shipton*[4]

[1]*Global Change Research Institute CAS, Brno, Czech Republic*
[2]*Aquaprocon s.r.o., Climate Change Adaptation Department, Prague, Czech Republic*
[3]*DHI a.s., Prague, Czech Republic*
[4]*DHI Water and Environment Ltd, Auckland, New Zealand*

ABSTRACT

This chapter presents the field of hydroinformatics as a game-changer in the water sector business. It reflects on Mike Abbott's original ideas and evaluates their impact from the perspective of water business owners, water consultants, clients, and communities. Mike Abbott promoted a novel concept, combining theoretical and practical aspects of modelling, towards 'industrial modelling systems'. The chapter traces the development of this concept, from its early adoption by government-funded research institutes engaged in scientific support to hydraulic engineering, through establishment of market for water-modelling software products, to its broad adoption by consultants. Mike Abbott also saw a significant opportunity for hydroinformatics in the developing world. New policies and businesses were easier to implement in countries without a mature market of ideas resisting a change. Many of his followers were eager to explore business paths, which were much faster and more agile than those in academic circles. Most large companies that are currently developing commercial applications and tools for water modelling matured during the so-called 'third modelling generation', when generic modelling tools started to be developed. Subsequently, the real growth was observed when the knowledge encapsulation and robustness of the applications expanded their use from specialists to a broad pool of water professionals and practitioners. The social impact of hydroinformatics was significant as it spread through a wide range of organisations impacting decision making, education, innovation and knowledge sharing. As modelling systems became a norm in the water business, the business itself has been transformed. Impacts of current technologies and new data sources on water modelling business are presented (such as the Cloud, High Performance Computing – HPC, virtualisation, internet of things – IoT, remote sensing, and others), and how they impact current needs of software producers and users. Diverse business models have emerged, which need to meet the future challenges of urban and environmental water management under challenges of climate change. With the large and continuously growing community of modelling experts and users, it can be argued that the Mike Abbott's vision of hydroinformatics has been well established.

Keywords: Hydroinformatics, water business, simulation models, GIS, cloud computing, collaboration

5.1 INTRODUCTION

Research institutes and universities, with their scientific and technical competencies formed a base in the traditional water business, providing the needed expert knowledge to contractors and consultants. This business was primarily associated with the need for water *engineering* works. Even though the complexity of natural processes and their interconnectedness in water environments were known for centuries, sophisticated tools were not in place to study important interactions in full complexity. Water business understood that, and accepted relatively simplified approaches and design methods based on established best practices and standards, with generous safety margins when designing and implementing water infrastructure. Most systems were considered static, ignoring rapid changes and trends. Furthermore, the lack of verified data sets and rudimentary monitoring systems, often based on indirect mechanical sensors, simply did not permit desired progress of new methods and holistic approaches as we know them today.

The engineering practice was also characterised with ignorance regarding environmental concerns, such as ecology and biodiversity. The minimal efforts to evaluate the coexistence of industrial and environmental worlds and their interconnections were insufficient for using a fully holistic approach. Further demand for maximum yields from any resource led to negative ecological consequences, sometimes even disasters in many regions in the past, such as polluting groundwater aquifers by chemical waste, inappropriate irrigation practices, excessive drainage, unsustainable withdrawal of water or ignoring sediment transport processes.

In the second half of 20th century, and especially since the 1970s, there has been increasing concern that the existing approach is non-sustainable and often detrimental to the natural environments, and that new, more holistic approaches are needed. The process of growing environmental awareness was in fact coinciding in time with important developments in information and communication technologies (ICTs), with increasing computing power and communication possibilities. Such ICT developments enabled the inclusion of more processes in the algorithms used in simulation models and tools, allowing for simulations of more complex water systems and aquatic environments.

To harness the emerging ICT power, some of the knowledge and technology leaders started developing their own computational centres, where they could execute extensive studies using novel ICT elements. In the beginning, the knowledge was guarded and not shared outside their inner circles. However, the growth of computational knowledge was increasingly being encapsulated within the tools developed for research or projects, paving the way for the emergence of a new discipline – *hydroinformatics*, and its application in a broad business environment.

An important advantage provided by the new modelling tools was that they enabled easier setup and execution of 'what-if-then'-based simulations, which help to determine in advance the consequences of different interventions. What-if-then simulations are unparalleled for understanding system consequences in situations that could cause severe damage to the physical system and cannot be therefore tested in reality (such tests would be too expensive, largely unsustainable and without possibilities for insights into long-term implications). Many types of such analyses, previously only possible with hydraulic scale models, started to become available with the new simulation tools (see Chapter 2 of this book). Using a hydraulic model, various applications of what-if-then simulations started to be used in water safety and early warning, design, planning and maintenance, and online training simulators for system operators.

Such advantages were initially known only in academic circles; however, the business environment recognised the potential advantage of using the new hydroinformatics tools in their consultancy. Standardisation of software tools and trained employees allowed consultants to use these new possibilities for their clients more often and with increased benefits.

This expansion of hydroinformatics tools into the water consultancy business was not accidental. Selected individuals and institutions were having the vision and leading the activities to achieve these goals. Mike Abbott, Jean Cunge and Roland Price could be regarded as the most influential computational hydraulics experts of their time and, at the same time, visionaries standing at the cradle of hydroinformatics. Other key experts joined them to spread the knowledge and development of applications, both in academic and consultancy fields. It is not easy to distinguish between the role of individuals and institutions in these developments, especially in consultancy related to water. The role of institutions was essential for the early development of computational hydraulics tools when investment in these pioneering trends was risky and uncertain. The spread of knowledge was important, but the financial support to stimulate progress could be obtained only by strong institutes with shrewd leaders.

As Price (1997) stated, hydroinformatics addressed the commercial market for information technology (IT) goods to manage water. He sees three categories of providers of such IT goods: (1) the system houses (such as LOGICA, CAP-GEMINI, etc.), (2) hydraulic laboratories and research institutes in water (Delft Hydraulics, DHI, HR Wallingford, SOGREAH and others) and (3) consultants in the water sector (such as SWECO, CDM, EDF, COWI) competing for contracts. Each category has own interests and provides related IT solutions, but for the emergence of hydroinformatics, relevant are the roles of the last two categories. The hydraulic laboratories and research institutes carried out research in computational modelling and eventually became providers of commercial software – modelling systems and tools, which were then used by the consultants. The tools eventually included both simulation engines and full graphical user interfaces (GUIs). Further, some universities accomplished significant software development activities toward full-scale experiments within their professional fields of expertise. The emerging fields of application were, according to Price (1997), Environmental Hydroinformatics and Urban Hydroinformatics. His prediction was quite accurate, and today we would add the current climate change impacts and needed adaptations as an additional important factor influencing further development in hydroinformatics.

Leading research institutes in Europe (DHI, HR Wallingford, SOGREAH) and academic institutes and universities (IHE Delft, Delft University of Technology – TU Delft, Technical University of Denmark – DTU, and others) created momentum at the right time. It produced ideas, tools and followers, and this exploding mixture spread hydroinformatics into practice very fast in the 1990s. There was an interesting development in the USA and partly in German academic circles, where the focus was mainly on developing engines with little or no interest in developing GUIs. In the USA, grants from governmental agencies such as the US Army Corps of Engineers, the US Environmental Agency and the US Geological Survey supported the development of very robust engines with simple GUIs to be used and downloaded free of charge by users mainly in the US, but later in the entire world. More sophisticated GUIs for these simulation engines were being developed by commercial companies. For quite some time, the developments in Europe and America were different. While European software vendors were relatively successful in the US market in the new millennium, only several US-made engines were applied in Europe (such as HEC family of tools or EPANET). This situation has changed over the years, and major US commercial companies operating globally have also become successful in Europe, also via mergers and acquisitions (e.g., Bentley Systems, Innovyze, etc.). Today, that difference has largely disappeared, and both professional and free modelling software and hydroinformatics tools are equally used in markets worldwide. Current software tools based on sophisticated ICT developments can solve various problems in very complex systems and find the optimal setup of a system even under dynamic conditions. The impossible of the past became possible, even expected, in recent times.

To understand how and why hydroinformatics became a game-changer, it is essential to understand its roots and evolution, as described by Abbott (1991) through the classification of five generations of models and modelling systems and their impact on the business markets. The possibility of using and/or developing 'industrial modelling systems' as a new engineering tool was only possible when hardware and software became accessible to users working in the open markets. Therefore, we can say that the starting line of the game was established by introducing the third generation of models when the key leading research institutes in the water sector started to offer tools with *encapsulated knowledge* (Abbott, 1993). The simulation tools were sold in a form that a user with access to a computer and sufficient skills could utilise the tool for solving particular water problem. The tools rapidly matured and became more sophisticated, while computers were getting more affordable. Actual *modelling systems* (eventually to become most successful commercial products) came with the fourth generation of models, covering multiple generic modules for different hydraulic processes and sophisticated GUIs for pre- and post-processing. The fifth-generation models, as predicted by Mike Abbott in 1991, would combine computational hydraulics, data processing and knowledge-based systems based on traditional artificial intelligence (AI). Aspects of such systems are still in development, but many of these predictions were already confirmed by clear evidence in hydroinformatics applications deployed throughout the water and environment segments. Mike Abbott's foresight on how information technology will develop and enable the development of hydroinformatics and its crucial role in the water business was visionary.

The remaining parts of this chapter aim to trace these developments in more detail, and to provide some ideas about accomplishments and future opportunities and challenges from the perspective of engineering practitioners and consultants. Section 5.2 presents the evolution of hydroinformatics, followed by Section 5.3 where its key technical and commercial aspects are presented. The ways in which hydroinformatics realises its potential in broader social and political contexts of water management are addressed in Section 5.4. Future business opportunities and challenges are presented in Sections 5.5 and 5.6, respectively.

5.2 EVOLUTION OF HYDROINFORMATICS

Abbott (1991) recognises five fundamental periods in the hydroinformatics development. While this categorisation is subjective and open-ended, it does provide truthful description of the hydroinformatics evolution over the last 50–60 years, under changing societal and technological environments.

5.2.1 Building the fundaments: 1960s and 1970s

An in-depth analysis of the evolution of hydroinformatics from computational hydraulics is presented in Chapter 2. The brief overview given here will highlight the main aspects of the early stages important for the business perspective.

The first generation can be traced back to the 1950s when the scientific community made the first advances in solving water equations with specific analytical solutions. The following stages of computational hydraulics (later hydroinformatics), during the 1960s and the 1970s, were still confined to a small number of academic and research institutions and had no significant impact on business. Computational hydraulics was primarily a tool in the hands of experts and focused on applying numerical solutions to practical problems at hand.

However, the pool of computational experts was growing, knowledge was embedded into tools and these results were increasingly shared within the scientific community. During this time, we see the emergence and strengthening of specialised research organisations such as DHI, Delft Hydraulics, HR Wallingford, and SOGREAH, all based in Europe. In parallel, the USA Corps of Engineers was working on developing their own tools (as well as other US federal agencies and research centres). In the early stages, the advances in the field were primarily in the development of computational methods, information technology (hardware and programming tools) and novel ways for data acquisition and storage, the 'inner' workings of hydraulic modelling.

The first computer models required expensive mainframe computers and were complex to develop and use. They were fed by a large number of punched cards with instructions and data. The outputs consisted of reams of printed pages with numbers, tables and crude graphics.

5.2.2 The rise of computational modelling business: 1980s

In the 1980s, the first desktop personal computers (PC) were introduced to the market. This technological advance invoked the development of simulation models based on the DOS operating system. The first type of the so-called generic simulation models was developed (known as industrial modelling systems), primarily for simulating one-dimensional (1D) flows in river- and urban water networks. Generic models were flexible and could be adapted to different project sites and boundary conditions. Soon after the release of the first modelling software packages to the market, this was recognised as a milestone in opening the first business opportunities. The first menu-driven software packages (e.g., MOUSE – see Figure 5.1) – simulation models with encapsulated knowledge, elementary graphics and first standardisation trials were released, and the first front runner water consultants invested in the purchase of sometimes costly licensed packages.

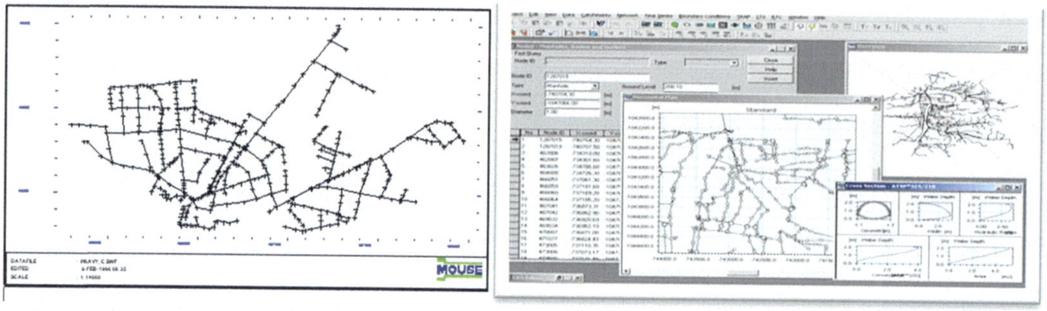

Figure 5.1 MOUSE modelling tool GUI development between early 1990s (left) and late 1990s (right) as an example of evolution of fourth generation of hydroinformatics tools (courtesy of DHI).

However, it took some time before competitive demand for modelling services appeared in the open market. The stakeholders during this stage consisted of a narrow group, mainly the governments and their research institutions, and the 'businesses' were publicly funded.

As the first modelling tools came out of large, specialised research institutions and universities, we observed different approaches. Some organisations saw commercial value in their tools and chose to develop fully functioning, off-the-shelf software products with development plans, regular updates and professional user support. Other players choose from the start to make their software free to use. They focused on developing core computational modules with little effort to make them easy to use and generally no support was offered. A toll was offered as-is, where-is, but it was free.

The price of the professional systems was high, and regardless of the encapsulation of the knowledge, extensive support was needed to use them. Only large engineering organisations could afford to purchase computers, own the tools, and develop expertise in using them commercially. However, the academic use was expanding through universities that have computing departments and were under no commercial pressure to turn a profit. An ecosystem of independent skilled users was starting to grow.

In parallel, the demand side for modelling services was starting to rise. Water was always recognised as being a facilitating or restricting factor of human activity. Rapid urbanisation and industrialisation, transport, food production, risk and damage reduction demanded better water management. The water business markets recognised modelling as a valuable but expensive activity. Independent engineers and scientists with entrepreneurial visions were facing an exciting emerging market. However, the investment requirements were high, demand uncertain, and projects' costs hard to control, creating an overall high-risk environment.

Many small companies that specialised in modelling started by offering their sub-consultancy to prominent consultants. Often, their skills were focused on developing their own add-ons and tools for data processing and modelling that enabled them to be more efficient in their consultancy projects.

Often, it was a case of 'learning while doing', with questionable data and no set standards or best practices available. Many early models were not very good, and the trust in modelling done outside scientific institutions was yet to be developed. Hydroinformatics faced a long road for maturing the technology and development of skills to create a solid base to be commercially successful.

The growth of the modelling business initiated by the releases of industrial software packages started the hunger for data. At the beginning of this period, very little data were available. Much effort was spent on data collection, digitisation, quality assurance, and so on. These went hand in hand with the issues of data quality, ownership and acquisition cost, causing problems for many modelling projects and bringing additional burden to the usually tight project budgets.

As a lecturer at IHE and a special advisor to DHI, Mike Abbott was uniquely positioned in the water business to develop and promote his vision of hydroinformatics. The combination of laboratory and numerical models together with field data provided good options for solving complex engineering tasks. His followers, mainly within the group of postgraduate students at IHE and DTU, created a nucleus of knowledge and saw early opportunities to apply it in the world of water consultancy in their countries, where the real and pressing need for using modelling was not yet being addressed. The new tools, when compared to standard studies or physical (scale) modelling, provided faster and cheaper ways to address engineering challenges.

During this time, many leading research institutes were gaining strength and expanding their physical and computational laboratories; earlier mentioned Delft Hydraulics in the Netherlands, DHI in Denmark, HR Wallingford in UK and SOGREAH in France were the largest in Europe. They were also becoming more agile and started offering their services commercially not only in Europe but also in America, Asia and Australia.

At the same time, the market in the USA was very cautious in trusting the tools developed by private or semi-private vendors. Instead, the regulators entrusted their own institutes, such as the United States Environmental Protection Agency (EPA) and USACE Hydrologic Engineering Center (HEC) and US Geological Survey (USGS) to produce quality tools to meet the needs of planning and engineering communities. These agencies focused on developing computational code and trusted that other commercial organisations on the market would invest in developing GUIs to interface their free simulation engines.

It could be argued that the strong commercial drive observed in European institutes was, in many ways, influenced by Mike Abbot and his quest and vision for hydroinformatics. He was not only a strategist but was directly involved in the development of novel methods and tools, mainly with DHI but also with some of the new business ventures started by his former students. One of the examples is his participation in the development of HYPRESS®, a modelling tool for Water Hammer with Hydroinform Ltd from Prague, the Czech Republic. The tool was based on a new, object-oriented approach, which was then a rather revolutionary idea (see Ingeduld *et al.*, 1996).

In parallel, new systems still adjacent to hydroinformatics were being developed in the same space of emerging information and computing technology, especially computer-aided design (CAD) and geographical information systems (GIS) applications that were starting to be commercialised. For example companies engaged in such developments are Bentley Systems, founded in 1984 and working on a stand-alone CAD system and ESRI with its ARC/INFO as the first spatial tool released in 1981.

5.2.3 Maturing and expanding: 1990s

In 1991 Mike Abbott published his famous book *'Hydroinformatics – Information Technology and the Aquatic Environment'* and, in many ways, started the period referred to as the 'Hydroinformatics era'.

Further development of personal computers during the 1990s, coupled with dramatic advances in computer usability through the introduction of graphically oriented operating systems and interfaces, facilitated the development of new generations of hydraulic modelling packages. Friendly GUIs provided for easier model development, simulations, and results analysis. Software development in relational databases (RDBS) introduced another technological progress for hydroinformatics modelling tools, simplifying data management and handling, storing, updating and error checking of data. Indeed, new technological advances further increased the interest of the water business in simulation modelling, and this period marks the start of the expansion of the hydroinformatics business. It was still risky to invest in hydroinformatics technology. However, those front-runners who took this challenge profited later from their risky decision (such as Ramboll, COWI, SWECO, CDM, ARCADIS, and HKV Consultants).

Hydroinformatics projects were growing worldwide based on the new model platforms that made the reuse of simulation models possible. The preparation of an urban drainage master plan for the city of Prague (the year 1998–2001) is one early example. Of the shelf 1D modelling tools, MOUSE and MIKE11 – developed and sold by DHI, were used to build the model of sewer and river networks and provide information about network capacity, overflows and impact on water quality. The same models were later expanded and used to optimise the operation of the waste water treatment plant. The reusability of the models opened new opportunities with clients and helped offset the high cost of the initial model development.

In the 1990s, hydraulic modelling projects were very different from what we are used to today. For example, in 1992, a new highway bridge across the Ohre River (in the Czech Republic) was being planned, and an assessment of the bridge's impact on the hydrodynamic flow pattern was needed. These types of analyses are very complex and usually required physical (scale) modelling. Zeman *et al.* (1992) proposed to use a 2D hydrodynamic simulation instead using a deterministic model FLUVIUS (Figure 5.2). It was a brave new world challenged by incomplete data, immature modelling tools and slow computers.

Abbott (1991) had foreseen this trajectory of hydroinformatics and indicated that its development would be dependent on new methods and tools, development of hardware and operating systems and available applications. He never believed that academia alone could drive his vision of hydroinformatics. Instead, he saw the development of the water business as a crucial move to fulfil the need of the society expressed through water regulations. The complexity of interactions between the natural and increasingly urbanised world calls for the integration of modelling, information and communication technologies and computer science. He recognised that such hybrids are not merely meeting of the knowledge and technologies but also seeds for sprouting new, higher learning. Climate change, urbanisation growth, energy and food are the main drivers of the need for predictive tools. Real-world experiments are either impossible or too slow and costly. Mistakes can be expensive, devastating and impacts irreversible. The only way to secure sustainable development in the water sector is to develop and use modelling tools. The tools need to cover all critical processes and answer the questions on the impacts of intentional or non-intentional changes in the environment, in other words, 'what-if-then' simulation, as Mike Abbott frequently repeated to his followers. The fourth and subsequent generations of hydroinformatics tools exhibit these aspects, and consultants in the water segment were under direct challenge to use these.

ICT advancements also contributed for significant improvements in data acquisition and management, which is crucial for successful modelling studies. Prominent consultants in the SCADA market (such as Honeywell, Schneider, Siemens or ABB) contributed to data-acquisition systems' development and data processing improvements. The operation of complicated plants and water works in real-time became more realistic and feasible. Integration of simulation models in SCADA systems started to be reasonable because it lowered the cost and reduced complexity, providing warning signals to dispatchers of plants when data from sensors differed from model outputs.

At the same time, Linux operating system was introduced to the market showing great potential for developments of cluster computing and high performance computing (HPC). This opportunity was taken by leading simulation model software vendors aiming to speed up the simulation time, especially for two-(2D) or three-dimensional (3D) simulation models.

The decade of the 1990s also brought in the data-driven models, as important hydroinformatics components. Contrary to the physically based models, data-driven models are connecting the system variables without much knowledge about the physical nature of the system. Following initial developments (see Chapter 3 of this book), usage of the so-called 'hybrid' models

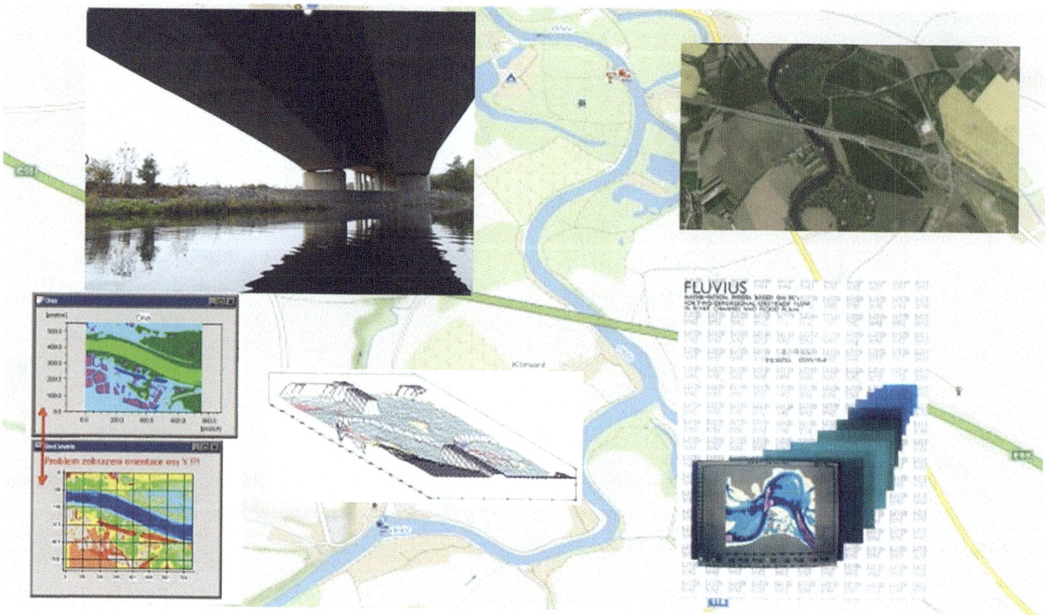

Figure 5.2 Highway bridge over the Ohre River close to Doksany, where the 2D model was used for the analysis of the position of the embankment of the designed bridge – first analysis in Central Europe using FLUVIUS 2D hydrodynamic unsteady model – results compared with scale air model (courtesy of DHI).

emerged, where physically based models and data-driven models are seen as complementary and are being combined to use their individual advantages (Solomatine & Ostfeld, 2008). For the feasibility studies and design analyses physically based models that enabled 'what-if-then' simulations would remain important in order to have the possibility to model alternatives with physical variations. Fast data-driven models would be used as emulators of physically based models, to be used in optimisation, uncertainty and sensitivity analyses, as models for discovering unknown knowledge hidden in data patterns, or for forecasting and other operational water management tasks (see again Chapter 3).

5.2.4 Water business in the new millennium

The development of hydroinformatics accelerated at the turn of the century, and so did the business opportunities. Advances were realised everywhere. Computational methods were being developed that take advantage of faster computers to solve problems in an integrated way. Engineering and scientific education was broadened to include computational methods and modelling, thus producing a growing pool of skilled modellers. Internet, wireless communications, and mobile telephony technologies emerged as new environments for developing and deploying hydroinformatics applications. Improvements in sensing and sensor technologies led to the concept of internet of things (IoT), and the ease of business-to-business interactions resulted in a proliferation of data and data providers. Lifelike visualisation techniques developed within the gaming industry started to be used in presenting outcomes of complex modelling simulations. These advances granted business opportunities that small and agile teams could monetise easier than large engineering organisations that often lacked the necessary skills in the emerging disciplines and technology.

All these developments led to an exponential increase in number of users of hydroinformatics tools. However, the number of key developers decreased, due to the concentration and integration of tool vendors. Further advantages of knowledge encapsulation in the hydroinformatics tools came via possibilities for integrating tools even from different vendors using agreed protocols such as the Open Modelling Interface – Open MI (Gregersen *et al.*, 2007; OGC, 2011). These features demonstrate significant advances in water systems modelling, especially with HPC frameworks, where widely used geospatial data and interfaces support integrated simulations on HPC infrastructure with the help of various interconnected tools.

Widespread use of the internet combined with spatial web applications such as ESRI, Google, Open Street Maps®, and so on., allows for projecting simulation results to maps provided by cloud-based software. Such maps can even be connected to external sources of observational data such as IoT platforms, which brings yet another perspective to presentations. Interactive result presentation tools allow users to explore outcomes of what-if simulations in a way superior to static charts in reports. Software development embraced spatially tagged data, while cloud-based advanced GIS solutions sufficiently matured to provide geospatial services for many application domains, including water. This focus on intuitive GIS-based interfaces also affected the next generation of hydroinformatics technology. Simulation models were migrated into a GIS environment that was a natural platform for model building and usage. For web-based

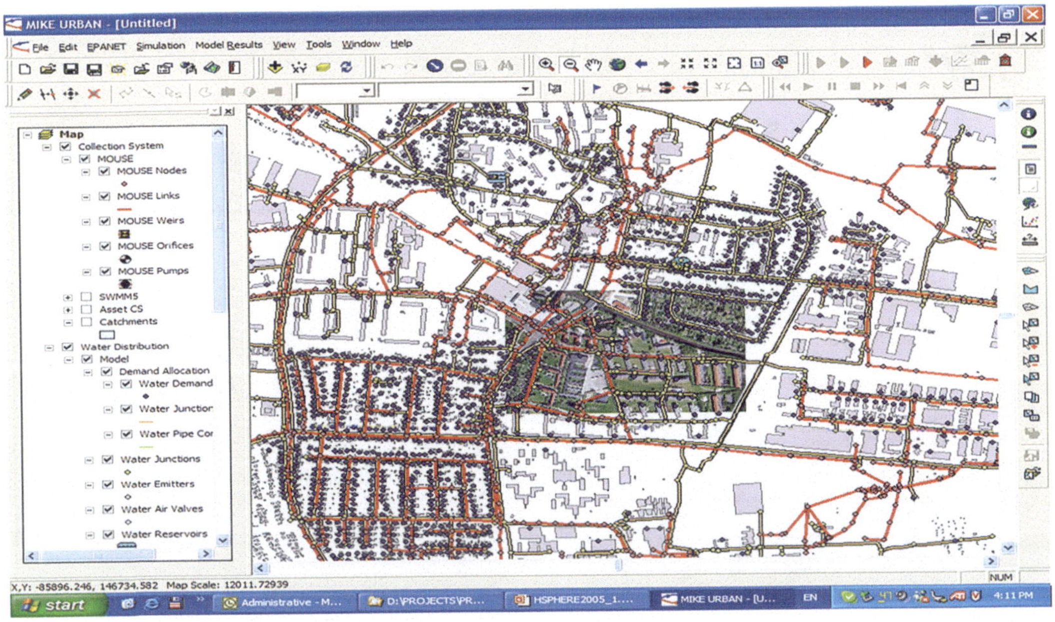

Figure 5.3 Example of fourth-generation tools – MIKE URBAN simulation model combined with ESRI ArcGIS (courtesy of DHI).

applications, the client–server architecture influenced simulation models' internal architecture, allowing for first trials in data sharing and remote access to data and models.

MIKE URBAN software package of DHI is presented here (Figure 5.3) as an example of such progress in technological development. The modelling package integrates two simulation domains, water supply and urban drainage, and encapsulates several commercial and free computational engines (MOUSE, SWMM, M1D, EPANET). Model configuration is stored in an RDBM database and is fully integrated into the environment of Esri ArcMap GIS, which is an industry leader in spatial applications.

The application of modelling technology also became recognised by international financial institutions and donors (European Investment Bank – EIB, World Bank – WB, Asian Development Bank – ADB, etc.). Many technical assistance projects mandated the use of simulation modelling technology as objective proof for any future investment, especially for large environmental projects. Just as one example, the 'Preparation of regional water and wastewater master plans for western region of Bulgaria' project is presented below. The Bulgarian urban water infrastructure assessment was elaborated, and development options were proposed utilising the MIKE URBAN simulation models. Almost one-third of the country was covered by simulation models providing information at various required scales, from the whole country down to a single street (as presented in Figure 5.4). Combining simulation models and GIS has proven to be a robust toolset for decision-making and long-term planning.

At the beginning of the millennium, the worldwide spread of internet technology reached the extent, capacity, and communication speed allowing for fast online exchange of large data sets. This development caused large migration of personal and industrial data to the internet. The developers of simulation models started considering to move from the desktop platform into the internet environment to start marketing the products as services (SAS – 'software as service' business model) rather than licensed modelling tools as desktop applications. Some simulation model providers even migrate their modelling tools to web-based applications with client–server architecture, back-end databases and remote simulation engines. These solutions are more oriented to specific applications in particular case studies (e.g., Kumar *et al.*, 2015; Almoradie *et al.*, 2015). With the advancements in virtualisation and cloud services offered by the tech giants such as Microsoft, Amazon, Google and others, even the traditional software products may become available via virtual platforms (operating systems). DHI has launched their MIKE Cloud in 2020 hosted on Microsoft Azure platform with the initial set of tools and engines to support model building and execution on the cloud infrastructure. Acceleration in the overall speed of the internet also allows for fast data collection from distant monitoring devices and introduces new potential for nowcasting and forecasting simulations. The 3Di interactive water management model (3Di Water Management, 2021) is presented here as an example of a modern emerged web-based integrated modelling technology of the 21st century.

In parallel, hydroinformatics experienced significant expansion outside of the hydrological and hydraulic simulation models into adjacent fields of cost–benefit analysis, multicriteria analysis and optimisation. Steps are also taken towards developing *generic* decision support systems (DSS) with their library, management, analysis, and communication functions, building on experiences from developing case-specific systems of that kind.

Similarly, the modelling on HPC platforms matures further, allowing for computationally intensive simulations to take place, such as climate change analyses or uncertainty modelling. The usage of HPC for such purposes is essential because of the enormous data fields and demands for parallel processing. Global or regional climate models provide boundary conditions for localised water balance/hydrological modelling. For uncertainty analysis, many runs of the models are required, which rely upon new algorithms and extensive computational power.

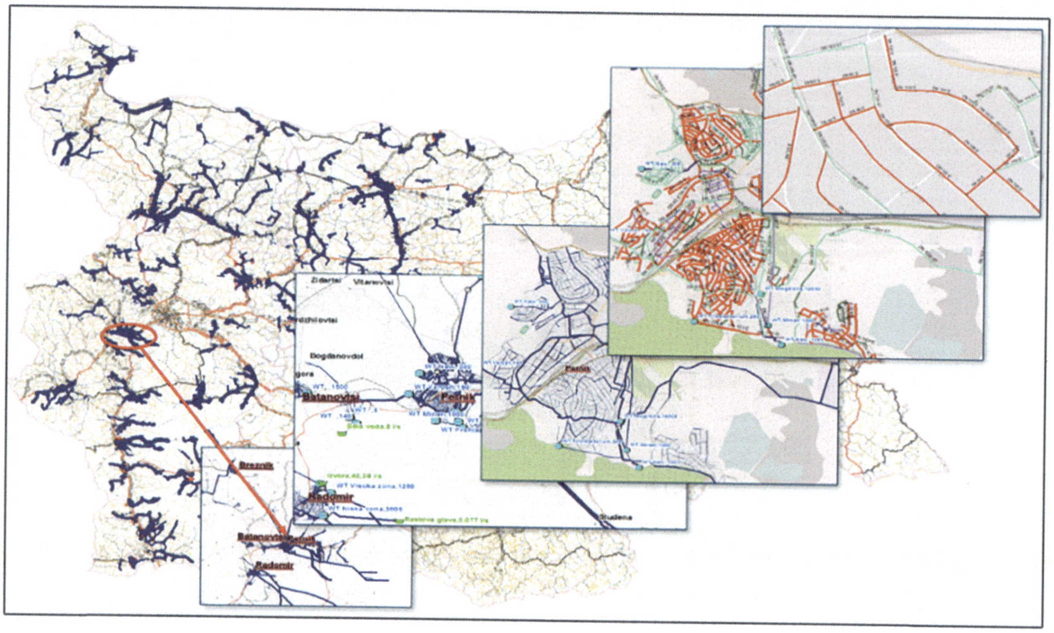

Figure 5.4 Example of fourth-generation tools – MIKE URBAN used in master plan studies in Bulgaria – simulation model combined with ESRI ArcGIS (courtesy of DHI).

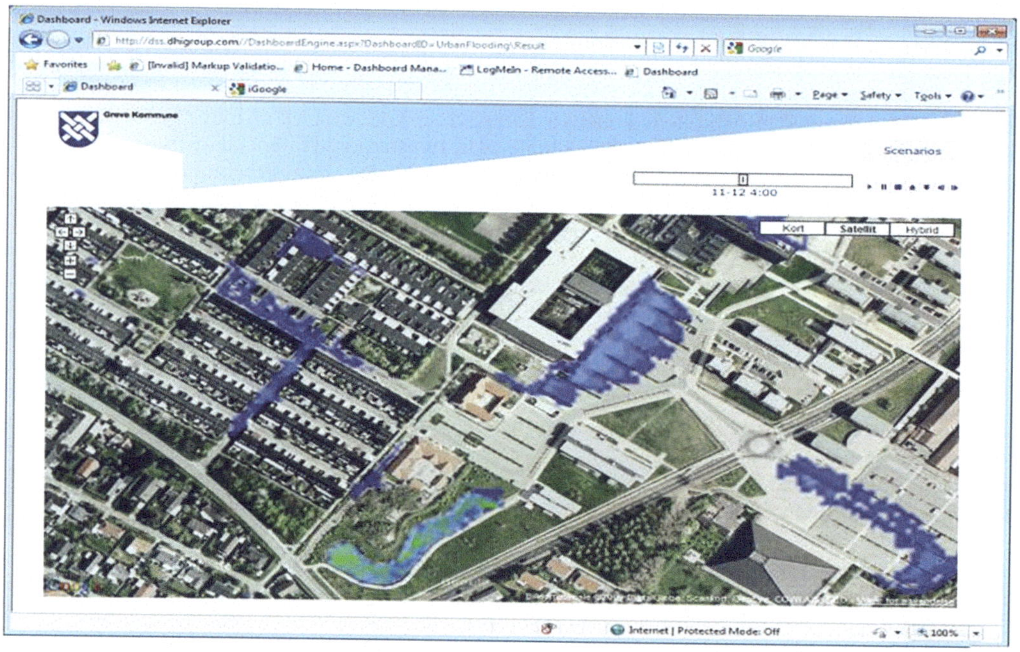

Figure 5.5 An example of the latest generation of simulation tools using internet technology – extend of flooding using orthophoto maps MikeFlood simulation tool (courtesy of DHI).

The advances in remote sensing, satellite-imaginary and similar techniques were also reflected in the development of modern GIS systems, with geospatial services becoming available over the internet. The introduction of map servers was combined with map browsers (e.g., GoogleEarth, StreatMaps, etc) and merging and combining geospatial information (maps, satellite images) in a variety of applications became possible (see for one example Figure 5.5). This technology enables provision of geospatial data together with other data attributes using standardised protocols such as Web Mapping Service (WMS) or Web Feature Service (WFS), and others, developed and managed by the Open Geospatial Consortium (OGC). Similarly, global repositories of terrain model data, such as. EarthExplorer.usgs.gov allow the selection of appropriate digital elevations across the world. These technologies make local project development and understanding the local sites easier than ever before. The combination of areal/satellite data and local data sets monitored by field sensors provides a comprehensive insight into a particular project location and helps eliminate errors or data miss-matches.

All these new technologies are impacting the hydroinformatics business actors. Traditional providers of software products are exploring opportunities for migrating their traditional solutions in the cloud, and integrating them with the new data sources. At the same time, new businesses, with small and emerging companies are taking specialised roles in the new business environment. These changes are discussed in more detail in the following section.

5.3 TECHNICAL AND COMMERCIAL ASPECTS OF HYDROINFORMATICS

Commercial and technical aspects of water management are interlinked through hydroinformatics. Growing challenges facing water practitioners are addressed by the 'knowledge owners', while new scientific and technological breakthroughs give rise to new emerging practices.

While outcomes of research projects often pave the way for more research, this is not what water management practitioners need. Mike Abbott understood that practice must meet research to make progress. The early hydroinformatics institutes, such as DHI, were heavily involved in practical project work where knowledge was not readily available commercially. Within these projects, based on physical and computational modelling, new knowledge was developed and encapsulated into a *product*. When a product was disseminated (sold) in the community of practitioners, it gave rise to new types of projects, providing a revenue stream for the institute that could be invested in encapsulating more knowledge into products. As Mike Abbott often said, 'projects follow products' and this symbiotic relationship has been of essential importance for many hydroinformatics institutes.

In parallel, new generations of hydroinformatics entrepreneurs have realised a growing emerging market perfectly suited for business start-ups. They were first in line to receive the new products, invest time in upskilling and service the water sector modelling needs. They were more agile, less expensive, and more adaptable to the end client's needs. Rapidly, hydroinformatics tools became understood and accepted as an integral part of water management projects and complex water environments.

5.3.1 Original view on hydroinformatics as part of the knowledge economy

The first issue of the *Journal of Hydroinformatics*, published in 1999, in Mike Abbott's words, 'marks the point where Hydroinformatics becomes accepted as a viable discipline, where it comes to presence in a more connected and homogeneous way, and where it attains to a certain permanence of form' (Abbott, 1999, p. 3).

This discipline of a 'new way of applying knowledge' paves the path for broad societal transformation. Knowledge becomes an entity to be produced, configured, brokered, marketed, transmitted, further transformed, and ultimately consumed in new ways. This opened the era of hydroinformatics expansion characterised by water-related 'knowledge consumption'. As stakeholder engaged in this process interacted, new hybrid knowledge was generated, larger than the sum of its parts.

Mike Abbott refers to two types of systems at play in knowledge management. One is a structured, advice serving chain of collecting, processing and dispersing the knowledge. Many engineering consulting businesses are developing their services in this structure. The other, more organic system is one that we may call 'knowledge farmers' market'. Here the cloud environment is a natural space for many small, niche businesses that generate knowledge in their own space and are 'found' by other knowledge providers, knowledge brokers and knowledge consumers.

5.3.2 State of technology

Digital computer performance speed has increased by orders of magnitude over the past decades, effectively doubling every two years (Moore's Law). The traditional models or typical engineering applications have likely reached a state where no more power is required, as the current average size models run sufficiently fast, and there is no need for a large model extent or a smaller grid size or a smaller time step.

However, the need for power is still growing, only shifting into solving operations or process optimisation problems. The real-time predictive control, machine learning and assessment of uncertainties represent a new and expanding need for HPC and cloud solutions. Many commercial modelling software packages have adopted their code to utilise HPC (such as MIKE 21).

Another important development is in the area of engineering project collaboration software suites, which have been designed for the architecture, engineering, and construction industries. They help project teams manage, share and distribute engineering project content and review in a single platform. Some of these systems provide integration of design tools and analysis vehicles with other elements of the project (see, e.g., Allplan, 2022).

Still, many organisations, especially infrastructure owners, have security concerns about adopting cloud-based solutions. Some legislations prohibit data of public interest to be stored outside of the region under their direct control and jurisdiction. Nevertheless, the expansion of technology giants such as Microsoft into the cloud, as well as 'native' cloud enterprises such as Google and Amazon, demonstrate their growing confidence that the cloud will be *the* space for business in the future. This confidence is also based on the fact that in many ways, in-house systems often suffer greater vulnerability due to a lack of investment in security and data encryption.

Supported by advances in technology, the way of working is also changing. It is possible nowadays to share work between teams, organisations and devices, which enables the decoupling of the technical work from a computer stationed in an office.

More recently, we can see the expansion of hybrid modelling where advantages of personal computers and cloud-based resources and services are utilised. The model development is often done on a personal computer; input data may be sourced via cloud-based data providers. The completed model may then be deployed on the cloud to run many scenario simulations, provide ongoing flood forecasts or be readily available via devices such as mobile phones, computers or tablets, for any day-to-day water management and operational tasks.

5.3.3 Changes in education of hydroinformatics and their reflections on business

Hydroinformatics in academia and in engineering practice took different tracks. As the knowledge encapsulated in hydroinformatics tools became more useful to engineers, infrastructure planners and operators, risk assessors and strategic decision makers, the need for skilled modellers grew. Very soon, the growth of software users vastly outpaced the growth of software developers. Modelling was first thought of through product training, but then the needs were also met at specialised educational institutions that highly invested in hydroinformatics education and research. Currently, modelling is integrated into most engineering and science degrees.

The introduction of hydroinformatics education for engineers not only gave them access to encapsulated expertise but also made the learning more interactive and engaging, with some additional observable benefits. Experimentation with the models facilitated the development of a deeper understanding of water movements, the kind of knowledge that was previously only gained with long experience working and observing in the field. Furthermore, in hydroinformatics-specific programmes, broader aspects have been introduced, such as the role of modelling in integrating knowledge for decision support (see Chapter 7 of this book). Confident in their skills, the generations of young modellers from universities were ready to enter the market. Their choices included joining large established engineering companies or governmental institutions – or starting their consultancy focused on modelling.

One of the critical steps for applying a simulation model in practice is selecting an adequate model for a particular project goal and objectives. Obviously, commercial organisations make use of modelling tools in their possession (often freeware tools) and try to adopt these models for project purposes, even when there are other tools better suited for the problem. This could lead to oversimplification of natural phenomena and processes and even obtaining misleading results, which is a challenge that education providers should address. Therefore, transparency of the inner workings of the modelling tools has become increasingly important. Although modelling has earned its place at the centre of the water business, some practitioners still see it as a 'black box' and therefore lack confidence in its outputs – another critical issue for educators of new modellers.

5.3.4 Business transformation

The 'knowledge supply chain' included such vehicles as animations, graphical expressions and stimulated imagination that led to innovation at all levels. The breadth of businesses entering the market and offering services in this knowledge market is evident. These, among others, include data gatherers and providers, scientists, engineers, spatial analysts, customisation specialists, animators, educators and game developers.

The innovation was not limited only to developing new domain knowledge but also innovative partnerships between previously disconnected disciplines. This new symbiotic way of thinking has produced awareness that all-inclusive knowledge promotes creativity and that mixing knowledge strata creates new kinds of knowledge. In practice, we have seen large multidisciplinary companies operating in the world of water trying to cover the breadth of the hydroinformatics field through business mergers, acquisitions or expanding organically within.

With technological advances such as the Cloud, apps, IoT, and social networks we see the proliferation of independent specialist data providers, services, analytical and numerical tools, AI and machine learning. The partnerships are forming, not by being alongside but being together, with seamless harvesting of knowledge on the 'farmers market', directly or through knowledge brokers. One such partnership example is Optimatics, a software company operating in the water systems market, established in 1996 and recently acquired by SUEZ. The Optimatics platform uses algorithms, intelligent automation, and the computational cloud to simultaneously balance cost, risk, and level of service against multitudes of criteria in an unbiased, transparent, and consistent way. Optimatics can utilise various numerical calculation engines, both open source (EPANET) and the two leading commercially produced packages from Innovyze and DHI, to evaluate the criteria in the urban water networks.

As technology started to work together, so did the professionals. We observe the integration of large teams of experts with different specialisations working together on analyses using hydroinformatics vehicles and tools. In administration and state authority sectors, hydroinformatics experts utilise modelling results for planning and project scoping activities. The monitoring systems and SCADA systems need experts for system design and maintenance, and other experts are required for ICT systems supporting local warning or forecasting systems.

5.3.5 Software providers and their business models

Unlike hardware, in the early stages of computing, software was not seen as tradeable commercial good. When in 1983, binary software became copyrightable, along with the growing availability of millions of computers based on the same microprocessor architecture, a mass-market ready for binary retail software commercialisation was created. Commercial software can be proprietary or free, or open source. Although not always recognised, open-source software and free software are not the same. Richard Stallman, the founder of GNU, states in his essay (Stallman, 2022) that the two terms describe almost the same category but adhere to fundamentally different values. In his words, open source is a development methodology; free software is a social movement.

In the hydroinformatics space, we can observe the first commercial applications and software products during the 'third generation' modelling systems. The 'fourth generation', led by organisations such as DHI (Denmark), Delft Hydraulics – Deltares (the Netherlands), and HR Wallingford (UK), was further adopted by many others, including Haestad Methods (US), MWH Soft (US), or BOSS International (US). Some commercial software market players have retained their original business model, while others have evolved over time.

The USACE Hydrologic Engineering Center (HEC) has been developing computer code for hydraulic and hydrologic engineering and planning analysis procedures since its inception in 1964. Although the software is designed to meet the needs of USACE's planning and engineering communities, it is made available for free to the public. However, the HEC software packages are not open source, and the license prohibits the modifications or reverse engineering. The fact that it is free software has made it a popular choice for many educational institutions and start-up modelling businesses. Technical support is provided through an annual subscription service, but the widespread adoption of the tools enabled a strong collaborative and self-supporting community.

DHI has also retained its original software business model, which was fully proprietary. The software branded 'MIKE Powered by DHI' has expanded into an impressive family of specialised tools covering all water environments, from urban infrastructure to water resources management and offshore energy applications, making it one of the most comprehensive software providers of hydroinformatics applications. As the applications shared the common code structure, the coupling of different engines enabled integrated hydraulic modelling for complex interlinked natural and man-made systems. DHI's ongoing applied research activities, supported by the government grants, ensure that technological and scientific advances are continually embedded into the tools. DHI software is fully documented, and ongoing software support is available through the annual service maintenance agreement (SMA). As mentioned earlier, DHI also has ongoing plans to continue and expand its software offerings as products and business applications, by making use of virtual platform services offered in the cloud.

Deltares, on the other hand, has morphed following the restructuring of the Dutch water research field, during which the former Delft Hydraulics also changed its name to Deltares. It is still a primarily publicly funded water research and development company focusing on large international projects and developing code to support its consultancy. Deltares has pivoted from the initial proprietary business model and has made most of its code open source. The exceptions include their real-time operation platform Delft-FEWS, which they describe as 'Open software but not open-source', stating that the decision is driven by the criticality of the systems and the need to have one organisation controlling the code. Similar to HEC, Deltares gets its software revenue mainly through paid support. As their code is open, it is often a platform of choice for other research institutes that may want to adjust the code themselves to suit their projects.

Innovyze is an example of a business transformation into a fully commercial organisation. The proprietary software, specialised in water infrastructure modelling, was recently sold to AutoDesk by their hedge fund owners. Over the past several decades, Innovyze has grown significantly through mergers and acquisitions. Their ancestry includes XP Solutions, Wallingford Software, Stantec, MWH Soft and MicroDrainage. Similar developments took place at Bentley Systems, which developed own software solutions for modelling and management of water infrastructure, but also expanded via acquisitions, most notably of Haestad Methods.

The business path of Sogreah and their TELEMAC-MASCARET software brand is another case of transition into the open-source model. Like Deltares, Sogreah promotes a range of software assistance fee-paying services to assist their software users.

We can also observe traditional engineering consultants integrating hydroinformatics into their services, primarily to strengthen their complex consultancy offerings. The recent example of Royal Haskoning DHV, a traditional Dutch consulting company, which acquired majority stake in Hydroinformatics Institute (Singapore) is a confirmation of such an evolution. Another example is a cloud-based collaboration platform 3Di Water Management, developed by several partners from the Netherlands, including consortium Nelen & Scuurmans, Deltares and TU Delft. 3Di WM is a tool for hydrodynamic simulation for pluvial, fluvial and coastal floods. It also aims to be a communication tool and bridge the gap between stakeholders with different backgrounds. These examples point out how the traditional boundaries between software producers and consultants as their main users become blurred, driven by increasing and continuously changing needs for knowledge provision with hydroinformatics tools and solutions.

5.3.6 Software users and their needs

The community of model users is diverse, large, and still growing. It covers a range of companies and physical entities, from large business enterprises to one-man offices, small NGOs to national authorities.

At the organisational level, hydroinformatics cannot be successfully implemented without specialised staff. The necessary skills range from hydraulics, computational fluid hydrodynamics, hydrology, data management, quality assessment, monitoring devices, GIS, CAD and, indeed, comprehensive understanding of a suite of simulation models. Such a broad spectrum of skills is not easy to find. A lack of qualified staff can lead to several problems during project execution. Without deep insight into data and data management processes, the simulation model can provide misleading results. There is also a strong distinction between skilled advanced modellers and 'button clickers', that is, model users capable of running the model but without understanding the background such as modelling assumptions and limitations, uncertainties, computational parameters, result interpretation, and so on. The phrase 'all models are wrong, but some are useful' acknowledges that models always fall short of the complex reality.

Large business enterprises benefit from their size and often develop internal skilled modelling teams. They can also use their size to procure proprietary modelling software at a lower cost and extract additional benefits from software vendors. However, due to the deep integration of the software into their corporate infrastructure, it is costly for them to change the modelling product provider.

Small and medium enterprises often cannot afford expensive proprietary modelling packages. Instead, they use free modelling tools and hence motivate further developments, especially in the open-source space. This approach fosters a competitive environment in the water modelling business as the free simulation modelling technology spreads in the water sector more easily and results in competitive project costs.

We observe increased prominence of other types of users, the engineers, planners, and operators with general computer knowledge but no skills in hydroinformatics. Their needs are very specific as they increasingly want to use a model for routine, repetitive tasks or results analysis without learning sophisticated modelling software. A new generation of software 'apps' focusing on ease of use is entering the market and is often cloud-based.

Based on the above facts, it is argued that adequate attention should be given to the model building process during project planning phase to ensure that good quality and trusted model is developed for further use.

5.3.7 Data collection, sharing and publishing

Data have always been enormously important in the water business, integrated with modelling, design and decision making. The earliest data of interest for hydroinformatics professionals came from monitoring and mapping the natural environment. These datasets were often kept in proprietary formats, bundled in the software packages, and thus not readily available. Many businesses quickly saw the value in data itself, which led to a proliferation of commercial data providers.

More recently, new data have been synthesised from overlapping natural and socio-technical information. The value of new knowledge being so created is higher than the value of the sum of its parts. Data are being generated at an ever-increasing rate, mainly as a result of the proliferation of sensors, both physical and virtual, produced by either synthesising the raw data or by models. Data, in order to be useful, have to be verified, validated, findable and readable by humans and, increasingly, machines. Technologies such as the IoT, digital twins and early warning systems depend on sharing and exchanging data between the outputs of a monitoring system and the outputs of modelling tools.

Democratic governments and societies recognise the role data and facts have in public engagement, so the investment is made in collecting data which is then freely accessible through national agencies. The data published for public consumption commonly include environmental monitoring data. Examples of 'discoverable' data in this way include rainfall, flows in rivers, sea levels, terrain (LiDAR), soil maps, buildings outlines, transport networks, modelled flood levels, maps of natural hazards or bathing water quality. Data can be either downloaded from central repositories or dynamically interfaced when needed. Sharing data ensure all stakeholders access the same validated data, and the maximum value is extracted from the public investments.

In parallel, technological progress in devices that can, along with other uses, capture data and information-sharing infrastructure has democratised creativity in data gathering. Crowdsourcing of data is rapidly becoming an integral part of data acquisition. For example, multi-functional devices such as smartphones can use the phone camera to record falling rain and then pass this to the cloud-based algorithm that transforms the video image into rainfall intensity (Allamano *et al.*, 2015). When such devices are connected, they form a dense network of sensors that can record spatial variability and improve early warning systems.

Automatic data discovery relies on sufficient metadata and well-structured datasets. The business opportunities in data warehousing started by providing secure large digital storage in data farms but are fast transforming into providing expanded data services. As hydraulic models, especially models of civil infrastructure, are being treated as assets, there is also an increasing need for them to be catalogued and managed with sufficient metadata so that models are used with confidence. Although the need for cataloguing is widely recognised, it appears to be the least attractive area for business, and no mature commercial solutions are currently available in the market.

5.4 SOCIETAL AND POLITICAL ASPECTS OF HYDROINFORMATICS

As a discipline of a new way of applying knowledge, coupled with rapidly evolving tools, hydroinformatics paves the path towards broad societal transformation. This was the vision of Mike Abbott from the very beginning (Abbott, 1991, see also Chapter 1 of this book). As with any transformation, the change is not always welcomed, and the resistance comes from many corners of society.

The academic and scientific community argued that only 'formal knowledge' and expert opinion matter due to the complexity of the issues. They were not accepting that the real stakeholders, people most impacted by the actions and interventions into the shared world of water, have anything to contribute. Diverse interests and traditional knowledge, accumulated from historical real-life trials and errors, cultures and beliefs, in their opinion, have no value as the rapid change in environment and the new 'acceptable' standards go beyond the accumulated 'traditional' knowledge. Such position was also accepted by institutions responsible for water- and environmental management. As Vojinovic and Abbott (2012) state, the fundamental problem of hydroinformatics, at the level of its application, is changing or bypassing or otherwise getting around the socio-institutional roadblocks erected in its path.

5.4.1 Social justice and hydroinformatics

Hydroinformatics must provide knowledge to as many persons as possible, utilising technology and information networks to achieve this purpose. It must give people the capabilities to consume knowledge and express their beliefs and fears. Across most of his recently published work, Mike Abbott stated the importance of including cultural and ethical considerations. Fairness is his new focus, as he starts from the premise that 'the things that really count cannot be counted' and ends with the problem of 'who is the one to decide if a risk is acceptable?'

Hydroinformatics institutes early on acknowledge the need to get involved in activities that promote their social role. One such example is the 1996 partnership between DHI and UNEP and, later, Danida to form the UNEP-DHI Centre on Water and Environment to serve as a centre of excellence and leadership on freshwater issues and promote sustainable water resources management and other SDG targets.

5.4.2 Learning by playing

The first premise of hydroinformatics is that all kinds of knowledge must be made accessible to all stakeholders. But how to translate complex interactions of the natural world into forms that all stakeholders can assimilate and interpret based on their abilities and interest? How to bring democracy into complex decision making?

Learning by playing, or 'serious games', was recognised early on as one of the concepts that can educate a broad spectrum of stakeholders (Jonoski, 2002). The technological advances in integrated modelling tools and visualisation techniques, maturing of the gaming industry coupled with widening accessibility of the internet provided a base for early attempts to develop engaging hydroinformatics knowledge vehicles. Two leading players, DHI-UNEP and Deltares, developed their first serious games in the late 2010s. DHI-UNEP's product 'Aqua Republica' focused on integrated water resources management in basins, while Deltares developed several gaming products such as Sustainable Delta and Climate Adaptation app tailored more towards the technical audience. The hydraulic models lay at the heart of the serious games, evaluating the impact of players' decisions. In the true sense of delivering hybrid products, the serious games included different types of knowledge. For the game design, the hydroinformatics players have partnered with companies operating in the gaming business to deliver the software product and design the game's flow. The delivery of the games to the stakeholders was also a collaborative effort, often involving professional workshop facilitators and educators. While Deltares' audience was mostly water professionals, Aqua Republica had the ambition to reach deeper into society, inviting stakeholders with little professional involvement in water.

An example of one of the more ambitious projects with societal engagement was 'Aqua Republica Eco-Challenge', an event held in New Zealand for school children aged 11–17. Youth around the country tried their hand at balancing the needs of human populations with those of the environment over a two-week online event. DHI's MIKE Hydro Basin, a science-based modelling engine for integrated water modelling, was used in this strategy game where players act as the catchment managers applying system-wide approaches to integrated water resource management (Figure 5.6).

Figure 5.6 AquaRepublica, New Zealand game, GUI for using included modules in the integrated approach to catchment management (courtesy of DHI).

The game was fully customised to local conditions, making it relevant for the players. The environment was built from an actual New Zealand catchment, and cultural considerations included indigenous Maori values and beliefs; negotiations and consultations were required and rewarded for certain activities. An online science hub was set up to prepare how-to-play videos and resources linked to the background science and concepts.

This is just one example of use of a serious game for dealing with water-related issues. This approach has in fact become quite widespread with potential for introducing the necessary learning and attitude changes needed for more transparent and inclusive water management (see, as examples, Armstrong *et al.*, 2021; Craven *et al.*, 2017; Medema *et al.*, 2019; Savic *et al.*, 2016; van der Wal *et al.*, 2016).

5.4.3 Transparency of modelling and ethical aspects of presentation materials

Computerisation of hydraulics knowledge is described by sets of numbers that are increasingly extensive and can hardly be studied without adequate presentation techniques. Decision-makers can better understand phenomena and, hopefully, make more informed decisions when modelling results are conveniently presented to them. Ethics is essential, especially if there is a need to simplify the presentation or render it less ambiguous or more convincing.

There is a belief that data and data visualisation, respectively, are apolitical or somehow ethically neutral. This is not the case, and the tendency to view them as mere structuring of objective fact is dangerous. Data and modelling results are not unbiased; they are always collected or processed by someone for some aim (Correll, 2019).

Presentation techniques have dramatically changed over the years. The early plots and graphs aimed at knowledgeable users were replaced by advanced 2D and 3D computer animations or even virtual reality techniques which place the user inside an experience.

Advances in modelling and presentation techniques can be seen in the two illustrations presented in Figures 5.7 and 5.8, both describing results of 2D flow in rivers/flood plains.

This ability to present modelling outputs, with all inherited assumptions and limitations as 'true to life', puts an enormous responsibility on data presenters. One way to address this issue is to apply some of the rigour seen in scientific research, especially with peer reviews that are becoming the standard approach in many markets. Indeed, we again see the business recognising this need as an opportunity. Many experienced users of hydroinformatics tools built their careers on being independent, trusted to check how modelling was applied and if presentations were accurate and transparent. In return, awareness of the likelihood that their work could be scrutinised forces providers of modelling studies to lift their game.

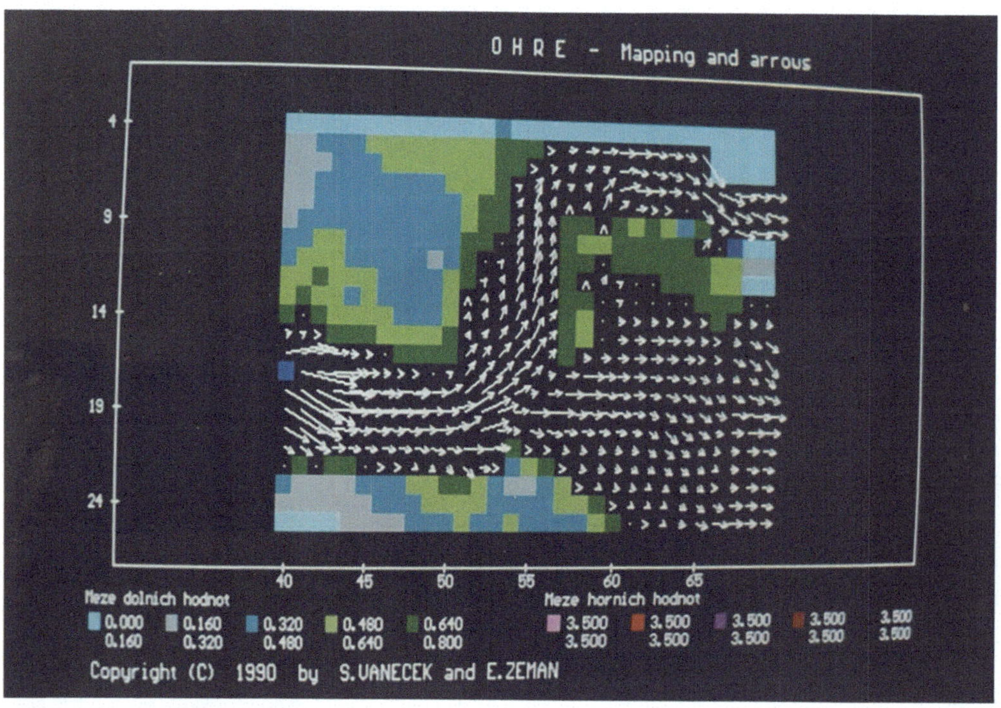

Figure 5.7 A photograph of a computer screen displaying results of a hydrodynamic simulation of unsteady flow in rivers, taken in 1990 when computer-based screen capturing was not yet technically possible (courtesy of DHI).

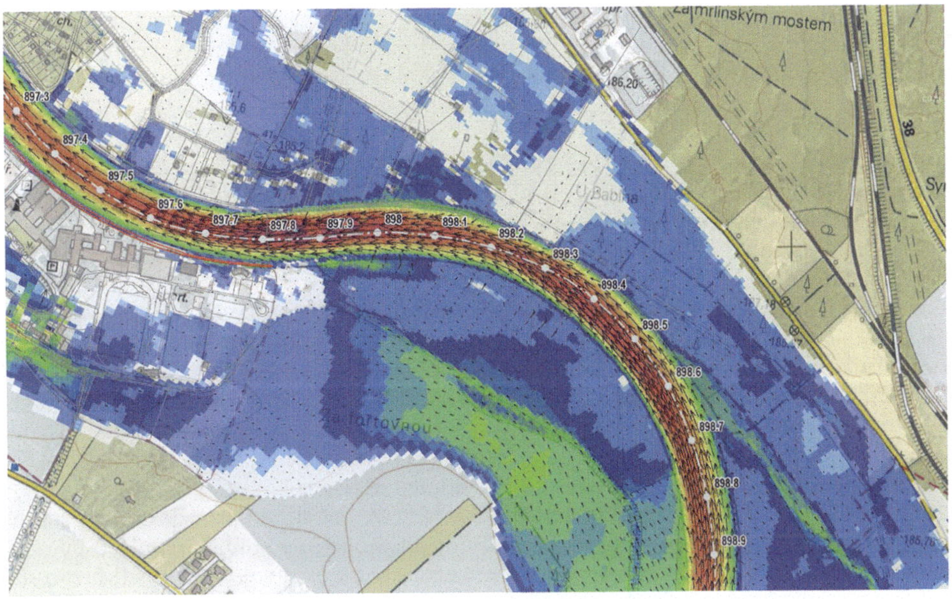

Figure 5.8 Computer-based animation displaying results of a hydrodynamic simulation of unsteady flow in rivers (velocity fields with direction and norm on depth mapping), 2021 (courtesy of DHI).

5.5 BUSINESS OPPORTUNITIES AS SEEN FROM 21ST CENTURY PERSPECTIVES

5.5.1 Water in cities market

Water has become an increasingly crucial issue facing many cities around the world today. Rapid population growth, urbanisation, and economic development lead to growing pressure on water resources in urban areas. The demand for water is constantly rising, and with the demand exceeding water supplies, the water shortage has become more prominent in many cities across the developed and developing world. Numerous cities have outgrown their local water supplies, particularly in semi-arid and arid regions. All close at hand surface water sources are being tapped, and groundwater resources are being exploited in an unsustainable way. In the future, more cities will need to rely on more distant water sources or inter-basin water transfers to cope with rising demands for water. Simultaneously, there is a strong drive for water reuse, leakage reduction, and overall improvement in efficiencies throughout the urban systems.

Digital transformation takes place in cities through the concept of 'smart cities' and 'resilient cities'. The availability of real-time data makes it possible to create a 'digital twin', a computer model of reality (Valverde-Pérez *et al.*, 2021). Water distribution digital twin, for example, will allow for simulation of any situation that may affect the network and water supply, anticipate problems, forecast future conditions, and optimise water delivery. A smart city is a perfect case for hydroinformatics, as it cannot be achieved without collaboration at all levels. Its complexity opens vast opportunities for new innovative businesses covering a whole spectrum of skills, from technical and commercial, to political and societal. Business opportunities will follow, as large investments in urban areas are foreseen, accompanied by requests for sustainable investments in the water sector on a long-term scale. Applications of formerly mentioned methods and tools of hydroinformatics would shorten the building duration and enable the optimal operation and maintenance of water systems in cities.

5.5.2 Water resources market

In recent decades, the water resources market has become more important for the water sector. Demand pressures and reduced supply of readily available good quality water create challenges. Further, there are clear signals that climate changes will worsen the situation regarding freshwater availability due to various reasons, leaving less water for the environment and human activities. This imbalance requires optimising usage and protection of freshwater sources.

Hydroinformatics offers some tools to address the integrated hydrological cycle, but there are significant uncertainties associated with the various processes and interactions within the cycle. To understand the water balance in water resources more accurately, better and faster modelling tools are required so that the impact of various uncertainties can be understood. Based on the gained understanding, optimising the use of the available water should take place. The water needs to be allocated fairly, balancing competing demands from different stakeholders in time and space within the boundary conditions, influenced by climate change impacts or other influences with known trends. Surplus runoff can be used to actively recharge the groundwater; losses can be reduced through smart irrigation and smart networks; water quality can be protected with science-based regulatory policies.

Currently, most of these challenges are solved using disconnected tools not dynamically linked. This level of simplification is increasingly recognised as unacceptable, and there is a need (and a business opportunity) for more sophisticated tools. The tools need to combine stochastic and deterministic approaches as we observe important changes in variables in time due to climate change. The variables, which were taken as semi-steady in the past, are now subject to specific trends. Optimising water resources in more general terms must also consider the demand part. For instance, water supply systems analysed by modelling systems and supported by decision support systems (DSSs) could identify water losses in water distribution systems. This activity integrates once again the urban and water resources market.

5.5.3 Marine market

The challenges today's world faces present opportunities for hydroinformatics in the marine environment, such as sustainable energy, climate change mitigation (blue carbon), and food production.

The World Bank predicts a rapid expansion of offshore wind from mainly Europe and China-based installations to emerging markets worldwide (Going Global, 2019). Unit prices have continued to drop thanks to technological improvements, economies of scale, maturation of supply chains, better procurement strategies, and the efforts of large project developers. Increased awareness and concern about the environmental impact from construction, operation, noise and vibrations, visual pollution in tourist areas, impact on shipping lanes, migratory patterns, and fisheries require sophisticated multidisciplinary approaches and a high level of local stakeholder engagement.

Aquaculture, a term used to define a fish farming activity, is practised at many levels, from small farmers in developing countries to multinational companies. The market is projected to continue to grow at a pace of over 5% per annum. Farming in water is a complex process that is often highly regulated in terms of allowed farm capacity, managing water quality impacts and overall impact on ecosystem and biodiversity. A holistic approach is required with increasingly demanding scientific evidence, continuous monitoring and impact modelling studies required by regulators. Another challenge (and opportunity) is in the improvement of the uncertainty assessment, which is difficult due to long model runs. Faster modelling tools are needed, to be sought in combining machine learning with deterministic models.

Climate change has aptly been described as humankind's most significant moral and technical challenge. An enormous amount of research is being conducted to try to slow down the Earth's warming caused by greenhouse emissions. 'Blue carbon', a term for carbon captured by the world's ocean and coastal ecosystems, is one of the more promising areas of research into carbon sequestration. Although the knowledge is still being developed, it is unlikely that the multi-faceted effort required for carbon reduction will be without the largest ecosystem – the oceans. Getting involved in this market opens many opportunities.

5.5.4 Other market segments related to water or the environment

Other related opportunities are in many ways interrelated with already mentioned segments. Inland waterways and navigation represent a vital activity in the transport segment. There are undertakings towards sustainability in any new development of European Waterways and new connections. Hydroinformatics tools have a prominent position in this segment to support sustainable solutions for the future.

Hydropower and new projects for the everyday use of renewable energy utilise innovative solutions, including online monitoring, SCADA and optimisation of resources using hydroinformatics tools. Other water-related segments are cooling in industry, water for the food industry, and water for sports activities (e.g., artificial snow for downhill skiing).

5.6 FUTURE CHALLENGES FOR HYDROINFORMATICS BUSINESS

5.6.1 Uncertainty and parametrisation of models

The models are burdened with a certain degree of uncertainty, manifested by the deviation of their outputs from the actual (measured) state. Uncertainty can be introduced into the model's output either by the degree of approximation of the modelled processes, the accuracy of its parameterisation, or the input data that the model uses. The parameterisation of the model could be refined using better measurements or with methods of inverse modelling and the implementation of a sophisticated calibration process.

A frequent question from the stakeholders looking at the modelling outputs is, 'how certain is this value'? For example, rather than saying that the simulated rainfall event will result in a particular flood level at a property, it would be more desirable and transparent to assign the probability of flooding being at a certain level or above. Using, for instance, the Monte Carlo method, it would be possible to achieve this and display the effect of the uncertainty simulation on the model output. The simulation can have considerable computational complexity, depending on the input probability distribution and sensitivity of the given input parameter. Such approaches are already widely investigated, but use of uncertainty in practice is still lacking. Further developments of usage of uncertainty in decision making are needed, where businesses can also identify new opportunities.

5.6.2 Climate change and its impacts

Most observable characteristics of climate change and global warming are manifested in meteorological and climatological changes (especially an increase in temperature averages and more frequent occurrence of temperature extremes, changes in precipitation patterns and increased frequencies of droughts and floods). Considering that the design of reservoir or flood protection schemes and their construction takes 3–4 decades, the forecast of the development of water balance under climate change is essential for most countries if they are to apply adaptation measures on time. Hydroinformatics tools will definitely play an important role in climate impact assessment in complex studies.

5.6.3 New business models in simulation modelling

Traditionally, purchasing software was the only, later even preferred option for businesses. Leasing software has become an option for desktop or on-premises (in-house) computing, and it is a common way of accessing remote cloud applications. Leased software gives the customer access to the program's latest version and is a cost-effective option for casual users.

Cloud computing is an emerging paradigm that conveniently allows users to access computing resources as pay-per-use services. Various cloud offerings are available, such as Amazon's Elastic Compute Cloud, Microsoft Azure or Google Apps, which are rapidly increasing their user base. However, engineering applications and enterprise software migration towards the cloud are still in very early stages due to significant changes to the software architecture and licensing policies that make designing viable business models challenging.

A new business will likely be created under the BIM (Building Information Modelling) umbrella. The BIM system for 3D framework and reinforcement planning enables integrated working between the team of engineers on the virtual support structure model. In conjunction with the workgroup manager, the project team can work simultaneously on a building model and accurately coordinate the various planning steps. An example project (led by Witteveen + Bos) is the bridge construction across the River Waal, which was an enormous challenge, especially as the timetable was tight. This was part of Europe's largest flood protection project, named 'Ruimte voor de Waal' (Space for the Waal). The whole construction of the bridge has been designed and executed under the BIM umbrella (see again Allplan, 2022).

5.6.4 Growth in hydroinformatics: consultancy demands

Growth in hydroinformatics is evident across many disciplines and sectors. The market for all variants of the knowledge providing systems has already expanded from capital projects financed by the governments to small communities trying to improve the resilience of their water infrastructure; from global production companies looking at the security of water supply to small aquaculture food producers wanting to maximise their crop and not to breach the environment carrying capacity.

The growth we have already seen is accelerating and covers various services combined into a new, holistic approach. The era of traditional consulting, in which a project is scoped, analysed, reported on and report shelved, is largely gone. The models are being recognised as assets, continually evolving, improving, refining, and complementing with new data to form not only broader but higher insights. They are used to test as many potential scenarios as practicable, assess the uncertainties, optimise new data collections and generate new knowledge. As Mike Abbott foresaw in 1999, hydroinformatics is here to stay, and the creative space has opened widely by social and technological advances.

5.6.5 Vision for future or business

The authors of this chapter are all engineers and practitioners, some also with academic background. Our collective vision reflects our current observations about knowledge providers and knowledge consumers, both structured and under 'farmers market' conditions.

The process and information modelling tools, deterministic and stochastic, are generally mature, and there is a variety of paid and free packages to choose from. The software development business has been consolidating over the last decade, and this is expected to continue through mergers and acquisitions. Hydroinformatics tools are also moving from being on the fringe to being in the centre of broader engineering disciplines and, consequently, corporations. However, easy-to-use interfaces and robust computational code create future knowledge users that rely on technology without sufficient understanding of the science to critically evaluate and interpret the outcomes. Recent acquisitions of computational hydraulics tools into larger packages, such as GIS or CAD systems, may lead to a loss of knowledge needed to maintain the current- or develop new numerical solvers.

We can envisage that the water segment will attract more private entities, especially in the urban sector, where significant investments will be needed. They will seek to use sophisticated automation and digital technology to compete for the business in running the city's infrastructure. New business models will use more cloud-computing technologies, more powerful tools (HPC) and BIM. Technicians need better sensor and control mechanisms; interacting with the systems requires smarter visualisation; model-based forecasts can be improved with data-based now-casts. Stakeholders require transparent assessment and communication of costs, benefits and impacts, including uncertainties. The Cloud, SaaS, IoT, AI and VR, real-time controls of the system's digital twins create a new base for the growth of the hydroinformatics business.

One can see that, in parallel and overlapping with hydroinformatics, other holistic digital technology-driven transformation concepts are maturing, such as Industry 4.0 and Water 4.0 (see Moore, 2019; Philbeck & Davis, 2018, or Sedlak, 2015), based on a seamless combination of real and virtual worlds or cyber-physical systems. Many experts believe that BIM is the future cooperative working methodology. According to EU estimates (EU BIM Task Group, 2022), BIM can save up to 21% on planning and construction costs and 17% on operations. A new type of business alliance between infrastructure owners and operators is already established in which the operators are paid according to the realised savings and key performance indicators.

Companies currently supplying the market with software packages will likely move to the Cloud and offer subscriptions to smaller, task-oriented web applications with slick GUIs and powerful 3D visualisation. This will be an especially attractive environment for building quick 'screening' models, sourcing data, solvers and computational resources from an ecosystem of cloud data and service providers. The Cloud enables 'finding' and combining the best tools for a problem at hand. Another opportunity that the Cloud has is readily available access to powerful computing resources that can be used for quantifying uncertainties, something that practitioners increasingly need.

Alongside the market consolidating around big players, the challenge of innovation will still be open to newcomers. The new business of hydroinformatics could evolve into a business of ideas and disruptors of established markets. One can envisage science and business entrepreneurs setting up their low-cost start-ups and innovation incubators. Many of these will incorporate new combinations of existing, currently 'siloed' fields and solutions. There is plenty of funding available in the world markets through angel investors looking to place their surplus profits. The new game is not to build and grow the business but to grow ideas and, as quickly as possible, sell them on the market. Successful innovations will be bought and assimilated (or killed) by big established players, and the inventors will move on to new ideas.

REFERENCES

3Di Water Management (2021). 3Di as state-of-the-art hydrodynamic simulation software for pluvial, fluvial and coastal floods. https://3diwatermanagement.com/ (accessed 15 July 2022).

Abbott M. B. (1991). Hydroinformatics: Information Technology and the Aquatic Environment. Ashgate, Aldershot, UK and Brookfield, USA, ISBN 1856288323.

Abbott M. B. (1993). The electronic encapsulation of knowledge in hydraulic, hydrology and water resources. *Advances in Water Resources*, **16**, 22–39.

Abbott M. B. (1999). Introducing hydroinformatics. Journal of Hydroinformatics, **01**(1), 3–19, Inaugural issue.

Allamano P., Croci A. and Laio F. (2015). Toward the camera rain gauge. *Water Resources Research*, **51**, 1744–1757, https://doi.org/10.1002/2014WR016298

Allplan (2022). https://blog.allplan.com/en/waal-bridge (accessed 15 July 2022).

Almoradie A., Cortes V. J. and Jonoski A. (2015). Web-based stakeholder collaboration in flood risk management. *Journal of Flood Risk Management*, **8**(1), 19–38. John Wiley & Sons Ltd and The Chartered Institution of Water and Environmental Management (CIWEM), https://doi.org/10.1111/jfr3.12076

Armstrong M., Kramer L., de Senerpont Domis L. N., van Wijk D., Gsell A. S., Mooij W. M. and Teurlincx S. (2021). Flipping lakes: explaining concepts of catchment-scale water management through a serious game. *Limnology and Oceanography: Methods*, **19**, 443–456, https://doi.org/10.1002/lom3.10436

Correll M. (2019). A Reader on Data Visualization, MSIS 2629 Spring.

Craven J., Angarita H., Corzo Perez G. A. and Vasquez D. (2017). Development and testing of a river basin management simulation game for integrated management of the Magdalena-Cauca river basin. *Environmental Modelling & Software*, **90**, 78–88, https://doi.org/10.1016/j.envsoft.2017.01.002

EU BIM Task Group (2022). [Online]. Available at: http://www.eubim.eu/ (accessed 15 July 2022).

Going Global (2019). Expanding Offshore Wind To Emerging Markets (English). World Bank Group, Washington, D.C. http://documents.worldbank.org/curated/en/716891572457609829/Going-Global-Expanding-Offshore-Wind-To-Emerging-Markets (accessed 15 July 2022).

Gregersen J. B., Gijsbers P. J. A. and Westen S. J. P. (2007). OpenMI: open modelling interface. *Journal of Hydroinformatics*, **9**(3), 175–191, https://doi.org/10.2166/hydro.2007.023

Ingeduld P., Havlik V., Vanecek S. and Jiroušek P. (1996). Object-oriented hydroinformatics tools for water distribution analysis. In: Water Supply Systems in NATO Environmental Series nr. 15, Springer Verlag, Berlin Heidelberg, Germany, pp. 400–418, ISBN-978-3-642-64739-0.

Jonoski A. (2002) Hydroinformatics as Sociotechnology: Promoting Individual Stakeholder participation by Using Network Distributed Decision Support Systems. PhD Thesis, Sweets & Zeitlinger B.V., Lisse, The Netherlands.

Kumar S., Godrej A. N. and Grizzard T. J. (2015). A web-based environmental decision support system for legacy models. *Journal of Hydroinformatics*, **17**(6), 874–890, https://doi.org/10.2166/hydro.2015.007

Medema W., Chew C., Adamowski J. F., Mayer I. and Wals A. (2019). Understanding Game-Based Approaches for Improving Sustainable Water Governance, Special Issue Published in *Water Journal*, MDPI, Basel, Switzerland.

Moore M. (November 2019). What is Industry 4.0? Everything you need to know. *TechRadar*. https://www.techradar.com/news/what-is-industry-40-everything-you-need-to-know (accessed 15 July 2022).

OGC (2011). Open Geospatial Consortium (OGC®) and the OpenMI Association (OpenMI), MOU between OGC and OpenMI about using to promote the development, use, management and maintenance of the OpenMI, a standard for the exchange of data between computer software in environmental management applications.

Philbeck T. and Davis N. (2018). The fourth industrial revolution: shaping a new era. *Journal of International Affairs*, **72**(1), 17–22, ISSN 0022-197X. JSTOR 26588339.

Price K. R. (1997). Hydroinformatics, Society and Market, inaugural address for Chair of Hydroinformatics at the IHE, Delft.

Savic D. A., Morley M. S. and Khoury M. (2016). Serious gaming for water systems planning and management. *Water*, **8**(10), 456, https://doi.org/10.3390/w8100456

Sedlak D. (2015). Water 4.0 – The Past, Present, and Future of World's Most Vital Resource. Yale University Press, Yale, ISBN 9780300212679.

Solomatine D. P. and Ostfeld A. (2008). Data-driven modelling: some past experience and new approaches. *Journal of Hydroinformatics*, **10**(1), 3–22, https://doi.org/10.2166/hydro.2008.015

Stallman R. (2022). Why Open Source Misses the Point of Free Software, GNU.Org. https://www.gnu.org/philosophy/open-source-misses-the-point.html (accessed 15 July 2022).

Valverde-Pérez B., Johnson B., Wärff C., Lumley D., Torfs E., Nopens I. and Townley L. (2021). Digital Water: Operational digital twins in the urban water sector: case studies. White paper, © 2021 International Water Association.

van der Wal M. M., De Kraker J., Kroeze C., Kirschner P. A. and Valkering P. (2016). Can computer models be used for social learning? A serious game in water management. *Environmental Modelling & Software*, **75**, 119–132, https://doi.org/10.1016/j.envsoft.2015.10.008

Vojinovic Z. and Abbott M. B. (2012). Flood Risk and Social Justice. IWA publishing Alliance House, London, UK, ISBN 9781843393870.

Zeman E., Vaněček S. and Ingeduld P. (1992). A Mathematical Model for Simulation of Unsteady Flow in Rivers and Flood Plains – An Element of Decision Support System in Aquatic Environment. HYDROCOMP '92 Proceedings, pp. 253–260.

doi: 10.2166/9781789062656_0075

Chapter 6

Hydroinformatics in China: overall developments and showcase of accomplishments in the Changjiang River basin

Y. Huang[1], H. H. Ba[2] and S. Yu[2]

[1]Changjiang Water Resources Commission, Ministry of Water Resources, Wuhan, Hubei Province, China
[2]Changjiang Survey, Planning, Design and Research CO., LTD, Wuhan, Hubei Province, China

ABSTRACT

Over the last few decades, China has experienced tremendous growth and socioeconomic development. This has also been a period of significant transformations in many social and economic sectors, including water management. Starting from revision of policies and regulatory frameworks, modifications of water engineering approaches have been introduced, together with broadening the overall management of river basins and water resources to encompass ecological and other nature services, next to considerations for economic benefits. These water management transformations have also been facilitated and enabled by introducing novel information and communication technologies, very much along the lines that Mike Abbott predicted and promoted as he was developing the field of hydroinformatics. In fact, he was always considering China as a country that would greatly benefit from hydroinformatics technologies as it has been going through rapid socioeconomic development, and, that the field itself would then make progress via novel Chinese applications. This chapter briefly introduces the main water management transformations and challenges that China has faced over the past few decades, and proceeds with presentation of hydroinformatics tools and technologies in the sequence of data→models→knowledge→decision-support applications, as they have been deployed in China. To showcase the use of such technologies in actual water management activities, the example of Changjiang River basin is presented as a case study, with demonstration of applications on flood management. These applications integrate monitored data with a suite of different models for flood forecasting and regulation of water infrastructure – particularly large scaled reservoir groups, where flood management is addressed in conjunction with other water resources functions. Decision support applications, such as to enable broader stakeholder and public participation – the platform serving for river/lake chief system – are also presented.

Keywords: China, water conservancy, hydroinformatics, river chief system, lake chief system, Changjiang River basin, flood management, joint regulation, Yangtze River

6.1 INTRODUCTION

Water management in China has a long history. Being a populous country that has some of the largest rivers in the world within its territory, it has faced challenges with floods and droughts since ancient times. Building on traditional knowledge, modern China has developed the concept of *shuili* (水利) that literally means 'water's benefits', and which is commonly being translated as 'water conservancy' in English language (Boxer, 2001, 2002). (This term is also used in this chapter.) Since mid-20th century, when modern China was established, this concept has been for several decades associated with the approach of 'hard engineering', dominated by construction of large water infrastructure, such as dams, reservoirs, irrigation schemes and inter-basin water transfers. This strategy was found to be necessary for early development of the country.

In late 1980s and 1990s, as China was introducing its shift to a market economy, it was recognized that water management in the country needs to become much more holistic. Fast-paced industrial and economic development quickly brought about serious concerns related to overall increase of water usage, poor water-use efficiency and water pollution, which contributed to the shift in thinking. Traditional focus on economic benefits from water supply, irrigation and protection from floods and droughts, needed to be expanded with considerations for non-tangible benefits from ecosystem services and clean aquatic environment, also in relation to its historical and cultural values. In 1999, a major shift in national policy regarding water management was introduced through the concept of *ziyuanshuili* (资源水利), that could be translated as 'resource-oriented water conservancy', where water is considered a *resource* that can have different functions, which need to be considered jointly in an integrated approach (GWP, 2015; Jiang, 2015; Liu *et al.*, 2013; Wang, 2006). In fact, the concept of integrated water resources management (IWRM), as it has been developed and promoted internationally, was also adopted in China, but with considerations for specific conditions and development needs of the country (Song *et al.*, 2010). With growth of many Chinese cities, water management in large urban areas gained special importance (The Nature Conservancy, 2016). This new approach has recently been named as 'soft-path' (Liu *et al.*, 2013), as the focus has shifted away from construction of large engineering water projects towards integrated water management that would result in increased water-use efficiency, reduced water pollution and effective management of existing water infrastructure for realizing multiple benefits.

It should be noted that in China comprehensive plans for managing each large river basin have been made and upgraded every two–three decades, which provide engineering and non-engineering measures to fulfill multiple objectives of flood control, integrated water resources utilization, protection of water–environment (aquatic ecology), in an integrated manner. With the master plan providing the top-level design, which specifies not only demands and purposes, but also boundaries and constraints of water management, the engineering and non-engineering measures developed should jointly serve as much as possible to realize actual integrated management. For instance, a reservoir should serve for meeting objectives of power generation, water supply, navigation and so on, alongside with a major purpose of flood management. Examples include the well-known Three Gorges Reservoir that provides 22.15 billion m^3 flood storage which is the backbone of flood management of Changjiang River, or, the Danjiangkou reservoir from which on average 9.5 billon m^3 of water is transferred from southern to northern China, with 19 billion m^3 storage of flood during the flooding season. Thus, in the 'post-engineering' era the focus has been shifted from *construction* of engineering works towards their *management* for improved resources utilization via regulation of numerous water projects. New challenges have been raised as how to make the best use of this existing water infrastructure to satisfy the multiple objectives as they were planned, which naturally resulted in the rapid development of hydroinformatics as a vital non-engineering measure to facilitate and optimize operational decision-making processes.

The period of these significant transformations in Chinese water management coincided with the developments of the field of hydroinformatics, led by Mike Abbott. Hydroinformatics involves the construction of information infrastructure, including data monitoring and management, and technologies of representing the physical water systems in the virtual world using all kinds of models and informatics technologies. Mike Abbot and his fellows have discussed the possibilities of applying such technologies in China at least since 2007, when he visited Changsha in China to attend the 2nd Yangtze (Changjiang) River Forum. Later on, as hydroinformatics was developing as core technology for construction of smart water management tools, he elaborated more extensively on possibilities to use technologies such as Software-as-a-Service (or SaaS) in China for developing platforms where the multi-dimensional and multi-temporal-scale physically based mathematical models including hydrology, hydraulics, sedimentation, water resources allocation, environment, water and soil maintaining and so on, are integrated and applied for better decision making (Abbott & Vojinovic, 2013). In addition, at a later stage in the past decade, with the accumulation of larger amounts of data, applications of data-driven approaches such as neural networks, genetic algorithms and so on (see Chapter 3 of this book), were widely introduced as supplementary or complementary means to perform either separately or coupled with the physically based models, with the ultimate goal of producing better models of the physical world of water and water-related phenomena, to be used in actual management.

Starting from 1990s, China began to promote construction of information and communication technology (ICT) infrastructure, and development of hydroinformatics, as computers gradually became common tools to facilitate decision making and operational management, such as flood forecasting, water allocation and distribution, and so on. Considerable research and applications have been made through national key research and development programs such as the '5-years research program'. These supported the development of numerical models for quantitative precipitation prediction, rainfall–runoff models, hydrodynamic models for flood routing, and other related modelling tools, which were then applied in flood forecasting systems that formed the very first generation of hydroinformatics applications in China. Significant developments came in 2000, after the big floods that happened in 1998 which brought large loss of life and economic damage across China (in Changjiang River basin and in several other river basins), which triggered the Chinese government to spend huge efforts and resources for improving both engineering and non-engineering measures. National-wide developments of decision support systems (DSS) for flood management took place, covering the entire process from data acquisition, to modelling and decision support, all with implementations that were following the new water management policies introduced earlier. Considerable progress on the implementation of hydroinformatics concepts and technology have been made, and in 2021, it was proposed to carry out the construction of digital twin river (DTR) as the core of smart water development to achieve the national *14th 5-Year Plan for Smart Water Conservancy Construction*. The digital twin technologies have been widely introduced and applied in many fields, particularly in manufacturing industries (Grieves, 2014), and lately they have entered the field of water management (e.g., Qiu *et al.*, 2022; Valverde-Pérez *et al.*, 2021). These technologies are in fact closely related to hydroinformatics systems, as introduced by Mike Abbott, even if they are named differently (Abbott, 1991).

For China, at this 'post-engineering' stage, where all rivers are mostly affected or rather controlled by water projects such as reservoirs, sluices and gates, latest developments of digital twin technologies enable automated operation (or regulation) of multiple water projects that are incorporated in the DSSs of river basin management. Such applications provide the 'intelligence' for the virtual river (or DTR) to simulate, analyse and visualize scenarios that may or may not happen in reality. From the guidelines for development of the DTR, it is expected that the new platform shall enable synchronous simulations, virtual-real interaction and optimization of the activities that could take place in the physical river. The ultimate objective is to provide computation and analysis capabilities for water conservancy applications with analyses and inter-comparison of scenarios, to support river basin management activities such as flood management, water resources reallocation, and other related management tasks.

Development in digital twins in Chinese water management has been implemented within the so-called 'state-system for flood management and drought relief'. Currently, a large effort for developing '2 + N' DTR applications and digital twin water conservancy projects is being promoted nation-wide. The meaning of '2 + N' is that the first focus is on two major application areas of flood management and water resources management, which are followed by ' + n' other applications such as the river/lake-chief system, the soil and water conservation, safety of infrastructures and so on. Among this 'n' applications, the river/lake-chief system is an innovative institutional arrangement for river basin management where different levels of government and stakeholders participate and identify river basin management tasks and priorities (Wang *et al.*, 2021). At present, all river commissions and all provinces are setting up plans on how to develop the corresponding DTR system.

All these developments in digital twin technology include components that can be identified as being part of hydroinformatics systems. In this chapter, applications of hydroinformatics in China are therefore presented following the information flow structure of 'data→models→decision support applications'. Changjiang River is taken as a case study to show how hydroinformatics is implemented and functioning in river basin management practices. Section 6.2 presents the decision support framework based on the DTR approach in China, followed by Section 6.2, where data types and data-acquisition techniques are briefly presented. Models, as core components of hydroinformatics systems are described in Section 6.2. Decision support applications, as used in the river/lake-chief systems are presented in Section 6.2. A more elaborated presentation of actual use of these applications in the case of Changjiang River basin, with the main focus on flood management, is given in Section 6.2. The chapter concludes with a short summary in Section 6.2.

6.2 DTR DECISION SUPPORT FRAMEWORK

A DSS is a typical non-structural measure for river basin and water management, which is developed based on a data management system, and integrating/coupling of multi-dimensional and multi-temporal–spatial-scale mathematical models, including meteorological model, hydrological model, reservoir regulation model, hydraulic model, and so on. These models then provide main information inputs to decision support components. This information flow direction (data→models→decision support applications) has also been followed in two technical specifications – Technical Guidelines for Digital Twin Water Conservancy Project Construction and Technical Outline of Digital Twin River Construction – recently (March 2022) issued by the Ministry of Water Resources (MWR) of China. The approach forms the core of the so-called intelligent water conservancy, which is both a framework and a platform of DTR and/or infrastructures. As presented in Figure 6.1, it consists of three major elements: *database*, *model-base* and *knowledge-base*.

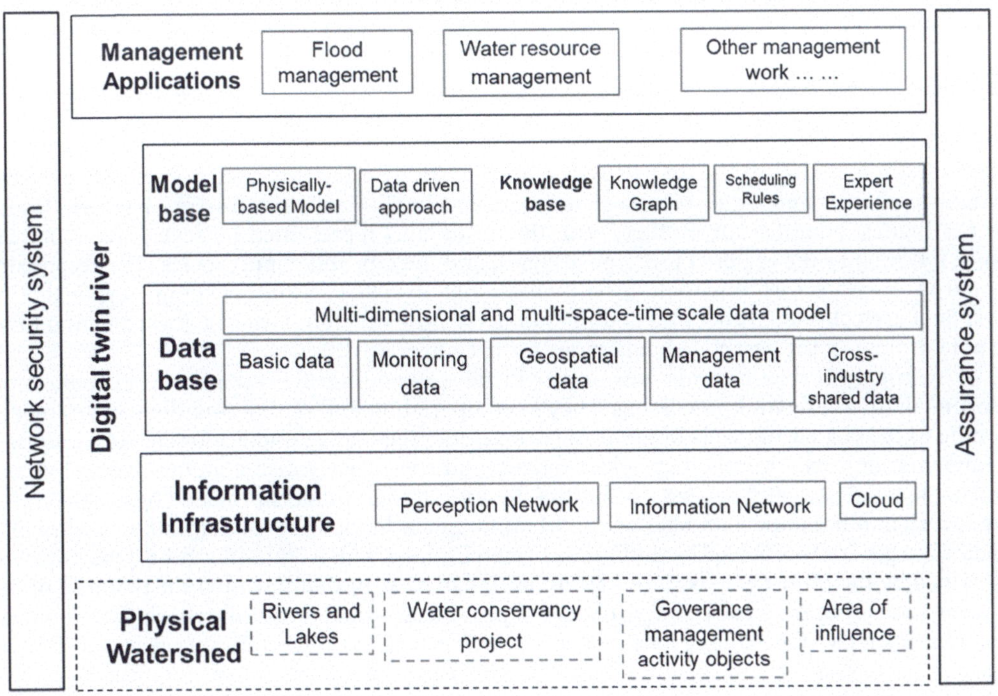

Figure 6.1 The framework of DSS of river basin management with the core of DTR (Technical Specification of Digital Twin River Construction, issued by MWR, March 2022).

In fact, the proposed structure is similar to most generic DSSs frameworks/platforms used in water resources elsewhere. The following sections present the key issues in the three major components of the framework, as they are being developed and incorporated in river basin management in China. The presentation is from the perspective of major river basin commission, therefore the main focus is on flood (and droughts) management, but, as mentioned earlier, multiple objectives are increasingly being included.

6.3 DATA ACQUISITION AND MANAGEMENT

6.3.1 Types of data

Data are the foundation for the development and application of hydroinformatics. The types can be classified under the following categories:

(1) **Basic data of water conservancy**: Basic water conservancy data refer to the attributes of objects such as rivers, lakes, reservoirs, hydropower plants, hydrological stations, dykes, flood storage and detention areas, sluices, pumping stations, and water diversion projects and so on. Actual data include: codes, names, location, topological relationships, characteristic values, characteristic curves and other attribute data of rivers, lakes, reservoirs, hydropower plants, hydrological stations, and water diversion projects.

(2) **Hydrological monitoring data**: Hydrological monitoring data refers to data such as water conditions, rain conditions obtained through online monitoring, as well as remote sensing data, video and online data provided by the public. Data could be historical monitoring data (i.e., time series), real-time monitoring data and future forecasted data, with different time intervals, as required. These data are collected on variables such as: water level, discharge, rainfall, river morphology, water quality, disasters, groundwater levels, water intake, soil moisture, and so on.

(3) **Geospatial data**: Geospatial data are geographic information that can be used in the computation and analysis of water issues, as well as the basic data of river basin management on which all elements and human activities are inter-connected. Different data types are used, such as: digital line graph (DLG), digital elevation model (DEM), digital raster graphic (DRG), digital orthophoto map (DOM), digital surface model (DSM), point cloud (PC) and other standardized GIS data.

(4) **Management data**: Management data refer to the data used- and generated by the management work during the operation and maintenance of the water conservancy information system. It includes management-relevant data for different objectives related to water resources, flooding, ecology, and so on, as well as management support data such as water supervision, water administration, and public services data. Together they provide comprehensive decision-making, operation and maintenance and other business-related data.

(5) **Sharing data from other industries (sectors)**: These are data from other industries that are not part of the water conservancy industry data but will have an impact on managing water conservancy objects, such as meteorological data, ecological and environmental protection data, natural resources, agriculture and rural areas information, socio-economic data and so on.

6.3.2 Data acquisition

Data acquisition is the foundation of the digital twin platform. Its working mechanism is highly associated with the institutional structure of river and water resources management in China. The Ministry of Water Resources (MWR) at state level is the highest administrative office, with the basin level represented by seven river commissions as the agencies of MWR, followed by the regional (provincial) level, and the city and county levels. The responsibility for river basin management is shared across this multi-level institutional structure. Data collection responsibility is similarly shared, and most data are collected, processed and maintained by the different levels of governmental offices, normally by the bureaus of hydrology at the associated data monitoring centres at all levels.

The bureaus of hydrology of river commissions or provincial water authorities are in general responsible for most of the data monitoring, maintaining and processing. They keep upgrading and optimizing their hydrological monitoring networks according to the change of needs in terms of parameters, frequencies and accuracies of water level, discharge, evaporation, sedimentation, river morphology, water quality and so on. They also conduct research and development work on monitoring technologies and devices, in addition to purchasing products provided by others.

Data sharing is also another important channel for obtaining data to facilitate management work. For instance, the meteorological data are generally provided by the bureau of meteorology, which is another state department next to MWR. GIS data are provided by the Ministry of Nature Resources. It can be seen that in China the majority of data producers and users are governmental offices. In recent years, some provinces attempted to purchase data monitoring services from the market. One example is the Guangdong province which is working on purchasing services of hydrological data monitoring from a private company that installs measuring devices and produces data as required by its users. However, this way of working is still in a trial stage, and is yet to be determined whether it will be used more broadly.

It needs to be noted, however, that as measuring, processing, maintaining and development of data are carried out by governmental agencies in China, this provides a consolidated basis for developing DTR/infrastructures following

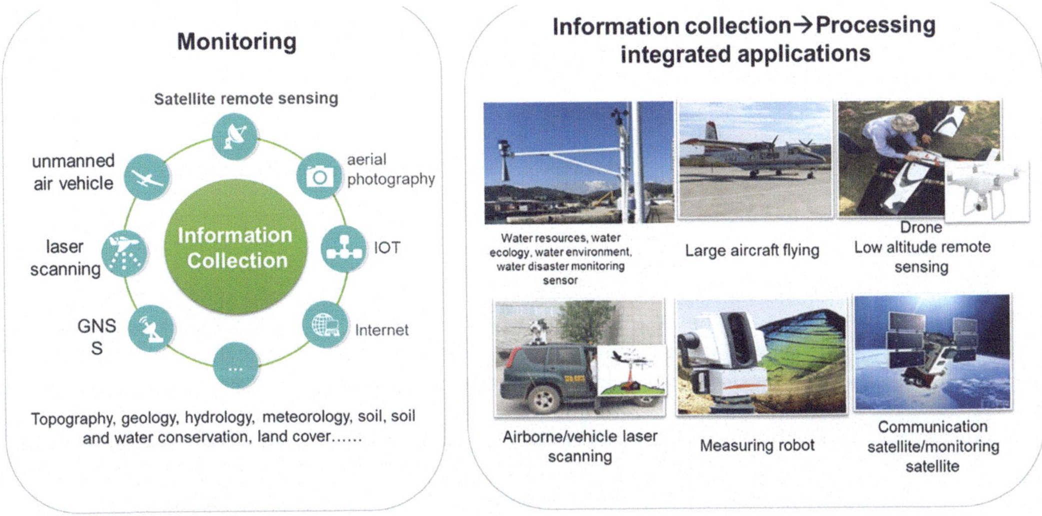

Figure 6.2 An integrated air-space-ground monitoring system for water management.

the same standards. This is particularly important for the sharing of data, models, across computational and communication infrastructures such as cloud and other networks.

6.3.3 Data monitoring technologies

Further to the traditional on-ground measurement monitoring, with the development of aerospace technology, sensor technology and information technology, a variety of remote sensing sensors are adopted to obtain multi-temporal and spatial resolution images of targets. These include sensors using visible light, infrared, microwave, synthetic aperture radar, lidar, and so on, based on aerospace and near-Earth platforms.

Known as intelligent monitoring technologies, the latest monitoring technologies such as smart video monitoring, high-precision sensors, micro-electro-mechanical systems (MEMS) smart sensors, smart acquisition equipment, global positioning system (GPS) equipment, drones, multi-beam detection systems, three-dimensional laser scanners, and interferometric synthetic aperture radar (InSAR) are receiving increasing attention. These can be applied to construct an air–space–ground integrated and comprehensive water engineering monitoring system for river management, particularly during emergency response processes. Figure 6.2 shows the basic information and monitoring methods applied for river data acquisition. These may also include information from mobile phones to obtain human real-time activities.

6.4 MODELS AND THEIR ROLES IN DSS

At present in China, as most water authorities have its own water expertise, water models such as hydrological/hydraulic models and reservoir regulation models, are developed as basin-specific and mostly by the experts of those water authorities. Unlike the commercial software providers by consultant companies such as DHI from Denmark, Deltares from the Netherlands, Innovyze or Bentley from USA, or the freeware providers such as US Army Corps of Engineers and others, there are no well-recognized providers of generic commercially or freely available software packages in China. Thus, most models and DSSs involved are developed/maintained either by the Bureau of Hydrology (BOH) at different levels, or by consulting companies engaged in a given project work.

Meanwhile, according to *Smart Water Planning* (developed as one of the planning components of the 14th 5-year plan) of the country, a model-base is proposed to be developed and equipped by MWR by the year 2025, which can provide model services to different DSS systems for river basin management at all governmental levels. After decades of development, a basin-specific model-base is gradually emerging, which can be shared as a service for free within river management organizations. For instance, in Changjiang Water Resources Commission (CWRC), the models and DSSs for flood management are developed and upgraded by the bureau of Hydrology of CWRC, and 90% of these models (including hydrological models and hydraulic models) have been gradually registered in the model platform to be used by other systems of the commission.

It needs to be mentioned that future expansion of such sharing of models and other DTR components will necessarily be in the cloud. Development and implementation of the digital twin systems or DSSs require large computation sources, so cloud computation is inevitable. In China, there are various commercial cloud platform providers, such as the *Ali Cloud* provided by Alibaba internet company, *Tencent Cloud* provided by Internet company of Tencent, and *Cloud* provided by Huawei – the largest mobile phone manufacturer in China. These clouds are commonly adopted by many cities or provinces for their governmental administrative work or public services, as well as by private users. However, in the water field the administrative water offices at all levels remain working on their own dedicated networks and servers using the so-called private cloud. This situation is a legacy from the intensive construction of special communication networks in the past, However, with the implementation of smart water planning, there are clear movements towards using commercial cloud platforms to build water-oriented cloud in the coming 2–3 years, which shall cover water authorities at all levels, including river commissions.

In the following sub-sections, the different types of models and their use in river basin management in China are presented.

6.4.1 Physically based water-related models

6.4.1.1 Meteorological modelling

At present, the meteorological numerical models applied for precipitation prediction in China are Global/Regional Assimilation Prediction System (GRAPES) models at the Chinese Meteorological Administration (CMA), known as CMA-GRAPES models. These include models with different spatial resolution, described as follows:

- The **CMA-GFS Global Forecast System** is an ensemble numerical forecast system developed by Earth system and Prediction Centre (CEMC) of CMA. It was officially put into operation on 5 August 2014. The system works on a spatial resolution of 25 km and a time interval of 3 h, and can provide meteorological prediction for a lead-time up to 10 days.
- The **CMA_GFS Global Ensemble Forecast System** is an ensemble numerical forecast system developed by CEMC. It was officially put into operation based on the same CMA-GRAPES model. The system works on a spatial resolution of 25 km, and can provide meteorological prediction for a lead-time up to 15 days.
- The **CMA_MESO regional numerical forecast product** works on a spatial resolution of 3 km and a time interval of 3 hours. It can provide prediction for a maximum lead-time of 72 h. The forecast elements include air pressure, temperature, wind speed, relative humidity, and precipitation.

6.4.1.2 Hydrological modelling and stochastic modelling

In China, as rivers are with different scales and geographic conditions, almost all kinds of hydrological models (or rainfall–runoff models) and flood routing models can be found in management practices and project research works. The hydrological models are used for various river basin management tasks (such as water resources assessment, planning, allocation, etc.), but their application in combination with flood routing models is of vital importance for flood management.

The current DSS on flood management for large rivers in China is set up to prevent or reduce damage when flood happens. In the system, the safety level of a given area is pre-determined with design standards. For instance, if a city is with 10-year return period flood safety level, it means such flood can be successfully managed without causing severe damage. However, once the flood exceeds the safety standard, referred to as 'over-standard flood', or a large basin-scale flood happens, mitigating flood risk using the extra capacity of engineering measures will be needed. Accordingly, regulation rules for implementing reservoirs or retention basins during emergencies should be developed or designed in advance. In order to develop those flood control regulation rules for real-time management, flood samples with certain design standard are needed. In view of the fact that the measured data of large flood events are of limited availability, a random simulation method aiming to generate data of large flood events using the Copula function has been developed, with a procedure depicted in Figure 6.3.

This stochastic flood simulation enables obtaining a large number of flood events through random simulation based on statistical laws of historical flood observation data. There is a certain correlation between the flood hydrograph, the flood peak and flood volume of those flood events. Copula function shows good performance to describe the correlation of multiple variables, and it has been successfully applied in flood simulation problems (Li *et al.*, 2020; Xu *et al.*,

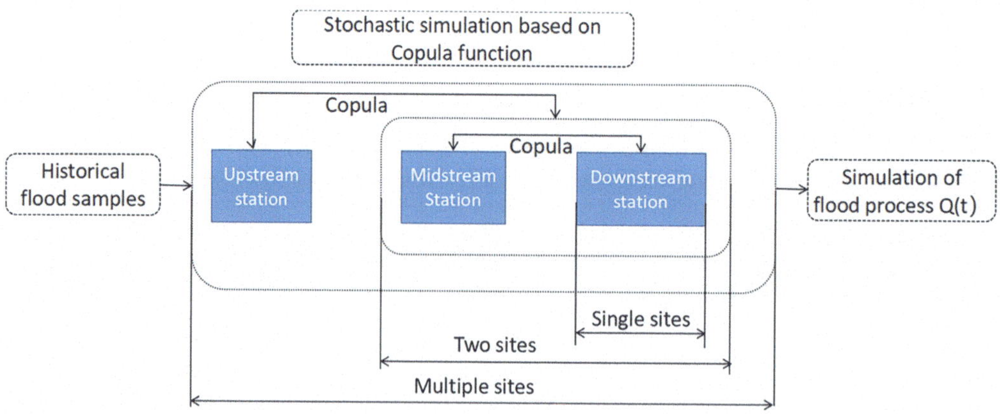

Figure 6.3 Structure of flood stochastic simulation model based on copula function.

2016). Thus, a multivariate joint distribution function is constructed using Copula distribution, which resulted in a stochastic simulation model of flood, taking into consideration the flood composition and its spatial–temporal characteristics. With this model, flood events can be generated by considering specific combinations of flood sources from different tributaries or regions associated with hydrological data coming from its gauge stations.

6.4.2 Regulation models

At present, there are 98 000 reservoirs with total regulatory storage of 93.23 billion m^3 constructed in the rivers of China. To achieve river management objectives as well as maintain a healthy river eco-system, *joint regulation* (or *integrated regulation*) of reservoirs and water diversion projects has become an important non-structural measure to coordinate multiple and contradictory objectives. In China, each river basin (at its commission level), or province or city (at their respective levels) has its own DSS developed and maintained by its technical support. Such a system could either be developed by their own offices such as the Bureau of Hydrology, or by consultant companies or research institutes hired through project work. Through nearly one decade of development, in the present DSSs used in many large river basins, reservoir regulation models have been developed at various stages – some remain focused on individual objectives mainly for flood management, and other, more advanced DSSs have taken multiple objectives into account. Objectives involved in regulation model development include (but not limited to) flood management, water resources utilization, sediment transport, ecological restoration, power generation, navigation and so on (Figure 6.4). In addition, *optimization models* are adopted as well to find appropriate regulation schemes particularly when contradictory objectives are involved.

The following sections present the main characteristics of regulation models for each individual objective as well as for the case when multiple objectives are considered.

6.4.2.1 Regulation model for flood management in a river basin

Regulation for flood is mainly applied in decision-making procedures during flooding seasons. In the current DSSs that are in operation at different levels of water authorities in China, real-time and multiple-lead-time forecasted hydro-meteorological data such as precipitation, water level and discharge and so on, are used to assess the flood risk of a flooding event. In the DSSs for flood management that are currently applied in MWR, different river basin commissions or provincial water authorities have their own systems with different interfaces and models, however with similar structures. Models for the regulation of floods include the following:

- **Reservoir regulation model**: which controls discharge (release) process of the reservoir according to the inflow and flood risk situation of the protection target downstream. This model usually consists of two parts: (1) the backwater computation which calculates the water profile in the reservoir area during a regulation process, and (2) the discharge control module which calculates reservoir releases to the downstream according to the regulation rules.
- **Regulation model of detention/retention basins**: to store water temporarily in order to reduce water level or discharge and maintain the protection target of cities or towns along the river, models of using detention or retention basins are being developed. Detention basins are generally small areas, constructed in the river (or lake area), between embankments, with

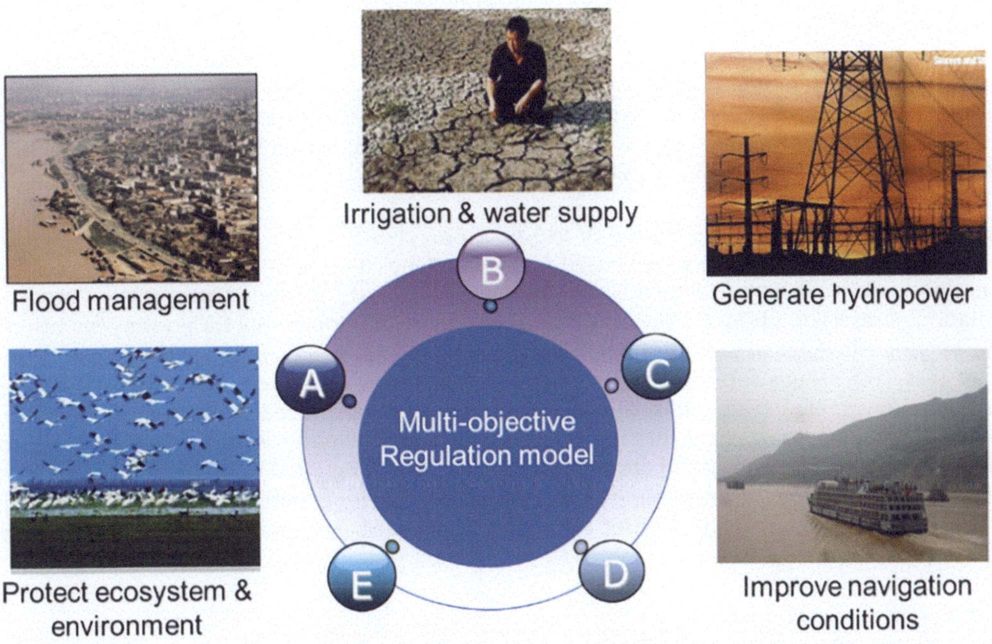

Figure 6.4 Regulation models for individual management objectives and multi-objectives.

low safety level and relatively lower dyke heights. These are normally not inhabited areas. Retention basins are constructed outside of the river channel or lake area, beyond embankments and with relatively higher dykes. They are larger areas, sometimes with significant population, and if they are used – evacuation is necessary. In any case, the required models take water level or discharge of the river reach as the initial and boundary conditions, and simulate diversion discharge and water stored in the retention or detention basin according to thresholds of the river, gates opening or flow from a dyke breach. This is actually a combination of different models, such as river hydrodynamic model, dyke breach model and 2D inundation model with closed physical boundaries.

- **Integrated regulation model using various types of infrastructures**: this model provides functionalities to simulate hydrodynamics in the river channel and floodplain areas, taking into consideration of contribution from drainage and pumping actions alongside the river (normally water is pumped from a city to the river). Such actions are triggered with threshold values at certain gauge station for the use of particular infrastructure.

The optimization procedure for the allocation of flood control storage of reservoir group has been developed to reallocate storage of each reservoir when the reservoirs behave as a group. The reallocation of storage is made considering spatial and temporal flood risk distribution. It takes flood control capacity of all reservoirs in the basin as a whole, and flood protection tasks of each reservoir as constraints, with minimizing the flood loss of the entire river basin and maximizing the overall flood management benefits as the optimization objectives. The ultimate goal of the procedure is to ensure the protection target with minimal losses and maximal integrated water resources utilization benefits such as power generation, navigation demands and so on. The results of this procedure present a reasonable division and allocation of storage for different flood control areas. With that achieved objective, the discharge hydrograph of each reservoir is derived accordingly.

6.4.2.2 Water resources regulation model

In China, water resources of river basins are unevenly distributed in time and space, with acute shortage in the arid regions. In the dry years and dry seasons, there are conflicts between tributaries, and between source areas and areas that receive water from inter-basin water diversion projects. Therefore, to minimize the losses caused by drought and safeguard water supply for domestic demands, while coordinating consumptive water and demands to maintain healthy environment and ecosystem, water resources regulation models are being researched and developed. In addition, it is currently planned in the *14th 5-year plan of hydroinformatics development* in China to develop such models to facilitate drought-resistant and water reallocation for different regions and water sectors.

6.4.2.3 Ecological regulation model

There are many different ecological regulation modelling objectives, but in China they may be exemplified by important applications for inhibiting algal blooms and protection of important fish species. The ecological regulation model for inhibiting algal blooms is through the control of reservoir discharge which aims to avoid the favourable hydrological and hydraulic conditions for algal bloom. Considerable research has been carried out to study the algae composition and the biological characteristics of the dominant species of the blooms in the tributaries of the reservoirs (Liu *et al.*, 2012, 2016). In addition, numerous field experiments were conducted to find the critical water velocity for different dominant algae species. To satisfy the requirement of flow velocity with which algal bloom maybe mitigated or reduced at the estuary areas, the joint regulation rules of reservoirs located at tributaries and reservoirs on the mainstream are derived. The reservoir ecological operation model for the protection of important fish species aims at reshaping the hydrological process to stimulate the natural reproduction, based on comprehensive studies on fish spawning and habitat factors, and ecological hydrology and hydraulics (Tao *et al.*, 2017).

6.4.2.4 Regulation model for power generation of group of reservoirs

The operation mode of cascade reservoir groups used for power generation is impacted by factors such as incomplete information and uncertainties in power demands. It is therefore necessary to study the operation mode of joint power generation regulation of reservoir groups in the power market. The regulation model for power generation of a reservoir group aims at improving efficiency of water use for power generation, taking advantage of large-scale cross-basin, cross-grid operation of reservoir groups for joint peak regulation (according to the characteristics of each power source (thermal, wind, hydropower, etc.). The output of each generator set is optimized and adjusted to reduce the peak-to-valley difference of the grid load, so as to ensure the stable operation of the power system). In the model, various aspects are considered, namely the excess power and absorption capacity of the regional power grids, the connection between different power grids line topology, transmission quota and grid security constraints requirements. This results in a regulation method considering joint regulation of cross-grid and inter-basin operation of reservoir groups for power generation is developed (Xie *et al.*, 2018).

6.4.2.5 Multi-objective regulation model

River management in China has shifted from the single-objective of flood control to multiple objectives involving water supply, shipping, ecology, power generation and others. It is therefore necessary to strike a balance timely and accurately against the competitive- and towards synergistic relations between the various stages and objectives. Tasks concerned include preparation and training before flooding season, flood management during flooding season, water impoundment at the end of flooding season, ecological restoration and environmental protection in dry seasons, forming a full cycle of reservoir operation in a given year (Huang *et al.*, 2021b).

6.4.3 Risk assessment models

Assessing risks and benefits of different regulation schemes during flood management generally involves models for one-dimensional (1D) and two-dimensional flood inundation, flood risk assessment, benefit assessment, and reservoir backwater calculation. The *flood risk assessment models* focus on inundation depth and duration, as well as their impacts, such as number of affected people and loss of properties and other values in the protected areas. The following models are supporting decision-making processes:

(1) **Dyke breach model**: When large floods occur, protective dykes can breach, either intentionally for flood diversion purpose, or unintentionally, by overtopping or damage of the dyke. The dyke breach model enables simulations of dyke breaching progress and of flood diversion flow. At present in China, dyke breach processes and the associated changes in flow and sediment movement are being studied and models are under development.

(2) **Backwater model in reservoir area**: The backwater calculation model of the reservoir area is applied to calculate the highest water profile along the reservoir area given the impounded water level and the inflow. At present, models for calculating backwater in the reservoir area of are well developed and mature, and are being applied for different reservoirs.

(3) **Flood risk loss assessment model**: The flood risk loss assessment model aims to quantitatively estimate the flood damage in the inundated areas such as the floodplain area, and the retention and detention basins that are to be flooded. Damage indicators include water level, discharge, number of affected people, the submerged area, submerged cultivated land, property loss, impact to important facilities, and so on. For reservoir areas, the flood risk is assessed using the backwater model combined with the socio-economic information of the affected area.

6.4.4 Data-driven approaches: artificial intelligence models

6.4.4.1 ML and AI models used for hydrological forecasting

Hydrological forecasting is an important basis for decision-making during flood control, drought management, water resources utilization, water ecology and environmental protection, and operation and management of water conservancy and hydropower projects. In recent decades, due to their powerful ability to deal with nonlinearity, ambiguity and uncertainty, machine learning (ML) and artificial intelligence (AI) technologies have been extensively researched and applied for hydrological forecasting (see also Chapter 3 of this book). For example, many scholars in China have widely used artificial neural network (ANN) models, long and short-term memory neural network models (LSTM), support vector machine (SVM) models and other AI and ML technologies to develop short-, medium- and long-term smart hydrological forecasting models and methodologies (Zhang *et al.*, 2021).

For flood forecasting of small- and medium-sized rivers, with short time of flood response, traditional hydrological forecasting models may have difficulties in producing accurate forecasts. In addition, due to complex geographic conditions, difficulty in data monitoring, and relatively scarce data in small and medium river regions, the alternative AI models show their advantages. They are fast-running models that work without a deep understanding of the physical mechanism of the hydrological flood generation process. With their strong ability for nonlinear fitting and relatively simple structure, these models are becoming more widely applied in flood forecasting of small and medium rivers (Kong *et al.*, 2018; Ma *et al.*, 2016).

6.4.4.2 Intelligent regulation technology based on a knowledge graph

Traditional regulation modelling technology is mainly based on customized development of a flood event, and it is time consuming and complex to apply to the large-scale, multi-level, high-dimensional, cross-regional water project operation and management needs, under rapidly changing environment.

The rule-base regulation has been developed using knowledge graph technology which connects the effects of using different combination of flood management engineering works such as dykes, reservoirs or retention basins, or their combinations, to information regarding flood risk, status of the engineering work and so on. Thus, a set of regulation–response relationships are developed, which can be used in real time when timely decision support or inter-comparison of scenarios of regulation schemes is needed (Huang *et al.*, 2022). This concept is presented in Figure 6.5.

6.4.4.3 ML and AI models for other water management tasks

In addition to hydrological forecasting and reservoir operation, the ML approaches have been applied in water quality prediction, flood routing, water resources allocation and other water management tasks. In the field of water quality prediction, Huang *et al.* (2021a) comprehensively reviewed the application and research progress of ML models in water quality prediction, and proposed the application and development prospects of ML models in water science. For flood routing, used in flood forecasting in China, physically based hydraulic model using numerical solutions is most common. Although this method can simulate the flood evolution accurately, the model establishment and computation are rather time consuming. With the development of new technologies in ML and AI models, these have also achieved reasonable success in flood routing (Wu *et al.*, 2021). In the field of water resources allocation, many scholars in China (e.g., Qiang, 2012) use ANN models for water demand prediction, and some scholars solve multi-objective optimization problems in water allocation based on genetic algorithms (e.g., Mou *et al.*, 2019). It should be noted however, that in China, the AI models are currently mostly studied and used for research purposes, and there are only few applications in river management practices.

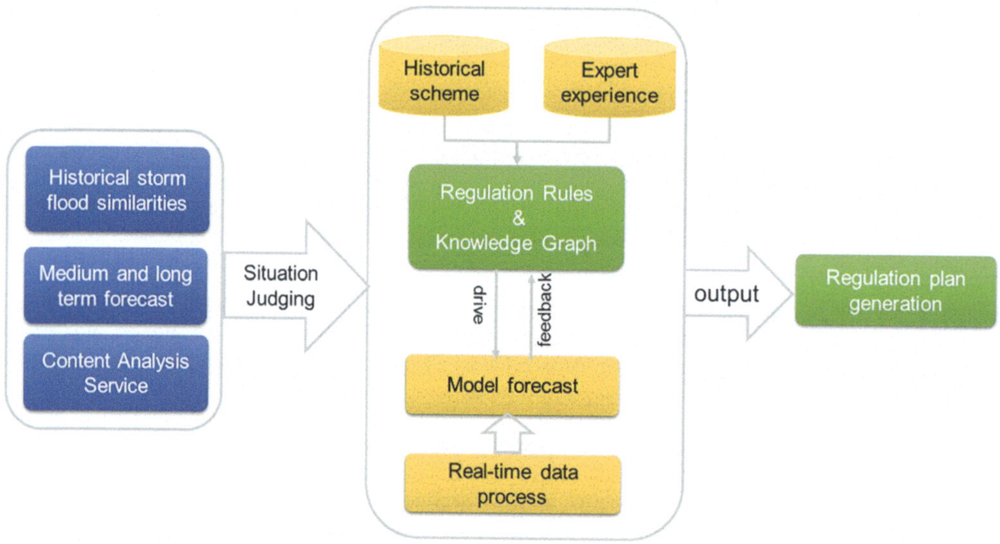

Figure 6.5 Intelligent regulation technology based on a knowledge graph.

6.5 IMPROVED GOVERNANCE AND PUBLIC INVOLVEMENT: THE RIVER/LAKE CHIEF SYSTEM

The *river/lake chief system* is the collective name of the river chief system (RCS) and the lake chief system. It is a national policy which established an institutional arrangement for all rivers and lakes in China, with which individual persons from responsible positions of government offices at all levels shall serve as river and lake chiefs, charged with the responsibility to organize and lead the corresponding river and lake management and protection, including collaboration with stakeholders and the public (Wang *et al.*, 2021).

'River Chief' is a special management system to supervise the improvement of river water quality and the aquatic environment. Each river in China has its river chief. The system originated in Taihu Lake Basin Management in Jiangsu Province in 2008, in response to the fact that the water quality of the rivers in the basin directly affects the Taihu Lake. Wuxi City determined that the leaders of the government at all levels in the city will fully implement the RCS for water function zones. The RCS is divided into four levels: the main leaders of the municipal party committee and the municipal government serve as the first-level river chiefs of the main rivers, the main leaders of the relevant departments serve as the second-level river chiefs, the main leaders of the relevant towns are the third-level river chiefs, and the village cadre of the administrative village is a fourth-level river chief. By 2014, more than 10 provinces including Jiangsu, Yunnan, Hebei, Anhui, Hubei, Sichuan, Heilongjiang, and others had implemented the RCS, and successively established the RCS in the Huaihe River, Haihe River, Dianchi Lake, Chaohu Lake, Songhua River and other key pollution control basins. The leadership responsibility system for water pollution prevention and control explores a new model for promoting the country's regional ecological environment restoration and river basin water environment governance.

In December 2016, the General Office of the Central Committee of the Communist Party of China and the General Office of the State Council issued the 'Opinions on the Comprehensive Implementation of RCS, and issued a notice requiring all regions and departments to conscientiously implement it in light of their actual conditions. Since 2018, the main leaders of the governments of 31 provinces (autonomous regions and municipalities directly under the Central Government) in China have served as provincial-level river chiefs. More than 900 000 village-level river and lake chiefs (including river patrolmen and river guards) guard the 'frontline' of rivers and lakes in China.

In order to support the RCS, the *river and lake information system* was developed at the level of policy and for interactions of different levels of government, and between government and public. The architecture of river and lake chief information system usually includes a transmission layer, a platform layer, an application layer, and a decision-making layer.

- The *transmission layer* includes hydrological monitoring equipment and sensing systems, and, water conservancy networks and transmission networks.
- The *platform layer* includes the river and lake chief cloud platform, the data resource pool and support platform.
- The *application layer* is the service monitoring platform, mobile work platform and public service platform together with other smart applications.
- The *decision-making layer* is the River/Lake Chief Working and Command Centre.

In accordance with the requirements of managing people, rivers and affairs, the river/lake information platform focuses on six tasks, including: *water pollution prevention, water resources protection, water environment governance,*

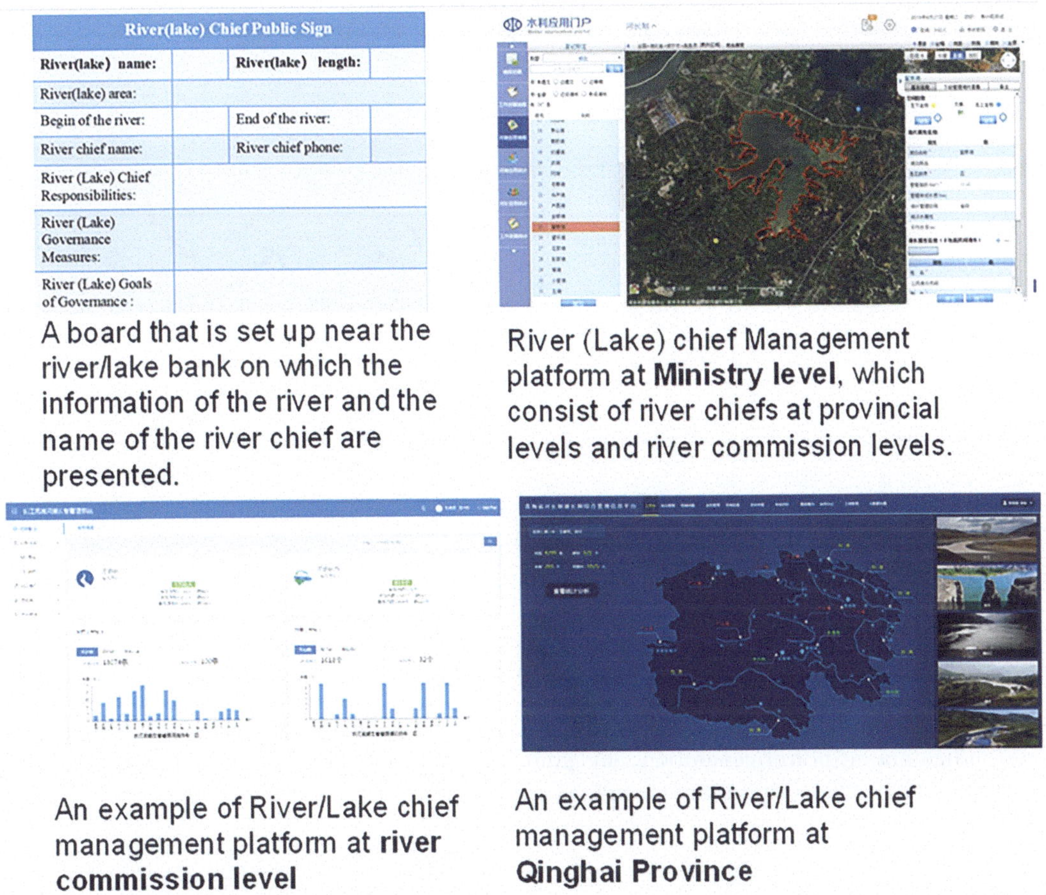

Figure 6.6 Schematic diagram of the River chief system bulletin board and applications at level of the ministry, river commission and province. Platform at provincial level and commission level can receive reports from the public: when people discover something unusual, they can report such information to the river/lake chiefs using a mobile phone app.

water shoreline management and protection, ecological restoration, and law enforcement supervision. 'Command, coordination, monitoring, handling, supervision, feedback, and assessment' is the main line of the operation and is currently being promoted and applied in China. Figure 6.6 shows the schematic diagram of the river/lake chief system bulletin board and application examples at levels of the ministry, river commission and provinces.

The River/Lake Chief System is relatively new in China and it is still in development. Further work on increased public participation (Wu *et al.*, 2020) or potentials for inclusion of citizen science (Wu *et al.*, 2022) are being considered, and are expected to be supported in future versions of the river chef system information platform.

6.6 APPLICATIONS IN THE CHANGJIANG RIVER BASIN

6.6.1 Flood management

China is one of the countries that suffers from frequent flood and drought disasters. After years of development, a flood and drought disaster prevention system that combines engineering and non-engineering measures has been formed. The Changjiang River (see Figure 6.7) – also known as Yangtze River – is the largest river in Asia and China, and the third largest river in the world. It originates from the Tanggula Mountains in Qinghai Province, with a total length of about 6300 km and a catchment area of 1.8 million km^2. It spans three major economic zones in eastern, central and western China, with a total number of 19 provinces, municipalities and autonomous regions (and 30% of China population), which produce about 30% of the national GDP.

As mentioned earlier, with the rapid construction and operation of water projects, particularly large-scale reservoirs, river management in China has shifted from the engineering construction period to the era of operation of water works. With the large number of engineering structures, the hydrological regime of the river basin has been affected. The regulation mode of these water projects has also transformed from individual operation to a joint operation mode, particularly for reservoirs. Their regulation rules were developed individually, but later (since 2012 when the very first joint regulation rules were developed), all reservoirs and retention basins in the river basins in China were gradually moved to joint regulation mode. In addition, with the use of flood forecasting, the way of passive response has been changed into anticipatory joint operation of water projects.

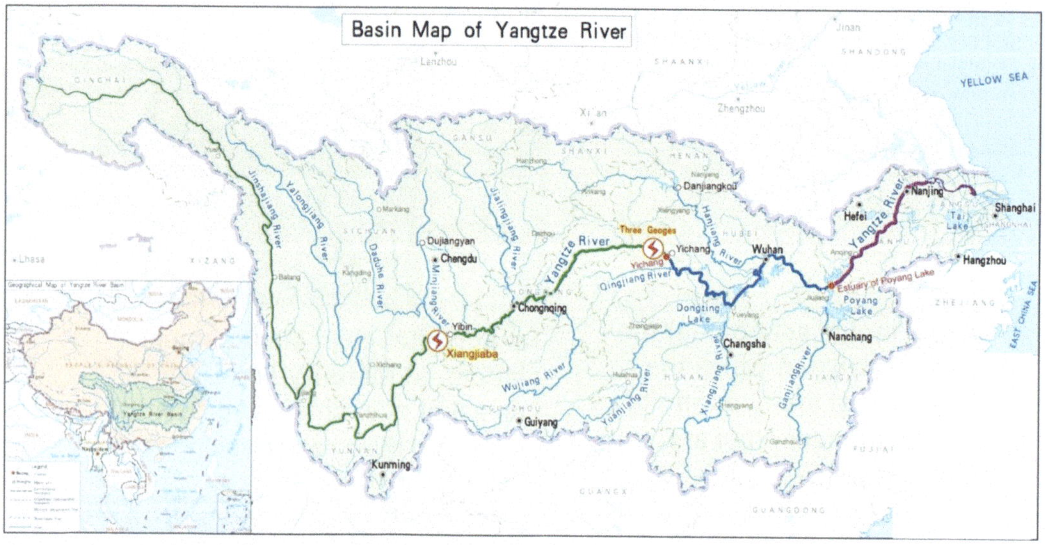

Figure 6.7 Map of Changjiang River (also known as Yangtze River) in China.

Consequently, in this post-engineering period, integrated river management introduces requirements for coordinated project regulation schemes and methods, and consideration for the trade-offs, the competition and the synergy between different objectives at all time during the year.

To improve the capability of constructed water works to realize the flood and drought management tasks, as well as integrated water utilization for power generation, navigation and other functions, it was of increasing significance to strengthen the developments of hydroinformatics support.

The construction of hydroinformatics support tools in the Changjiang River basin involves the following components:

(1) Collection and fusion of multi-source *data* such as hydrology and meteorology, environment, ecology, water engineering, geography, and social-economy in the river basin, and development of databases made available via a data platform, to facilitate the development of modelling processes and water applications.

(2) Development of *mathematical models* based on physical processes of hydrology, hydraulics, water quality, water ecology, reservoir regulation, risk assessment, and so on, to facilitate the development of water applications to support management activities with computation, simulation and analysis capabilities.

(3) Development of *joint regulation rule-base* and its *kick-off engine* to meet demands for flood control and water resources management purpose, as well as to enable trade-off among contradictory objectives with intelligence represented by the regulation rule-base and its kick-off engine. Similar to Alpha-Go (the AI computer game that plays the strategy game 'Go'), the joint regulation rule-base and its kick-off engine translate complex descriptive regulation scheme into digital and logical rules, which is to be executed for particular water management tasks. This is the key intelligent hydroinformatics component for river basin management.

(4) Development of *DTR* using IT technologies such as cloud computing, Internet of Things, big data, AI, mobile Internet, GIS, BIM, 3R, 'blockchain' and so on, to analyse and mine data relationships and to form the knowledge base or knowledge map of flood management, which can facilitate the construction of the inter-connection and response between natural elements and management activities. The DTR enables the complete integration of processes of water flow, information flow, business flow, and value flow.

(5) Development of *DSSs* for river management. In fact, DSS has been developed and updated regularly, since 1990s when first hydroinformatics developments were introduced, primarily oriented to the most urgent management need – flood management. Flood-related applications were the first generation of hydroinformatics tools for Changjiang River. At present, with the promotion of DTR, more management applications are under construction, such as water resources allocation and management, river (lake) chief monitoring system and others.

6.6.1.1 Data acquisition in the Changjiang River basin

Through long-term development, a comprehensive network of monitoring stations integrating hydrology, water environment, water ecology, river and lake shorelines, and water and soil conservation, has been well developed. Consequently, a full chain of information flow has been enabled, such as monitoring, collection, fusion processing, reorganization, and application.

At the same time, after decades of accumulation, the Changjiang River Water Resources Commission has collected and sorted out basic information on hydrology and sedimentation, geological survey data, river observations, wading engineering, geographic information, ecology and environment, social economy, and so on. This systematic monitoring has formed a comprehensive basic monitoring data set for the river basin. A batch of management databases has also been established, covering different tasks, such as: planning, flood and drought disaster prevention, water resources management, river and lake management, conservation and protection, sand mining, soil and water conservation management, supervision and law enforcement. Among these, management databases for flood management are comparatively better developed because of its importance. Such long-term large-scale multi-disciplinary data accumulation is an important basis for the development and application of hydroinformatics in the Changjiang River basin.

6.6.1.2 Models and DSSs for the Changjiang River basin

Physically based mathematic modelling is the core technology for river management. At present, China's domestic physically based models, including hydrological models and hydrodynamic models are well developed regarding their theory and technology, but their service capabilities in implementing technical standardization and versatility are limited. Therefore, in the construction process of hydroinformatics support, according to the needs of flexible and rapid construction and iteration, a general model library (or model-base) has been developed, which consists of 16 rainfall–runoff hydrological models, six river routing models, two water resources allocation models, eight water project regulation models, and four flood risk assessment models. This current model base is regularly maintained and upgraded as new models are being developed and applied.

6.6.1.3 Joint regulation of water projects for flood management

Starting from the approval and application of the 'Three Gorges Optimal Regulation scheme' in 2009, the Changjiang River Basin has experienced three stages of regulation development: (stage 1) – the Three Gorges as the core single reservoir (2008–2011), (stage 2) – a group of reservoirs (2012–2018), and (stage 3) – a variety of water projects (reservoir + flood storage and detention area + drainage pumping station + water diversion project, etc.) (2019–present). After more than a decade of development, a joint regulation scheme has gradually developed with flood control as the main objective of the engineering system.

As shown in Figure 6.8, after more than 10 years of development, in 2021, a total number of 107 water projects are involved in the joint regulation scheme, including 47 reservoirs with a total regulatory storage of 106.6 billion m^3 and a flood control storage of 69.5 billion m^3.

The characteristic periods in a year are used in management of the Changjiang River basin: (1) May–September is the main flood control period, (2) September–October is the main water impoundment period for most of the reservoirs, and (3) November–May is the reservoir drawdown period, with an average use of about 80 billion m^3 water resources. Real-time river basin management covers flood and drought disaster prevention, integrated utilization of water resources (e.g., power generation, water supply, shipping), ecological restoration and environmental protection, at the mainstream and major tributaries of the river basin. Therefore, operation and use of these large storage water projects for meeting the demands of different stakeholders, in different time/space, is of great significance for the sustainable socio-economic development, particularly in the economic belt of the Changjiang River.

Hydroinformatics support tools at the Changjiang River basin aim to solve the difficulties of joint regulation of large-scale reservoirs and water diversion projects based on real-time and predicted hydro-meteorological conditions. In accordance with the principle of maximum benefit and minimum risk, the appropriate regulation scheme is proposed for decision making, considering multi-objectives of flood control, water impoundment, power generation, shipping, ecology and sedimentation, and other objectives. To this end, with flood risk mitigation as the main goal, the regulation model with the rule-base for flood control in Changjiang River basin has been developed, which can be considered as symbolic (or knowledge-based) AI system (for details of this system, see Huang *et al.*, 2021b).

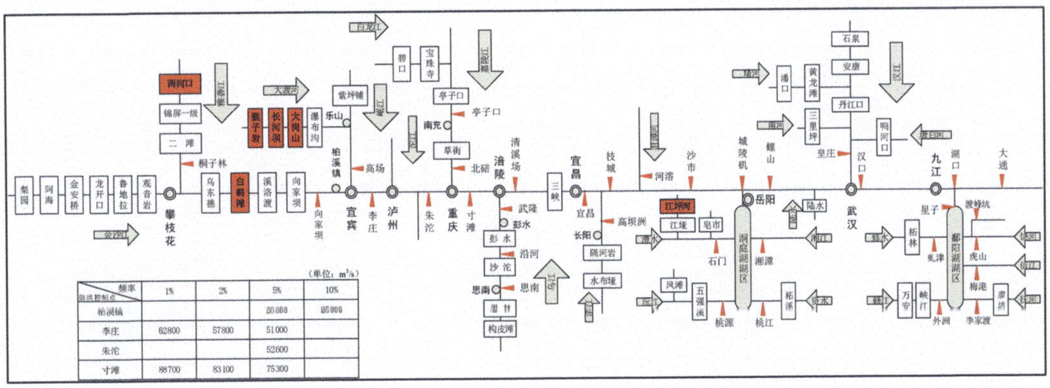

Figure 6.8 Schematic diagram of the joint regulation scheme of water projects in the Changjiang River Basin in 2021. Each box shows a reservoir along tributaries at each side of the main channel as the central line. The red box indicates a reservoir that remains under construction, therefore cannot be used for normal flood control purpose but only when extra-large flood happens. The grey arrows represent names of the rivers.

The established DSS for disaster prevention and utilization of the water resources in the Changjiang River basin includes the management process of joint intelligent regulation of water projects. The system provides functionalities of flood forecasting using precipitation forecast. Based on this, flood risk is assessed and the combination of water projects is proposed according to the regulation rule base. Then, through automatic analysis and human–computer intervention, joint regulation of various water projects is optimized and recommended.

During such analyses and joint operations evacuation is also considered, as one of the most important non-engineering measures for flood management. To better support flood management at all stages, the location based services (LBS) and big data technology is adopted to facilitate personnel evacuation, and to assist resettling of people threatened by the flood. This system facilitates guiding people to be evacuated from areas in danger of flooding – to safe areas. It combines hydrodynamic modelling results of a flood event, flood risk assessment model and an evacuation guidance system using LBS technology that is provided by mobile phone. In this system, analyses are carried out regarding population distribution and its characteristics, connection with the management groups responsible for evacuation, the traffic condition of roads, resettlement areas and other relevant information of the inundated area. For areas where an evacuation plan is available (for instance, the retention basins which are designed as temporary flood storage areas), the evacuation plan and the optimal path of transfer and resettlement can be provided to all people affected. When an evacuation plan is not available, the system provides functionality to formulate evacuation plans and monitor the evacuation dynamics in real time, and adjust the transfer plan in time when necessary. The evacuation plan can be developed using the following steps: 1 – determining the number of people to be evacuated and their locations; 2 – identifying appropriate or suitable resettlement sites for people to be resettled; 3 – mapping the transportation routes, scope, expected time and so on and 4 – sending messages to inform people to follow the evacuation plan in real time.

6.6.1.4 Application examples

As shown in Figure 6.9, downstream of the Three Gorges reservoir is the most important flood protection area, where large cities are located, such as Wuhan with more than 10 million citizens and vital economic importance (e.g., location of large car production industries). The two lakes of Dongting and Poyang which contribute to flood at the middle-downstream section of Changjiang River have an earlier flooding season compared to the rest of the areas in the river basin. Furthermore, the part downstream of the Three Gorges reservoir is relatively flat, which makes flood management difficult. Thus, once floods happen at the middle-downstream section of Changjiang River, it is necessary to make the best use of reservoirs upstream and the available retention/detention basins. Sometimes when the water level in the river channel is higher than the safety level (which is also called 'guarantee level'), detention basins may be used to store water temporally. Such a decision requires evacuation of people living the detention basin that is going to be used.

The system briefly described in previous sections has been well applied in recent years during flood management in this area. In the summer of 2020, the Changjiang River experienced a basin-scale flood which was third large on record (after the floods of 1954 and 1998). The following cases exemplify how the system was used during the flood in 2020.

Case 1 – Analysis and simulation of joint operation of water projects: This case considers the first flood of 2020, that occurred in early and mid-July in the basin upstream of Poyang Lake, which has five tributaries at its upstream that join the Poyang Lake before entering the main stream of Changiiang River. It was predicted that the flood would exceed the highest safety level at the gauge station of Hukou. It was crucial whether or not to use the detention basins in the

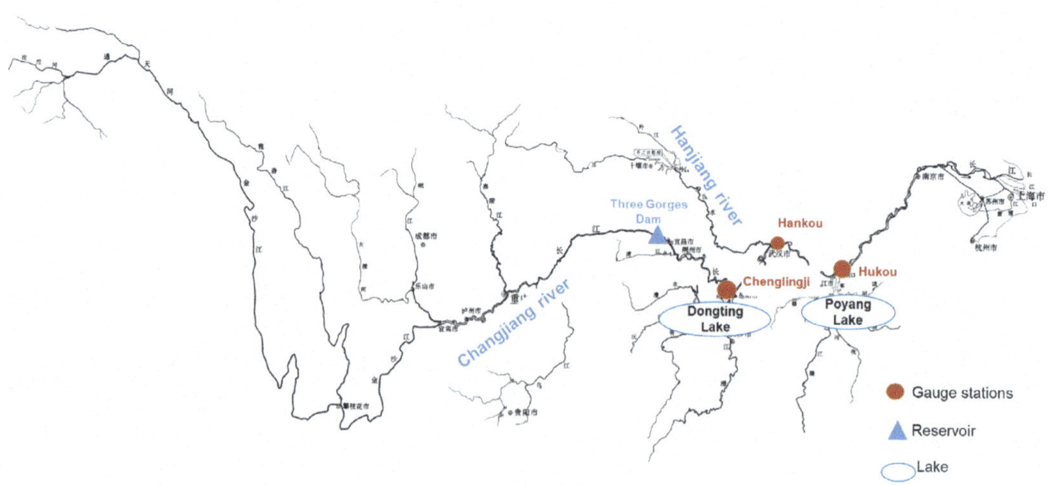

Figure 6.9 Schematic map of Changjiang river, with key gauging stations of Chenglingji, Hankou and Hukou, the most important Three Gorges reservoir, and two big lakes of Dongting Lake and Poyang Lake downstream of the Three Gorges reservoir.

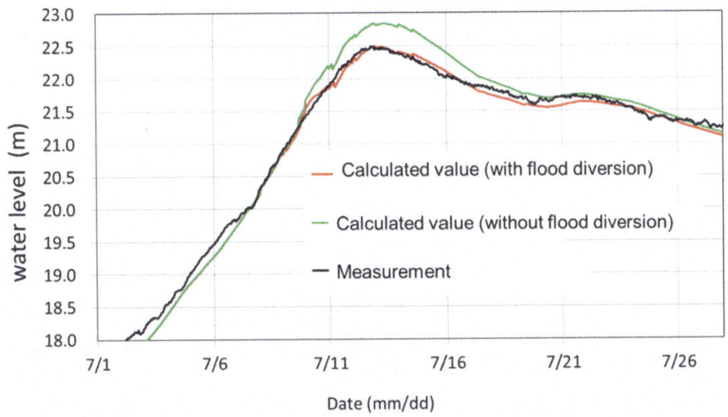

Figure 6.10 Water level at Hukou station during first flood in 2020: the green line is 'natural' flood without any regulation; the red line is the simulated water level using detention basins for regulation; the black line is the actual water level as measured.

lake. Using the system, with combination of short-term (3–5 days) and medium-term (7–10 days) lead time forecasts, the system provided scenarios of regulation schemes involving the groups of reservoirs located upstream of the lake, and detention basins located downstream of the tributaries and in the lake areas, as well as use of the embankments. It was accurately assessed that the water level at the gauge station of Hukou would be lower by about 20 cm if the detention basins are to be used (which meant that the dykes of those basins should be breached, naturally by overtopping, or deliberately). Decisions have been made to use those detention basins. As shown in Figure 6.10, the results show that the use of 185 detention basins has lowered the water level to 22.49 m, which is just 1 cm below the guaranteed water level of 22.5 m. As mentioned earlier, for these uninhabited detention areas there was no need for execution of complex evacuation plans.

Case 2 – Intelligent regulation based on rule base and interactive optimization: For the second flood that occurred in 2020, the flood situation in the middle and lower reaches of the Changjiang River was forecasted using quantitative precipitation forecasts as inputs. At the beginning stage of the second flood, it was predicted that the water level in the gauge station of Chenglingji (the most important gauge station for the downstream flood protection) might exceed the guarantee level of 34.4 m within 7 days after 16 July 2020. According to the regulation rule base, the Three Gorges reservoir should stop storing flood water (by which flood at Chenglingji station is controlled), when the water level of the Three Gorges reservoir reaches 158 m. If the reservoir keeps storing flood water beyond this reservoir water level, it will result in a rising risk of flooding in the reservoir area. Thus, it was necessary to trade off the risk between downstream areas and the reservoir area.

During such situation, the system provided functionality to adjust the regulation process of the Three Gorges reservoir through human–computer interaction, which can adjust the reservoir discharge process according to flood control needs downstream, regardless of the regulatory constraints. It allows the operators to adjust the reservoir discharge for the forecasting period, or set up the target water level of the reservoir and optimize the regulation scheme of upstream reservoirs accordingly. As a result, a joint regulation scheme involving various upstream reservoirs is obtained, which was applied in this case and reduced the flood peak of inflow of the Three Gorges reservoir by 9000 m^3/s. Consequently, through reducing the discharge of the Three Gorges reservoir, the water level at Chenglingji was reduced by 1.71 m.

As shown in Figure 6.11, the optimized regulation scheme was applied and has reduced water level of the second flood at Chenglingji to 34.39 m, which is lower than the safety level of 34.4 m. This regulation continued for the third, fourth and fifth flood (as seen in Figure 6.11), with some small exceedance of the level of 34.4 for a short period of time. Most importantly, however, the regulation avoided the use of detention basins along the river reach near Chenglingji station which would have caused large amount of flood damage to properties, and avoided complex evacuation of thousands of people living in those retention basins.

6.6.2 River/Lake chief system and public participation

This section takes the smart river chief system (RCS) in Chongqing city (a large city located at the upstream of Changjiang River) as an example to introduce the implementation of the RCS management and its public participation supported by hydroinformatics tools. The RCS consists of four subsystems, including: management, supervision, one river and one policy implementation supervision, and comprehensive display. The application terminals include PC, mobile devices, large-screen and WeChat (a social media APP). Based on massive multi-source data collected at all time, including those submitted by public participants, the system supports river chiefs at all levels (provincial level, city level and county level) to implement precise supervision and management actions, which significantly improves the performance of river chiefs in Chongqing city. The following sections present how the RCS works.

6.6.2.1 Functionalities of the RCS

Every river in Chongqing city has its river chief – a responsible person. Using mobile Internet, more than 17 000 river chiefs in Chongqing have performed their duties online, including: custom river patrol plan, one-click river patrol, analysis of river patrol effectiveness, problem reporting and handling, and reminders.

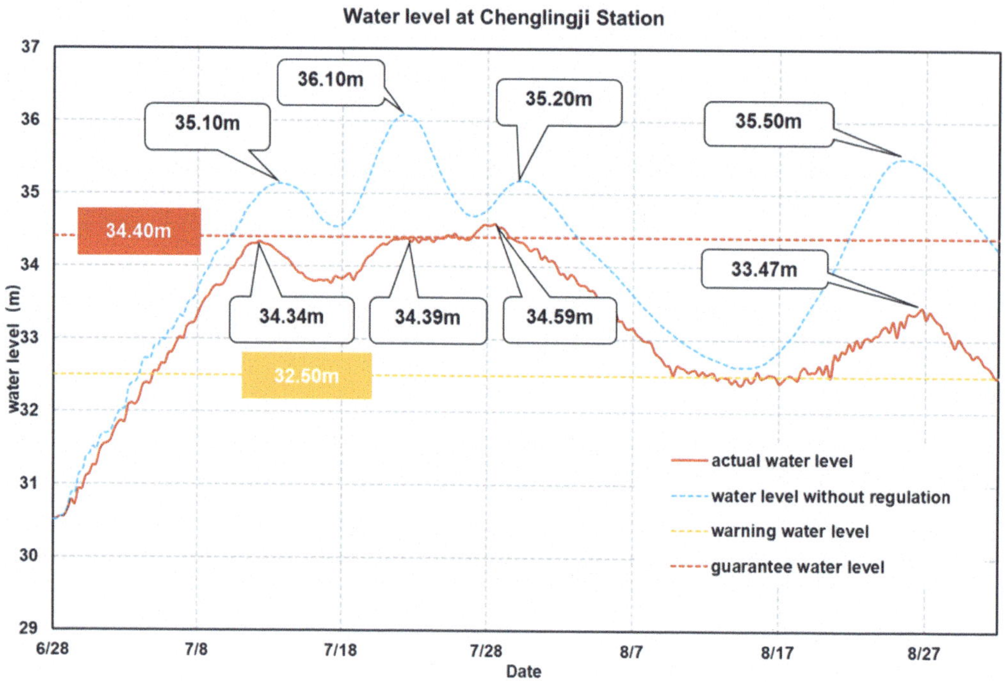

Figure 6.11 Comparison between regulated water level and unregulated water level at the Chenglingji station during flood 2020 in Changjiang River.

The RCS integrates information from different organizations such as the water conservancy bureau, ecological and environment bureau, and other units, and the various businesses into a GIS map, and presents a comprehensive display and statistical analysis of basic data, thematic data, and real-time monitoring and early warning data. Through this platform, all kinds of information needed for the supervision and assessment work are retrieved in real time, such as the performance of the river chiefs, the daily work of the river chief office, the implementation of various tasks of the RCS, and so on.. The system can provide technical support to the decision making or consultation meeting between river chiefs at municipal level and district/county level.

To support the operation of the system, the Chongqing Municipal Water Conservancy Bureau has installed 33 sets of automatic monitoring equipment for water quality and quantity, 189 sets of video equipment, as well as drones and unmanned boats on the Changjiang River, Jialing River, Wujiang River, Liangtan River, Longxi River and Longhe River, which constructs a kind of aerial-ground stereo perception network. The instruments and equipment of these monitoring stations use technical means such as AI identification, remote sensing analysis, pollution source tracing, and water pollution diffusion deduction, which effectively improve the ability of river chiefs to find problems and improve the overall supervision level of the RCS. Some examples are given in Table 6.1.

6.6.2.2 Public participation

The WeChat application is an effective channel for the public to participate in the health supervision of rivers, lakes and reservoirs. It is also an information platform for disclosing the progress of the RCS, policies and regulations. Through the WeChat application, the public can check the information about the RCS and its management, make complaints and provide advice or suggestions on river/Lake/reservoir management issues. It also plays the role as a platform for the public to supervise and monitor the management of the RCS.

6.6.2.3 Performance of Chongqing RCS

The Chongqing RCS has four management levels: city, district (county), township, and village. The users include river chiefs at all levels, staff of river chief offices, responsible members of member units, and the general public. Up to now, the number of users of the system in Chongqing has reached 30 000, the number of effective river patrols has reached more than 2.83 million instances, more than 40 000 problems have been reported, and the problem resolution rate has reached 90%, creating good ecological and economic benefits.

The river/lake chief system is an institutional innovation to ensure overall national water security. Through the measure of the river/lake chief system, the ecological environment of rivers and lakes across the country are expected to be continuously improved.

Table 6.1 Typical smart functionalities using AI technologies.

AI Functionality	System Visualizations
Problem identification using AI technology: Using video monitoring information, combined with human image recognition, deep learning and other technical means, it can identify and warn activities of swimming, fishing, floating objects on the water and other phenomena in video images.	
Deduction of pollution diffusion: For the Liangtan River Basin, a 1D hydrodynamic and water quality diffusion model is adopted, combined with the simultaneous monitoring of water quality and quantity, the deduction of water pollutants in time, space and concentration is assessed.	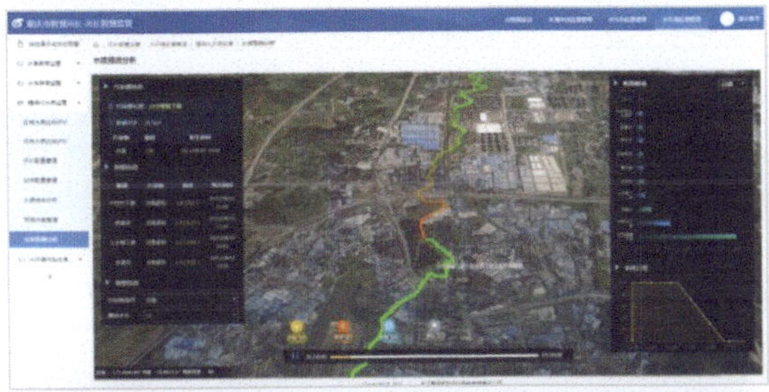
Pollution tracing: Through the combination of online water quality map fingerprint comparison and offline laboratory testing, to achieve accurate traceability of water pollution incidents, and provide technical support for the supervision of polluting enterprises' pollutant discharge behaviour.	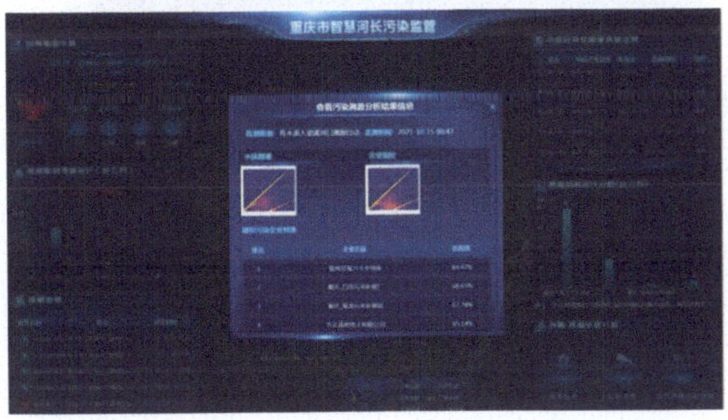

6.7 SUMMARY

As predicted by Mike Abbott, by making full use of modern information technology, in-depth development and utilization of water conservancy information resources, through the integration of data and models, including AI, hydroinformatics has become an indispensable part of modern river and water management. China has had its own path of developing hydroinformatics, but like in other Asian countries such as, Myanmar, Japan and others, such developments came from continuously updated requirements for water disaster prevention and water resources management. Over the years, hydroinformatics has expanded from a technology-centric approach to a sociotechnical discipline that supports stakeholders to participate in multi-level decision-making in the water sector, such as the river chief management system applied in China.

Following the logical information flow starting with data as the foundation, models as the computation core, and decision support as the ultimate goal of hydroinformatics, this chapter introduced the present concept of hydroinformatics in China, and the application status and technical development of hydroinformatics in the water industry, using flood management in Changjiang River as an example. Again, as promoted and predicted for a long time by Mike Abbott, public participation is becoming increasingly important and will be the driving force of water and river management. The RCS that is currently applied in all rivers/lakes of China at all governance levels (covering state level, provincial level, city level and county level), is relying on dedicated platforms for sharing visualization and use of data and modelling results. By including public participation through the use of WeChat application, RCS introduced different angles and scopes of monitoring and analysis, much broadened, and hence, led to improved river and water management capability of the government authorities.

In personal communications with the first author of this chapter, Mike Abbott used to repeatedly state that 'the future is in China', and that hydroinformatics, as it has been developed since 1990s, does have its place in China. This vision was doubtful at the time of early developments of hydroinformatics. As Chinese developments took place over the last three decades, the country developed its capabilities for dealing with all kinds of challenges and natural disasters (not only water-related, but, e.g., the dreadful earthquake in Sichuan in 2008, the two large barrier-lake floods in November 2018 in Upper Yangtze River, or, most recently the coronavirus pandemic). These developments also show that gradually hydroinformatics has found its place, and perhaps its future – in China, as predicted by Mike Abbott.

REFERENCES

Abbott M. B. (1991). Hydroinformatics: Information Technology and the Aquatic Environment. Ashgate, Aldershot, UK/Brookfield, USA.

Abbott M. B. and Vojinovic Z. (2013). Towards a hydroinformatics for China. *Journal of Hydroinformatics*, **15**(4), 1189–1202, https://doi.org/10.2166/hydro.2012.178

Boxer B. (2001). Contradictions and challenges in China's water policy development. *Water International*, **26**(3), 335–341, https://doi.org/10.1080/02508060108686925

Boxer B. (2002). Global water management dilemmas: lessons from China. *Resources*, **146**, 5–9.

Grieves M. (2014). Digital twin: manufacturing excellence through virtual factory replication. White Paper, Copyright © Michael W. Grieves, LLC 2014, available from https://www.researchgate.net/publication/275211047_Digital_Twin_Manufacturing_Excellence_through_Virtual_Factory_Replication [last accessed 15 July 2022].

GWP. (2015). China's water resources management challenge: the 'three red lines'. Technical focus paper, © Global Water Partnership.

Huang R., Ma C., Ma J., Huangfu X. and He Q. (2021a). Machine learning in natural and engineered water systems. *Water Research*, **205**, 117666, https://doi.org/10.1016/J.WATRES.2021.117666

Huang Y., Yu S., Wang X. M. and Yi L. (2021b). Integrated and intelligent reservoir regulation for flood management. Hydropower & Dams, **28**(1), 70–75.

Huang Y., Yu S., Luo B., Li R., Li C. and Huang W. (2022). Development of the digital twin Changjiang River with the pilot system of joint and intelligent regulation of water projects for flood management. *Journal of Hydraulic Engineering*, **53**(3), 253–269, CNKI, https://doi.org/10.13243/j.cnki.slxb.20210865 (in Chinese).

Jiang Y. (2015). China's water security: current status, emerging challenges and future prospects, *Environmental Science & Policy*, **54**, 106–125, Elsevier, https://doi.org/10.1016/j.envsci.2015.06.006

Kong J., Li S. and Zhu Y. (2018). Flood forecasting for small- and medium-sized rivers by ensemble extreme learning machine. *Journal of Hydrology*, **38**(1), 67–72, CNKI (in Chinese).

Li R., Huang Y., Li A., Li W. and Liu Y. (2020). Stochastic simulation of Yangtze River flood based on Copula function. *China Flood & Drought Management*, **30**(12), 25–28, CNKI (in Chinese).

Liu L., Liu D., Johnson M. D., Yi Z. and Huang Y. (2012) Effects of vertical mixing phytoplankton blooms in Xiangxi Bay of Three Gorges Reservoir: implications for management. *Water Research*, **46**, 2121–2130, Elsevier, https://doi.org/10.1016/j.watres.2012.01.029

Liu J., Zang C., Tian S., Liu J., Yang H., Jia S., You L., Liu B. and Zhang M. (2013). Water conservancy projects in China: achievements, challenges and way forward. *Global Environmental Change*, **23**(3), 633–643, Elsevier, https://doi.org/10.1016/j.gloenvcha.2013.02.002

Liu D., Yang Z., Ji D., Ma J., Cui Y. and Song L. (2016). A review on the mechanism and its controlling methods of the algal blooms in the tributaries of Three Gorges Reservoir. *Journal of Hydraulic Engineering*, **47**(3), 443–454, CNKI (in Chinese).

Ma K., Li S., Wang J. and Yu Y. (2016). Comparative study of data-driven intelligent flood forecasting methods for small- and medium-sized rivers. *Journal of University of Science and Technology of China*, **46**(9), 774–779, CNKI (in Chinese).

Mou M., Lin D., Zhou Y., Zheng W., Ruan J. and Ke D. (2019). An optimal allocation strategy for multi-energy networks based on double-layer nondominated sorting genetic algorithms. *Complexity*, **2019**, 11, Article ID 5367403, https://doi.org/10.1155/2019/5367403

Qiang X. (2012) Forecasting model of urban daily water demand based on artificial neural network. *Water Sciences and Engineering Technology*, **2012**(05), 58–60, CNKI (in Chinese).

Qiu Y., Duan H., Xie H., Ding X. and Jiao Y. (2022). Design and development of a web-based interactive twin platform for watershed management. *Transactions in GIS*, **00**, 1–19, https://doi.org/10.1111/tgis.12904

Song X., Ravesteijn W., Frostell B. and Wennersten R. (2010). Managing water resources for sustainable development: the case of integrated river basin management in China. *Water Science and Technology*, **61**(2), 499–506, https://doi.org/10.2166/wst.2010.848

Tao J., Yang Z., Cai Y., Wang X. and Chang J. (2017). Spatiotemporal response of pelagic fish aggregations in their spawning grounds of middle Yangtze to the flood process optimized by the Three Gorges Reservoir operation. *Ecological Engineering*, **103**, 86–94, https://doi.org/10.1016/j.ecoleng.2017.03.002

The Nature Conservancy. (2016). China's Urban Water Blueprint. The Nature Conservancy, Beijing, China. Available from https://www.nature.org/en-us/what-we-do/our-insights/reports/ [last accessed 15 July 2022]

Valverde-Pérez B., Johnson B., Wärff C., Lumley D., Torfs E., Nopens I. and Townley L. (2021). Digital Water: Operational Digital Twins in the Urban Water Sector: Case Studies. International Water Association, London, UK, White paper, © 2021.

Wang S. (2006). Resource-Oriented Water Management: Towards Harmonious Coexistence Between Man and Nature, 2nd ed. China WaterPower Press, Beijing.

Wang B., Wan J. and Zhu Y. (2021). River chief system: an institutional analysis to address watershed governance in China. *Water Policy*, **23**(6), 1435–1444, https://doi.org/10.2166/wp.2021.113

Wu C., Ju M., Wang L., Gu X. and Jiang C. (2020). Public participation of the river chief system in China: current trends, problems, and perspectives. *Water*, **12**, 3496, https://doi.org/10.3390/w12123496

Wu X., Zhao X., Ding X., Xu Z. and Qiu C. (2021). CNN Flood routing method based on data-driven training. *Journal of Hydroelectric Engineering*, **40**(5), 79–86, CNKI (in Chinese).

Wu Y., Washbourne C. and Haklay M. (2022). Citizen science in China's water resources monitoring: current status and future prospects. *International Journal of Sustainable Development & World Ecology*, **29**(3), 277–290, https://doi.org/10.1080/13504509.2021.2013973

Xie M., Zhou J., Ouyang W., Yan L. and Zhang H. (2018). Research on joint generation scheduling of cascade hydro plants in the dry season[J]. *Water Science and Technology: Water Supply*, **18**(1), 193–202, https://doi.org/10.2166/ws.2017.108

Xu C., Yin J., Guo S., Liu Z. and Hong X. (2016). Deriving design flood hydrograph based on conditional distribution: a case study of Danjiangkou Reservoir in Hanjiang Basin. *Mathematical Problems in Engineering*, **2016**, 16, Article ID 4319646, https://doi.org/10.1155/2016/4319646

Zhang J., Chen X., Khan A., Zhang Y., Kuang X., Liang X., Taccari M. and Nuttall J. (2021). Daily runoff forecasting by deep recursive neural network. *Journal of Hydrology*, **596**, 126067, https://doi.org/10.1016/j.jhydrol.2021.126067

doi: 10.2166/9781789062656_0093

Chapter 7

Hydroinformatics education at IHE Delft: past and future

D. P. Solomatine, R. K. Price, A. Jonoski, I. Popescu, B. Bhattacharya, L. Alfonso Segura, G. Corzo Perez, S. J. van Andel and C. Bertini

IHE Delft Institute for Water Education, Hydroinformatics and Socio-Technical Innovation, Delft, the Netherlands

ABSTRACT

From the very beginning of introducing hydroinformatics as a new discipline of research and practice, Mike Abbott worked together with his collaborators on establishing a supporting educational programme at post-graduate level. This chapter presents the establishment of the first hydroinformatics post-graduate programme at IHE Delft, in the Netherlands, later offered as an MSc degree in water science and engineering with specialization in hydroinformatics. The curriculum and the set-up of the programme are described, together with their developments over the last three decades. This programme has significantly benefited from the unique characteristics of IHE Delft: The institute is specifically oriented to providing water education to participants from the Global South, and, at the same time, it operates within a network of diverse organizations engaged in research and practice of water management in the Netherlands and Europe. The hydroinformatics programme has therefore benefited from contributions of top experts from such institutions who have taken roles as invited lecturers, as well as MSc and PhD mentors and supervisors, very often on water-related topics originating from-, or relevant for students' home countries. Challenges for maintaining relevant attractive hydroinformatics curriculum have come from continuous developments in information and communication technologies (ICTs), need for balancing ICT-related and water-related subjects, while still imparting knowledge regarding the primary role of hydroinformatics as an *integrating* discipline needed for supporting efficient-, but also transparent and inclusive water-related decision making. The associated PhD programme in hydroinformatics and some related Master programmes at IHE Delft are also briefly introduced. The chapter also presents an overview of hydroinformatics-related education at other universities around the world. The concluding section presents an outlook for future hydroinformatics education in which next to its water and informatics components, its social dimension and integrative purpose will become even more important.

Keywords: Hydroinformatics, education, post-graduate, Masters, PhD, integration technology, modelling, decision support, systems analysis, social dimension

7.1 INTRODUCTION

Mike Abbott's vision of hydroinformatics included not only its development as a new technology for dealing with the challenges of our water systems and the aquatic environment, but was also very much associated with a supporting educational and training programme. Such a programme was envisaged to provide generations of new engineers with much broader and deeper knowledge that interweaves informatics and water sciences. These 'hybrid' engineers – the 'hydroinformaticians', would then be much better equipped to employ such knowledge and associated tools for more innovative and efficient water management, but also one that will be more inclusive and just. This vision led Mike Abbott to establish a postgraduate course in hydroinformatics at IHE Delft in the early 1990s, which has educated and trained hydroinformatics experts ever since. Although initiated as a post-graduate diploma course, hydroinformatics has, in fact, been offered as a specialization of the Master's programmes at IHE Delft, first as part of its hydraulic engineering MSc and later as part of its water science and engineering MSc. This specialization provided foundations for subsequent development of education and research in hydroinformatics as part of the IHE's PhD programme.

Since its establishment, the hydroinformatics MSc specialization at IHE Delft has aimed to maintain some rather unique characteristics, stemming not only from the requirements of novel developments in hydroinformatics, but also from its location in the Netherlands, and in particular at IHE Delft. The Netherlands, with its vast expertise in water science, engineering and management, and its rich network of research organizations, consultancies and contractors, has provided a highly beneficial environment for education and training in any water-related discipline, and hydroinformatics has been no exception. IHE Delft Institute for water education, although rather small compared to Dutch Universities, has had a unique focus on educating and training international post-graduate students (primarily practitioners with some working experience, therefore called 'participants'). These have been primarily from the Global South (or the 'developing countries', a term mainly used in the past), and mostly funded by scholarships from the Dutch Government. The unique student body of IHE Delft has had an impact on the development and implementation of all IHE's water-related courses and programmes, including hydroinformatics.

During the last 30 years, as the discipline developed, other educational institutions have also embarked on delivering education and training in hydroinformatics through courses and even full post-graduate programmes. This has been primarily a development in Europe, where EuroAqua, the Erasmus Mundus MSc Programme offered by a consortium of European Universities, has been the most important hydroinformatics educational programme at MSc level, next to the programme of IHE Delft (see Gourbesville & Cunge, 2010). A number of universities in USA, Asia, Latin America and Australia have also introduced hydroinformatics or hydroinformatics-related courses in their post-graduate curricula. Overall, such courses have included the numerous developments of hydroinformatics on the technical side, but the adoption of the social and ecological dimensions of hydroinformatics, as promoted by Mike Abbott (1991a) has been very limited. Furthermore, many of these educational courses have been developed independently, resulting in a diverse understanding and follow-up design of hydroinformatics courses at post-graduate level. This situation is perhaps natural for such a young and rapidly developing field, but also presents significant challenges in analysing and even recognizing the state of the art in hydroinformatics education, needed for further development and delivery of relevant and attractive post-graduate curricula to future generations of hydroinformaticians.

In support of a better understanding of the past, and, hopefully leading to improvements in future hydroinformatics education and training, this chapter describes the postgraduate hydroinformatics specialization provided by IHE Delft, in relation to its development of Mike Abbott's ideas and goals, and exploring the extent to which these have been attained. In Section 7.2, it first introduces the original ideas and implementation of the courses, together with conditioning factors coming from the Dutch water sector and the unique orientation of IHE Delft to education of international students. Given that this was the first hydroinformatics course at post-graduate level, these initial decisions were highly important for future educational developments, not only at IHE, but elsewhere as well. Section 7.3 then presents some key developments of the hydroinformatics specialization over the last 30 years, and the numerous challenges and difficult decisions regarding the design and implementation of the curriculum, especially in view of the explosive growth of new informatics technologies. The current status of the hydroinformatics specialization is then presented in Section 7.4, demonstrating the way in which components of the programme are interconnected, so that graduates obtain not only key hydroinformatics competencies, but also a higher-level comprehension of the integrative nature of hydroinformatics. IHE's PhD programme in hydroinformatics is also briefly presented. Section 7.5 gives a brief presentation of couple of more recent IHE MSc specializations/programmes, closely associated with hydroinformatics. Hydroinformatics education outside of IHE Delft is then investigated in Section 7.6, where an attempt is made to demonstrate to some extent both the similarities and differences in understanding of the nature of hydroinformatics elsewhere in the world, leading to particular hydroinformatics educational events and courses. The chapter ends with Section 7.7, which provides an outlook for the future development of hydroinformatics education generally, as well as at IHE Delft.

7.2 MOTIVATION AND ORIGINS

The emergence of hydroinformatics as a discipline, required necessary and adequate changes in educational programmes to ensure the robustness of the knowledge background of researchers in hydroinformatics, of developers of hydroinformatics technologies and tools and of the users of these tools. Abbott (1993b) wrote about the changing requirements in education, and indicated that there would be new opportunities for educational institutions. It would no longer be necessary, or even possible, to teach all the details of modelling and other design processes. New opportunities would be created to lighten the curriculum related to detailed, scientific principles, thus providing adequate time for a much more critical, incisive and deeper view of the principles of design and other methodologies associated with hydroinformatics in practice. Mike Abbott's vision was that the training of scientists and engineers needed to be at the interface of informatics and water science, which were traditionally separate disciplines. In the beginning of the 1990s not too many people shared that vision, and the course in hydroinformatics given at IHE Delft was the first attempt to provide this kind of education.

Abbott (1991a) defined hydroinformatics as information and communication technology (ICT) applied to problems of the aquatic environment. As the notion has expanded, a more detailed definition has emerged, which is in use today: the study of the flow of information and the generation of knowledge related to the dynamics of water in the real world, through the integration of modelling, information technologies, optimization and artificial intelligence (AI), taking into account sustainability and social implications for decision support and smart management of water-based systems.

In 1990, Mike Abbot invited a panel of experts from Delft Hydraulics, DHI, HR Wallingford, RAND Corporation and other technological and knowledge institutes to discuss the syllabus of the future programme in hydroinformatics. During these discussions it became clear that such a course should start at the Masters and PhD level, assuming the undergraduate (Bachelor-level) education would provide the necessary foundations (Abbott *et al.*, 1994). That meeting resulted in the drafting of the course 'skeleton' that has been further detailed and developed into the syllabus of the post-graduate course and later Master specialization in hydroinformatics. Given the foundations of computational hydraulics for the discipline of hydroinformatics (see Chapter 2 of this book), the first hydroinformatics course was in fact largely based on the already established post-graduate course in computational hydraulics during 1970s and 1980s at IHE Delft. The new course 'arose as response to the observed need on the side of practice to educate and train a new cadre of engineers and scientists to meet the challenge of applying advanced information technology to the even more demanding problems of the aquatic environment' (Abbott *et al.*, 1994, pp. 204, see also Price *et al.*, 1998).

Interestingly, the first years gave rise to an ongoing debate at IHE Delft, and, to some extent, at the Delft University of Technology, about the significance and importance of hydroinformatics as an academic subject and as a professional discipline. Some professors were somewhat uneasy about hydroinformatics because it appeared to cut across the traditional subjects, in other words, there was no reason for it to maintain a distinction vis-à-vis traditional engineering. Mike Abbott was very good at arguing his case for hydroinformatics, along with support from the senior staff of IHE Delft, DHI and Delft Hydraulics. However, the division between those engineers and mathematicians who supported hydroinformatics as a unique academic subject consisting of an integrated collection of subsidiary academic subjects, and those academics who reject the need for such an academic subject continues to the present day, albeit with much less intensity. Mike Abbott worked unceasingly in writing papers and articles, addressing academics, practitioners and students at all levels to persuade them of the importance and value of hydroinformatics. It is his legacy that he was successful through his books and papers in bringing together a large number of academics and practising engineers from all over the world under the banner of hydroinformatics, in academic conferences, in peer-reviewed scientific journals, in academic and professional courses, and in a major group of the IAHR and IWA membership, with its distinctive integrative approach to observing, modelling and managing water.

The actual hydroinformatics course curriculum (as elaborated in Abbott *et al.*, 1994) consisted of three main parts: (1) foundational subjects, (2) core subjects and (3) applications. The first part on foundational subjects was expanded beyond topics related to hydraulics and hydrology, to include revisions of particular aspects of chemistry and biology, enabling the basis for water quality and ecological modelling applications to be covered later in the programme. A broader set of mathematical topics was also introduced. This included mathematical logic, set theory, numerical methods, partial differential equations, Fourier analysis and optimization techniques. For this first part, covering foundational subjects, it was mostly expected that large sections of these subjects would be already covered in undergraduate studies, and that these topics would be mainly introduced as relatively brief 'revisions'.

The second part of the curriculum (core subjects) was of crucial importance, because here water-related subjects were linked to informatics subjects. After introductory topics in ICT and software engineering, further programming subjects were devoted to problems of computational hydraulics and hydrology. The students learned how to actually program codes for small computational hydraulics and hydrological problems and associated 'modelling systems' that linked computational codes with user interfaces. Next to these physically based modelling topics, the core component of the course also covered subjects on AI, including both knowledge-based systems from the symbolic paradigm, at that time still dominant but already losing popularity, and data-driven modelling from the sub-symbolic paradigm that was emerging as novel and promising AI direction. The core component was completed with necessary supporting subjects on databases and geographical information systems (GIS), as well as some specific additional topics such as water quality and ecosystem dynamics modelling. The third part of the syllabus concentrated on 'applications', which was followed by integration subjects, with more realistic river, coastal and catchment simulations, linked to related water quality and ecological simulations, systems analysis, decision support, and more sophisticated AI systems applied to water problems.

This brief overview of the first hydroinformatics course is re-introduced here because the basic setup of IHE's hydroinformatics MSc specialization with these three components has in fact been maintained throughout the years. The principle has been to start with foundational subjects (some of which could appear quite separate and unrelated to each other to students/learners in the beginning of the course), and then to gradually introduce their interlinking by informatics technologies resulting in new tools for modelling and management of water systems and the aquatic environment. Once the possibilities of such interlinking and tools development have been opened to students, the integration subjects and their real-world applications would become more comprehensible, and different strands of the course introduced earlier could be said to 'come together', to reveal the true nature and potential of hydroinformatics tools and approaches. Individual subjects and topics in each of these components have been modified and adjusted over the years (as elaborated below in Section 3), but this basic principle of starting with rather different topics and weaving them together towards integrative hydroinformatics applications has remained.

One important aspect in this setup was the requirement that students entering the programme needed to have good knowledge of the foundational subjects. Given that IHE has been receiving international students from many different countries of the Global South, with different educational systems, standards and cultures, this requirement was not always easily fulfilled in practice (even with careful selection of students after they had applied for the course). Therefore, the first part of the course, with the 'revision' subjects, was, in fact, of vital importance for bringing the diverse students to a common environment level, so that they could continue well equipped for the next phases of the programme. Challenges of educating and training such diverse group of students from different cultures in topics already developed and well established in Europe and some other developed countries, have in fact been faced throughout the programme duration, and the long experience of IHE in this regard has also been utilized in the hydroinformatics programme.

The three parts of the curriculum presented above, however, cover only the 12-month taught part of the course, which in the very beginning was sufficient to obtain a post-graduate diploma. A short 'individual study' was introduced at the end of the taught part of the course, lasting about 4–6 weeks. This study was regarded as a project through which the students could demonstrate their abilities to independently carry out a small hydroinformatics project – leading them to a post-degree graduate diploma, which was later abandoned and replaced by a Master of Engineering (MEng) degree. Only students with sufficiently high average marks were allowed to continue with a longer, 6-month MSc research study, leading them to a Master of Science Degree. This setup continued for about 15 years, until the mid-2000s when the MEng degree was cancelled and IHE started offering only full Master of Science Programmes with 18-months duration, with the same setup of 12-month of formal teaching, and a 6-month MSc thesis research part. This was also the time when IHE introduced the modular composition of the MSc Programmes, in which a set of subjects from a same theme was to be delivered in a three-week period – a 'module', and the whole taught part of the programme became a series of connected modules.

It was in the setup and arrangements for students' individual MSc research that the unique opportunities of linking with the Dutch water sector brought most benefits for IHE students. The Netherlands, through its history of successful water management has developed a rich and complex network of water-related governance and management institutions – enabled and supported by research, consultancy and contracting organizations. Water management is institutionalized at all levels of government, from the *'Rijkswaterstaat'*, which is a part of the Ministry of Infrastructure and Water Management, responsible for design, construction, management and maintenance of main water infrastructure in the country, to the lower Provincial and Municipal levels, and the unique regional water management structure of the Water Boards – elected bodies for water management organized at catchment level. IHE has benefited significantly from long-term relationships with these institutions through which MSc research problems and topics of high relevance were offered to its MSc students. These institutions have often provided funding for particular MSc research topics of their interest. Such relationships have been even more developed with research and consultancy organizations, such as Delft Hydraulics, TNO, Royal Haskonning, HBV and others. Good collaborations were also maintained by engaging a number of invited lecturers from such organizations and from Dutch universities who offered contributions to the programme in regard to engineering practice and the latest research results.

Of particular interest and significance for the hydroinformatics programme, have been the collaborations on joint MSc research with companies and organizations that have been developing software packages for modelling water systems. Here, the primary partners have been Delft Hydraulics (currently Deltares) and DHI from Denmark. Mike Abbott has worked with both of these organizations (especially with DHI in the beginning of the programme) to maintain these close relationships so that hydroinformatics students would be fully aware of the potential of these latest modelling tools and understand their limitations. Some students have also been in a position to contribute to the further development of some of these products through their MSc research. Over the past 30 years similar relations have been established with other European partners (e.g., HR Wallingford), especially after Professor Roland Price took over the leadership of the hydroinformatics core group at IHE, and started developing broader collaborations with European research partners by joining consortia to carry out research projects funded by the different research programmes of the European Commission.

It should be noted, however, that the MSc research topics of the hydroinformatics students were not always coming from these relations and collaborations. All participants were encouraged to bring issues from their own countries, especially those that required approaches relying on novel modelling and analytical tools, or other ICT support. In many cases such problems also originated from operations of Dutch consultants in the participants' countries. In internal IHE discussions and documents, Mike Abbott always insisted that this kind of linkages between IHE participants and the rich network of organizations from the Dutch and European water sector is of vital importance. The arguments that he was using to support this claim, however, were beyond the immediately obvious. Of course, IHE hydroinformatics graduates needed to have knowledge and skills for applying modern modelling and ICT tools to address water problems in their own countries. More importantly, however, they were to become integrators of people with different kinds of practice experience and knowledge using innovative tools so that the design and management of water systems would be more inclusive, and therefore more integrated in the sense of addressing interests and concerns of various stakeholders. Mike Abbott often referred to the need of hydroinformatics graduates as people who could be in the position to persuade other people to take particular course of action. He was insisting that the 'state ethics' of international aid to the water sector of countries 'in development', including the work of foreign consultants and contractors and the 'philanthropic' mission of IHE itself, would sometimes clash with local values and value systems in recipients' countries. Hydroinformatics graduates from IHE would therefore need to be capable of recognizing such issues and be in a position to mobilize their acquired knowledge to modify the intended courses of action. There is no doubt that Mike Abbott considered such aspects as most important in hydroinformatics education (see Abbott, 1991b, and especially the appendices of Abbott *et al.*, 1994).

The extent to which such goals be attained by hydroinformatics education and training alone is certainly debatable. The limited duration of the programme and the need to introduce many different topics in a coherent curriculum were (and still are) a big challenge. About 10 years after the programme introduction, Mike Abbott, although already officially retired, was contributing to the programme with his 'Professorial Lectures'. These were provided towards the end of the taught part of the programme, aiming to introduce the comprehensive nature of hydroinformatics. His initial ideas regarding the proposed syllabus of such lectures included a long list of broad and diverse topics covering numerical modelling, data and information, software development (including object- and agent-orientation approaches), but also topics from philosophy, semiotics and postmodern critical theory. Due to time limitations, such overwhelming list of topics could not be included in the curriculum, and the study handbook eventually included a reduced list of topics. One can argue that the above competences that Mike Abbott was aspiring to for the hydroinformatics graduates can only be developed with experience, and development of right *attitudes*. The educational programme could nevertheless at least instil such thinking that could later be developed by practice and experience.

7.3 DEVELOPMENTS OF IHE'S HYDROINFORMATICS EDUCATION OVER LAST 30 YEARS

To understand the developments and changes of hydroinformatics education at IHE, it is worth mentioning some of its key principles:

- The large part of the course is devoted to using the modern computer-based modelling and analytical tools (known as knowledge processors), so it is '*tool-based*'.
- Special attention is given to *integration* – of data, modelling, systems analysis and decision-support technologies, which would naturally follow the information flow – from data to information to knowledge and to decisions based on optimal choices.

In the development of an educational programme, Mike Abbott's intention was to address a change that he described as follows (Abbott *et al.*, 2001):

> …the device that was previously a computer, as a means of making computations, now becomes a knowledge processor, as a means of manipulating encapsulated knowledge. Similarly, what was previously a data network, as a means merely of accessing data, now becomes an intranet, or even an extranet, or more generally an internet (in the generic sense and so with a lower-case 'i') as a device for communicating knowledge in the first place and data only in the second place. The new era, in which the engineer no longer works with symbols in the capacity of a knower, but instead works with signs in the capacity of a consumer of knowledge, is called quite generally the post-symbolic era… The equipment whereby knowledge is transformed and transferred belongs to the category of tools, so that we identify here a division between toolmakers and tool users. Correspondingly the most immediate and direct consequence of this division is that the most advanced hydraulics knowledge has become useable to far more persons.

This is what brought one of the main principles into the foundation of this course – to make it a *tool-based course*, so that the participants would be able to consume the knowledge encapsulated in the modelling tools (Abbott, 1993a). The amount of this knowledge is much larger than can be given in the form of lectures and traditional exercises, and, most importantly, it is put in the form of software which allows for its immediate and effective application to real problems.

However, to achieve these goals, the course needed to undergo significant modifications, as it developed over the years, primarily because the technologies for tool development, and the tools themselves have evolved rapidly. Associated with this was the accelerated spread of application of such tools in different areas that needed to be addressed in the hydroinformatics curriculum.

Significant impact on the curriculum has come from the emergence of the internet, and the technologies for developing web and later mobile phone applications. Such technologies were non-existent when hydroinformatics post-graduate course was launched. For several years, while these technologies were still emergent, a small number of introductory lectures for informative purposes were brought into the curriculum. From the mid-2000s, when these technologies became more established, decisions had to be made about the ways in which they could play an integral part in the curriculum. Eventually, they were introduced in two parts. First, they have been demonstrated in the introductory ICT-related subjects, mainly regarding the possibilities that have opened up for the acquisition of new data and their easy integration in applications. Towards the end of the programme, technologies for developing web and mobile phone applications were introduced as 'software technologies for integration' with small exercises of actual software development. The expectations have not been that the graduates would become full-fledged web and mobile phone apps developers, but that they should be sufficiently aware of the possibilities so that in their subsequent working practice they could engage in discussions with professional developers as they would be fulfilling their role of integrators of knowledge.

Similar transformations have also occurred in the software engineering subjects regarding programming languages used in teaching. Initially, the choice was for compiled languages such as Pascal/Delphi, with Java being introduced in some transitional years, but eventually decisions were made to move to scripting languages (with a higher-level of abstraction), first MATLAB and later Python. The popularity of the different languages was not the only reason for such transitions. In the beginning the course was oriented more towards actual software development, with the vision that hydroinformatics graduates could contribute to development of water-related modelling software. As such systems became more sophisticated, these tasks were to be fulfilled by professional software developers, and hydroinformatics graduates needed to develop much more skills in programming for scientific research and experimentation and prototyping, for which the scripting languages were more suitable.

The explosive growth of different methods for data-driven modelling and machine-learning (ML) techniques had to be accommodated as well. The original hydroinformatics post-graduate course introduced the methods of artificial neural networks and genetic algorithms (as an alternative to classical optimization methods), but as new methods have emerged they needed to be accommodated in the curriculum in a way that would offer a meta-perspective to them. This meant that more emphasis was put on more rigid introduction of ML, and also focussing on assessing their suitability for particular problem class and data availability (e.g., regression vs. classification). 'Classical' AI with knowledge-based systems has been completely removed from the curriculum.

As modelling, optimization and decision support have developed and became indispensable methods for so many different application areas, introducing all of these in the 'applications' part of the programme became a significant challenge. The only way of dealing with this challenge was to introduce elective modules that could run in parallel, and the students would be selecting them according to their interests and future career needs. For several years the programme had three blocks of elective modules (consisting of two modules each), covering the areas of: (1) river, catchment and flood modelling, (2) urban water systems and (3) environmental/ecosystems modelling. In recent years, all IHE programmes have moved towards increased flexibility and many such modules have become available to students to choose from, offered by other MSc programmes and departments.

The last area of significant change has been in the integration subjects, such as systems analysis and decision support. Topics on more sophisticated and more relevant optimization techniques (including model-based optimization) have been introduced, as well as methods for assessing and quantifying uncertainty in modelling and its use in decision support applications. Additionally, networked applications (again, using the web and mobile phones) have been introduced for enabling participatory decision support of multiple stakeholders. To exemplify such possibilities some multi-player gaming applications (which used water models as supporting calculation engines), have been developed and introduced in the regular programme and in other hydroinformatics training activities, such as short courses (Craven *et al.*, 2017).

One area that has seen less modifications is in the core subjects related to more classical numerical methods and their applications to problems of hydraulics and hydrology. Exercises with software implementation have been converted from one programming language to another, and some new methods have been introduced, but the topics through which the links between the 'hydro' and 'informatics' have been brought into the curriculum have remained those of river (free surface) hydraulics, and rainfall–runoff and groundwater hydrology.

Finally, it should be obvious that a number of topics from the curriculum of the original hydroinformatics post-graduate course (see again Abbott *et al.*, 1994) needed to be removed in order to make room for the new topics mentioned above. Some have been completely removed (such as already mentioned 'knowledge-based systems', and some of the mathematics topics), and others have been shifted to electives, offered by other departments (e.g., coastal engineering and simulations, ecosystems modelling and related subjects). Reductions in contact hours and study loads for students, as well as merging of some topics have also been necessary to accommodate new topics.

An important component of the programme was the introduction of a special fieldtrip to Florida, USA, in 2005. The collaboration with South Florida Water Management District (SFWMD) and USGS was enabled by Florida Earth Foundation who facilitated the organization of the field trip. Students could get acquainted with actual technical, research and organizational activities in the field of water management, hydraulic engineering and hydrology in the area of South Florida, as they have been implemented within the multi-billion US dollars Everglades Restoration project (Loucks, 2000). Since 2014, Florida Atlantic University took the lead from Florida Earth Foundation and organized these fieldtrips.

Currently, hydroinformatics is recognized as an *integrating* technology, which combines data monitoring and remote sensing technologies, with modelling, forecasting, optimization and decision support. This creates a challenge for educators who should find a balance between the breadth and depth to be able to teach all these aspects in a limited time.

7.4 CURRENT STATUS AND THE COURSE CONTENT OF THE HYDROINFORMATICS MSC SPECIALIZATION

The hydroinformatics MSc specialization is designed for hydraulic, environmental and water resources engineers from universities, consulting firms, research institutes, water boards and other government agencies. Education at IHE Delft is aimed at training young and mid-career water professionals, traditionally mostly engineers, who had graduated with a class honours degree in civil engineering or other similar engineering subjects, who will be well prepared to contribute to the development of their country. Therefore, participants of a course have a good understanding of the fundamentals of a range of physical processes, and those following hydroinformatics specialization will acquire knowledge of advanced modelling techniques, AI and information technology for water management. At the end of the course, participants should be able to select and use simulation models applied to water-based systems in a wide variety of hydraulic, hydrologic and environmental engineering situations. They must be able to use current software tools; and know how to design, develop and integrate decision-support systems and tools in order to provide advice to managers and stakeholders. For the latter, it is expected that the participants understand and practice collaborative work, making use of internet-based platforms.

The modern practice of hydroinformatics calls for a deepening knowledge of a number of fundamental subjects related to water, computer sciences, IT and AI, and directing further education to ensure that the knowledge and information arising from the corresponding subject areas can be integrated, and finally brought into hydroinformatics products and services.

Given today's needs for a hydroinformatics professional, compared with the graduates of the hydroinformatics courses given in 1990s, the current hydroinformatics programme has been extended over the years, to include additional

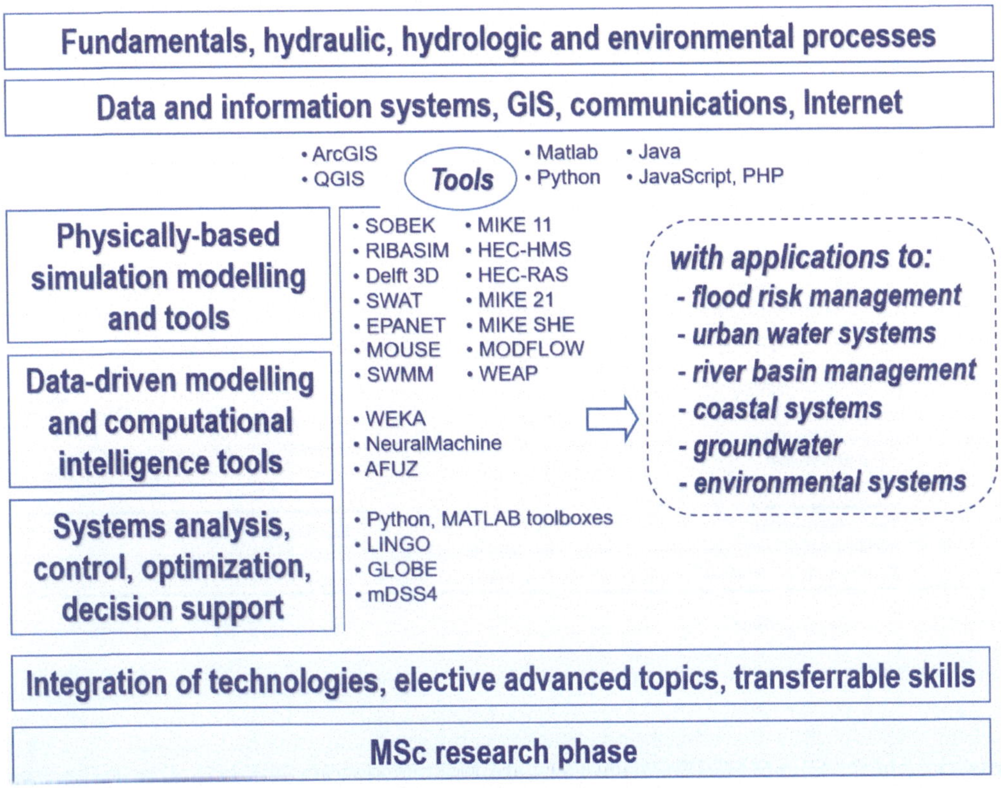

Figure 7.1 Structure of the IHE hydroinformatics course/specialization.

subjects on theory of modelling, probabilistic forecasting, uncertainty, optimization techniques and decision support systems. Furthermore, applied subjects like flood risk management, and urban water modelling and management are now included. In addition, data management, together with AI and ML techniques are employed to build data-driven models, which complement more traditional physically based hydrological and hydraulic models. The course also introduces high-performance and web-based computing, including modern programming languages with libraries for scientific computing.

The overall structure of the hydroinformatics course at IHE Delft is presented in Figure 7.1.

As with most of the programmes at IHE Delft, the Hydroinformatics specialization in the first year is divided into 14 modules, each of which lasts for three weeks. After every two modules there is one week reserved for examinations. This means that during a particular module the participants are focusing on a group of thematically interrelated subjects. The last six months of the specialization are dedicated to the MSc thesis research.

The modular structure of the course at IHE Delft allowed for the development of short and online courses in hydroinformatics, covering different application areas, for which professionals from all over the world could join for two to three weeks (see Jonoski & Popescu, 2012; Price *et al.*, 2007). The advantage of professionals joining the class with students studying for their MSc is that practitioners can share their country's experiences and ideas for a solution. The drawback of this approach is that short-course participants joining a regular module have different backgrounds and prior knowledge and it may take time to catch up with the workload.

An example of the sequencing of subjects in 'modules' is given in Figure 7.2.

This design of the programme reinforces the students' 'problem solving' skills through investigation of applications involving those skills, and gives students the opportunity to develop projects and assignments for later use in their career as water professionals. In addition to the regular academic staff of IHE, invited speakers and guest lecturers address issues relevant to a broad understanding of hydroinformatics and its applications in society.

The research phase of the Master's programme, and the ongoing research projects of the associated PhD programme, make room for the involvement of the participants in the real-life projects run in various countries, bringing them in contact with the practising engineers and managers. To ensure that, as mentioned earlier, important links are maintained with leading technological institutes and consultants. Experts from these institutions give lectures, and also guide the participants during their research projects in the last six months of the course.

During the research phase, participants are able to make a qualified judgement when choosing a particular theme for research, and to deepen substantially their knowledge on the narrower topic addressing a particular specific problem of interest and of social relevance. A quick search on the IHE library catalogue, on the hydroinformatics MSc theses shows how diverse the topics are. Research carried out by the members of staff feeds into the syllabus of the programme, but also research results achieved during the MSc research project contribute to the overall research output and publications of the chair group and of IHE as a whole.

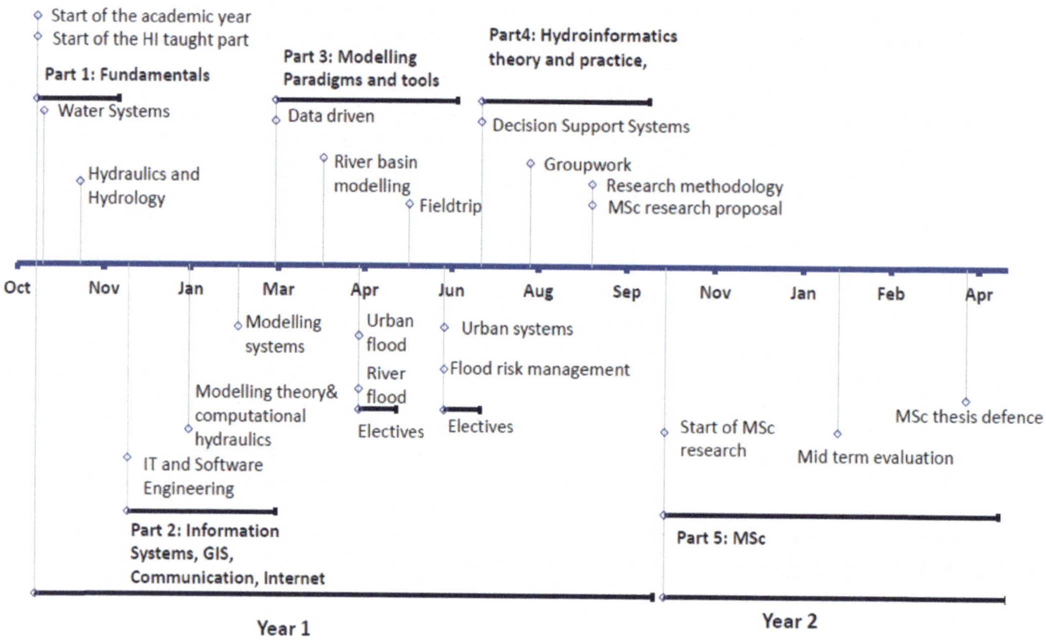

Figure 7.2 Sequence of study modules in the hydroinformatics course/specialization.

In 30 years, from 1992 to 2022 there have been 808 hydroinformaticians trained, of which 191 obtained an IHE diploma, and 617–the Master's degree (IHE started to award Master's degrees in 1994). A large majority of graduates continued their career in water, and many now occupy important positions in the public and private water sector.

Associated with MSc programme was the process of 'supplying' PhD fellows to the PhD programme of IHE and of universities – approximately a quarter of all hydroinformatics graduates continued with a PhD study. This shows that the level of the Master programme is at a sufficiently high level, and in fact is seen by some participants as an important stage on their way towards a PhD. Since 1993, when the first IHE PhD title in hydroinformatics was awarded to Ms. Khen Ni Ni Thein from Myanmar, there have been 54 PhD research fellows supervised by the members of the hydroinformatics chair group who have defended their thesis successfully.

7.5 ASSOCIATED MSC PROGRAMMES

The rise in need for hydroinformatics professionals around the world, due to evolution of computational models and data availability led to the development of hydroinformatics education outside of IHE Delft. Education in these technologies and their integration in water management research and practice, around the world was promoted by IHE Delft through development of curricula and associated MSc programmes, which are presented below.

7.5.1 Erasmus + Flood Risk Management Master's programme

In recent years, hydroinformatics education has taken up a new initiative in offering its core educational material and tools to a new pan-European master's programme in flood risk management, which is fully funded by the European Commission (Erasmus + Programme). Flooding is a serious threat to societal prosperity, which is being exacerbated by climate change, population growth and urbanization. Scientists have called for the adoption to integrated flood risk management, which aims at reducing the human and socio-economic losses caused by flooding while receiving the benefits from floods and the use of flood plains or coastal zones. The need for the adoption of a holistic integrated approach to managing flood risks has been reflected in the Flood Directive of the European Parliament.

Hydroinformatics offers its core technologies to the programme so that graduates attain the skills needed for implementing flood risk management solutions in diverse river basins, coastal zones and urban areas. These skills are specifically those of understanding and managing the flow of information from data to models, models to knowledge, knowledge to decisions and decisions to dissemination with the overall objective of reducing flood risk.

The joint master's programme is offered by a consortium consisting of IHE Delft, Technical University of Dresden, UPC Barcelona (BarcelonaTech) and University of Ljubljana. The associated partners include European hydraulics laboratories, namely, DHI (Denmark), Deltares (the Netherlands) and HR Wallingford (UK), key national organizations responsible for flood management, including Rijkswaterstaat, Rijnland Waterboard (the Netherlands) and National Research and Development Institute for Marine Geology and Geoecology (GeoEcoMar, Romania). Other associated partners include ICHARM (Japan), Institute of Water and Flood Management (Bangladesh), HydroLogic (the

Netherlands), Center for Environmental Studies of the Florida Atlantic University (USA) and International Association of Hydrological Sciences. All these partners bring their specific complementary expertise in flood risk management to the joint programme, which educates flood risk professionals with a broad vision of the processes occurring in river basins and in coastal zones at different spatial and temporal scales, and who will be able to master the links between systems, processes and natural and socio-economic constraints for all the aspects of the water cycle.

7.5.2 International Master's in Hydroinformatics

In recent years IHE Delft has developed cooperation in the area of hydroinformatics with Universities in Asia and in Latin America in teaching hydroinformatics. In turn, this led to the proposal and development of the International Master's in Hydroinformatics (IMHI). These initiatives have been motivated by MSc students, PhD interns, and PhD research fellows, who took part in the educational and research programmes of IHE.

IMHI China has started in 2005 by agreement with Hohai University in Nanjing, China. The IMHI China ran until 2018 and was oriented mainly at the Chinese participants. During the first block of 3 months of this programme, the participants were trained by staff of Hohai University, after which they came to Delft to continue their study. In total about 70 Chinese participants followed that programme.

The concept of IMHI continued in 2019 by an agreement to form a partnership in Colombia. A prestigious technical university in Bogota, the Escuela Colombiana de Ingeniería Julio Garavito (ECIJG), started the joint program for the region. This programme has been initially set-up for participants from Colombia, but it is also open to participants from other Latin American countries. In the first semester, participants follow courses in Bogota, and then come to IHE Delft to continue with Module 6, together with the Flood Risk Management participants, and joining the rest of the group that started the study in Delft. In this two-year Joint Programme the participants receive two diplomas, one from IHE and one from ECIJG. So far, each year four students have participated in this programme.

7.6 HYDROINFORMATICS EDUCATION WORLDWIDE

With the European origins of hydroinformatics, and its first educational programme at IHE Delft, it is not surprising that additional developments in hydroinformatics education also took place in Europe. As discussed above, the most significant of these is EuroAquae, the Erasmus Mundus MSc Programme with duration of two years (four semesters), offered by a consortium of European Universities (coordinated by University of Côte d'Azur, formerly by University of Nice Sophia Antipolis). Its setup is similar to the Erasmus Mundus in Flood Risk Management described earlier. EuroAquae was in fact one of the first Erasmus Mundus MSc Programmes, supported by the European Commission with fellowships for international students, in which students spend part of their study in each of the partner institutions (usually for about one semester). The different parts are still designed to form a unified programme, leading to multiple degrees or a joint degree, depending on the arrangements by the participating institutions.

EuroAquae was initiated in the early 2000s and it has been offered for about 20 years. An overview of the programme and its historical developments is provided by Gourbesville (2021), (see also resources at the programme web site https://master.euroaquae.eu/). Similar to other Erasmus Mundus Master Programmes, EuroAquae needed to develop and adjust existing curricula mainly developed for students from a particular European country, to become sufficiently relevant and attractive to international students, coming from Europe as well as from the Global South through the provided fellowships. Close collaboration and coordination of participating partners is a necessary requirement in these processes.

The curriculum of EuroAquae is similar to the hydroinformatics programme of IHE Delft, and roughly follows the concept of fundamentals, core subjects and applications introduced earlier. A rather unique component regarding hydroinformatics education of EuroAquae is the HydroEurope component (also supported by the WaterEurope project, see Gourbesville & Molkenthin, 2020), where the students learn to work collaboratively with students from other MSc programmes and Universities on a real-life engineering problem (flood-related problem around the city of Nice in southern France) that requires different kinds of hydrological and hydraulic modelling activities. This collaborative work is initiated early on in the programme with remote online collaboration, followed by face-to-face groupwork later in the year when all students meet in the city of Nice. This 'collaborative engineering' component was in fact initiated together with IHE Delft about 20 years ago, and for several years IHE hydroinformatics students participated in such activities with students from Universities that later set up EuroAquae.

An important characteristic of the programme is the partnership with different industrial partners, offering possibilities for students MSc research through internships, in setups similar to those for the IHE Delft Programme. Finally, EuroAquae has expanded its collaboration with associated partners from Asia and Latin America, especially in involving students from universities in the HydroEurope activities described above.

Currently, IHE Delft and EuroAqua are the only two European hydroinformatics MSc Programmes (New Castle University in UK, did have an independent hydroinformatics MSc programme, but currently they offer only EuroAquae, where they are core partners.) These two programmes are offering training and education to future hydroinformaticians that is closest to the vision of comprehensive and integrated hydroinformatics education, put forward by Mike Abbott and his collaborators. Of course, there are many European Universities that offer MSc programmes with somewhat similar content, but they are with different titles, and, consequently, somewhat different overall orientation. Some universities offer individual hydroinformatics courses (topics) or modules, as part of larger water-related MSc Programmes. Examples include University of Exeter in UK, Norwegian University of Science and Technology (NTNU), University of Stuttgart (Germany), and several others. These courses mainly cover technical aspects of data management, modelling and AI topics. From this situation, it can be observed that in

European Universities with water-related MSc programmes the need and relevance of hydroinformatics technologies is becoming more recognized, but that the comprehensive education of hydroinformaticians is still limited to the two main MSc programmes of IHE Delft and EuroAquae.

The situation in USA is in fact rather similar. Complete academic programmes at MSc level in the field of hydroinformatics cannot be identified, but there are a number of similar programmes (with different names), and a number of universities offer courses (topics) in hydroinformatics. These courses are often provided in support of research, for graduate students (at MSc and PhD level), in some occasions associated with dedicated hydroinformatics research laboratories (e.g. the 'Hydroinformatics lab' of Brigham Young University – https://hydroinformatics.byu.edu/, and the University of Iowa Hydroinformatics lab (UIHI) – https://hydroinformatics.uiowa.edu/). Different understandings of what constitutes hydroinformatics (also addressed in this book in Chapter 1), has also played a role regarding introducing hydroinformatics education in US universities. For example, the important book on hydroinformatics by Kumar *et al.* (2005) introduces it as a field for water data processing, analysis and communication, and the follow-up 'soft computing' – data-driven modelling. Relevant origins and links with computational hydraulics and other physically based modelling are not included, nor are aspects related to ecological and social dimensions.

An important initiative regarding technological advantages offered by hydroinformatics tools for advancing water sciences is the Consortium of Universities for the Advancement of Hydrologic Science (CUAHSI), consisting of 130 members (mainly universities, but also other partners), in operation for the last 20 years. CUAHSI has led several projects closely related to hydroinformatics with involvement of students and researchers, examples of which include Hydrologic Information Systems, CyberWater (on model and data integration on the web), Data Science and Analytics for Water, CrowdHydrology (on hydrological data using citizen science and social platforms). They also initiated the Hydroinformatics Innovation Fellowship programme where they offer grants to young innovators for novel water-related services and products. Universities that are closely engaged in CUHASI activities would also be those offering relevant hydroinformatics-related courses, even if predominantly on the side of technological enablement.

It should be clear, though, that many educational and research programmes at US universities in fact address other aspects of hydroinformatics, including its ecological and social dimensions, just that the term 'hydroinformatics' often is not used. This was already elaborated by Goodwin (2000), indicating how a number of US organizations (federal and state agencies engaged in water management, the National Science Foundation – NSF, or the US Army Corps of Engineers with their activities around HEC software suite development and worldwide usage) are in fact engaged in activities related to hydroinformatics. In the words of D. L. James, then Program Director of Hydrologic Sciences at the NSF (in Goodwin, 2000):

> (In) 'Introducing Hydroinformatics' Michael Abbot presents the difficult task of persuading 'one another in more equitable ways' in a world where water approaches the 'level of the religious'. He concludes with words that strongly echo the experience of the hydrology program at NSF in that the 'aim of 'building community' remains as a beacon towards which we must constantly steer'. Even though we seldom use the word "hydroinformatics", this is what water research at NSF is all about.

Naturally, many US universities have educational and research programmes related to these aspects of hydroinformatics, even though they would not be introducing them under that umbrella term. One clear example is the enormous body of research and education in systems analysis applied to water and environmental resources, which overlap significantly with hydroinformatics (see, e.g. Loucks, 2000).

Similarly, in Asia hydroinformatics courses (topics) have been provided in some countries with universities and research centres active in hydroinformatics research. Example institutions include Asian Institute of Technology (AIT) in Thailand and National University of Singapore. Some universities in India provided Master (MTech) educational programmes in hydroinformatics, but at the moment of this writing it seems that only the national Institute of Technology (NIT) in Agartala provides such programme ('MTech in hydro-informatics engineering'). Some collaborative efforts together with European centres who offer hydroinformatics education have been put in place (e.g., through HydroEurope of EuroAquae and through training and capacity development projects of IHE Delft), but without establishing a full hydroinformatics MSc educational programme, although many centres are engaged in hydroinformatics research at PhD level.

China, as a large country, with numerous water-related challenges as well as universities has somewhat different situation. Due to the fact that several of these universities and research centres have been engaged in hydroinformatics research, also with alumni educated in hydroinformatics over the past decades, currently there are several universities that provide hydroinformatics or hydroinformatics-related courses in their civil engineering and water-related programmes, as summarized in Table 7.1.

Notable from Table 7.1 is the fact that in several universities in China, hydroinformatics-related courses are offered at undergraduate level. In fact, as from 2021, one undergraduate major (lasting four years) titled 'Smart water conservancy' (closely related to hydroinformatics) has been officially approved by the Chinese Ministry of Education. This is rather unique, as hydroinformatics has not been on offer at undergraduate level elsewhere. To the best of our knowledge, at

Table 7.1 Hydroinformatics-related courses offered by some Chinese universities.

Name of University	Courses Offered
Hohai University (https://en.hhu.edu.cn/)	• Undergraduate courses (focus on hydrometry and hydrological data): (a) hydroinformatics (hydrological information technology), (b) water information and data processing • Graduate courses: (c) modern hydrological information technology, (d) hydrological modelling and forecasting
Wuhan University (https://en.whu.edu.cn/)	• Undergraduate course: hydrological information technology
Tsinghua University (https://www.tsinghua.edu.cn/en/)	• Undergraduate course: principles and practice of big data in water conservancy
Tianjin University (http://www.tju.edu.cn/english/)	• Undergraduate courses: smart water, building information modelling (BIM) and three-dimensional design for water conservancy, software for computing in water conservancy, Flood control in emergency and intelligent disaster mitigation • BIM and its application in water conservancy and hydropower engineering
Sichuan University (https://en.scu.edu.cn/)	• Graduate courses (in the PhD programme): turbulence numerical modelling, hydrological information modelling and coastal dynamics modelling
Huazhong University of Science and Technology (http://english.hust.edu.cn/)	• Undergraduate courses: databases, object-oriented design and programming, GIS and remote sensing, ML, modern optimization algorithms, water management information system, fundamentals of data science
Zhejiang University (https://www.zju.edu.cn/english/)	• Undergraduate course: smart water conservancy engineering
Sun Yat-sen University (https://www.sysu.edu.cn/en/)	• Undergraduate courses: water conservancy, big data and informatization, GIS and remote sensing

graduate level there are no full hydroinformatics programmes at Chinese universities, except individual subjects linked to water-related MSc and PhD programmes, again mainly in support of research activities.

In Latin American countries hydroinformatics education is being promoted through the activities of IHE Delft and EuroAquae consortium, as mentioned earlier. There are still no established hydroinformatics programmes in Africa, although there have been several past and ongoing initiatives with European partners.

Apart from ongoing educational programmes in hydroinformatics, designing of curricula for hydroinformatics was carried out through different donors. European Union Tempus and Horizon 2020 programmes supported the development for curricula in hydroinformatics through twinning projects like Technology Management & Integrated Modelling in Natural Resources: A University-Enterprise Win-Win Partnership (Ain Shams University, Egypt), 2008–2010; and Data4Water (Politehnica University Bucharest, Romania), 2016–2019. The latter, an EU Horizon 2020 funded project looked at the development of hydroinformatics curricula at Computer Science faculty at this university. A series of three summer schools in hydroinformatics were conducted during the project time, along with two courses for academic staff. Academic staff are now carrying out research and organizing workshops in hydroinformatics, however, the educational programme is not yet implemented in full. The main reason is that demand for IT solutions in other industries is so high that potential students of this computer science faculty are seeking more opportunities when selecting application areas in industry or finance. The above overview of educational offerings indicates that the need for hydroinformaticians is growing, even if with different approaches and at different levels. In most cases hydroinformatics is brought at graduate level, primarily because of the need of integrating knowledge from different disciplines; it would be interesting to see how the Chinese undergraduate major will develop. Particular need for hydroinformatics is recognized at PhD level, as can be shown in the following example: Recently, the UK government has funded the Water Informatics in Science and Engineering Centre for Doctoral Training, which aims to fill the skills gap discussed above. It offers a postgraduate programme that fosters enhanced levels of innovation and collaboration to train a cohort of engineers and scientists at the boundary of water informatics, science and engineering. Hydroinformatics is the main keyword in presenting this PhD programme (Wagener *et al.*, 2021).

> the requirement for hybrid skills and expertise in water and informatics will grow with the need for increasingly intelligent systems. New water industry roles such as digital e-service delivery, smart water networks and big data analytics are misaligned with established single-discipline post-graduate programmes and require new hybrid skill sets not normally taught as a package. For society to take full advantage of leading-edge technologies we need to provide training for hydroinformaticians, i.e. scientists and engineers capable of working at the interface of traditionally separate disciplines of informatics, science, and engineering to manage information and water cycles effectively.

The authors are writing about the need for a new type of professional, and this resonates very much with what Mike was advocating 30 years ago. This means actually, that this problem has become more acute, and more efforts are needed to expand and deepen education in hydroinformatics at various levels.

Finally, it also needs to be recognized that providing most recent and most relevant hydroinformatics content to undergraduate or graduate students is posing very serious challenges to the educators themselves. The fast pace of changes in enabling technologies, and the continuously increasing and diversified demands put forward for such education are not easy to meet (see Makropoulos & Savic, 2019; Popescu *et al.*, 2012). Assembling and organizing teams of educators with different expertise and skills will become increasingly important to meet these challenges.

7.7 OUTLOOK

The course on hydroinformatics at IHE Delft was the first one of its kind, and enjoys a worldwide reputation. Due to the growing applications of hydroinformatics in the water-professional world, several other universities have initiated courses or specializations in hydroinformatics, at Master and PhD levels (see again Wagener *et al.*, 2021), and introduced professorship positions in hydroinformatics around the world.

From the content viewpoint, and to continue ensuring the relevance of hydroinformatics education in the future, it would be useful to look at it from the perspective and evolution of its core components, about which Mike Abbott has already written 25 years ago (Abbott, 1996) (see also Abbott, 2008). As of now, we see these components to include:

- *hydro* (including models),
- *informatics* (including AI),
- the *social* dimension, and
- their *integration*/interface.

The *hydro* component includes the fundamental knowledge about water as well as its mathematical and computer-based representation (e.g., models). This is a more-or-less traditional part, which is also taught in the courses of hydrology and hydraulics, but which in the context of hydroinformatics should be associated with the most innovative approaches to increasingly difficult water-related challenges. The modern analytical, modelling, and forecasting tools allow for detailed studies of the possible consequences for society of various scenarios of human interventions, global and climatic changes, and environmental degradation. Global meteorological and hydrological models increase their resolution by an order of magnitude every five years, and allow for more and more accurate (and practically sound) predictions of the environmental variables globally. Digital twins, real-time digital duplicates of real systems, and their multi-objective optimization are being actively developed in the water sector for purposes ranging from operator training to optimal design and operation of systems, to public awareness raising.

The *informatics* component is making use of significant technological advances that are becoming more and more readily available. The computer-literate population in the world is growing fast, albeit with serious geographical distribution gaps. Computational power, in terms of floating-point operations per second, has increased 350 million percent between 1993 and 2020. In the same period, computer storage has increased by five orders of magnitude (Roser & Ritchie, 2022). Vast amounts of data are collected by satellites and other remote sensing instruments. ML-based methods to process, analyse and visualize huge amounts of information (big data) are increasingly available. With the help of freely available scientific software libraries, sophisticated computer codes can be written in a matter of days. Widely available inexpensive supercomputing facilities, and easily accessible tools for their use, allow for solving complex model-based optimization problems, or applying Monte-Carlo framework for uncertainty analysis of distributed hydrological models where thousands of runs can be done in a matter of hours, or carrying out generation and model-based analysis of thousands of future scenarios.

There is also an important *social* component that has been emerging rapidly for many years. Water-related problems such as floods, droughts and pollution are to a large extent social problems, and therefore solutions require a larger involvement of citizens and communities, participation of different sectors such as energy, environment and health, and a dialogue of experts from social sciences to geosciences. Online discussion platforms that distil information produced by predictive models are being made available to facilitate multilateral interactions. Model-based serious gaming is becoming an important instrument in supporting stakeholder discussions and helping to make decisions that are important for society. An important development is the citizen observatories, which complement the traditional monitoring networks instruments and gauges, and involve the citizens and communities, via smartphones and social media.

Hydroinformatics covers all of the above components, and its main role is *integration* of all of these components. We are aiming at teaching participants to be able to build innovative integrated hydroinformatics systems encompassing the full information flows, along the chain: 'data – models – knowledge – optimal choice – decisions – impact'. The graduates should be able to communicate with the other specialists in related areas, in water and IT, and be familiar with their language. For example, IWA reports that water utilities, which historically have been very slow in adopting technologies compared to other sectors, are now pursuing digital transformation because it is a must, not an option (Sarni *et al.*, 2019). However, they are facing the lack of workforce with the ability to bring the two worlds closer. The Action Plan for a Digital Single Market for Water Services, makes this even clearer, with one of its actions being defined as follows: 'Define, develop and promote new digital water skills, labour markets and training and education systems to fit for the digital age (in line with the need to invest in digital skills)' ICT4Water (2018, p. 23).

With the intensified development of all of the above-mentioned components, education in hydroinformatics has a lot to reflect on. These developments open wonderful opportunities, but there is a need for training specialists who will be able to integrate all the components and build systems aiding water managers. The further development of hydroinformatics education should be seen within the deeper integration of hydroinformatics technologies into the mainstream digital platforms and practices taking place in all areas of human activity. This includes a broader use of AI, big data analytics, supercomputing for simulation modelling and forecasting, as well as digital platforms empowering citizens by participation in water governance.

IHE Delft is also changing. Several individual Master's programmes have now been aggregated into a single Master programme called 'Water and Sustainable Development', which starts in 2023. It has four tracks and several profiles (specializations) across a number of tracks. In particular, it includes the profile '*Digital innovation and hydroinformatics*'. Its name has been chosen to denote and stress the perspectives for the further development of hydroinformatics education, broadening its coverage and appeal. We hope that it will also resonate in the minds (and souls) of young, enthusiastic people thinking of a digital future, who are eager to be involved in the *real and useful* digital innovation that change society. The new programme is seen as an opportunity to broaden the coverage of hydroinformatics, bringing in more developments related to the digital era, which are important for effective socially responsible, technologically advanced, and optimal water management.

As a concluding note, we would like to return to a typical Mike Abbott's lecture. Mike often asked to open windows in the classroom during his lectures. (He always enjoyed fresh air, in a wide sense of this word.) At the end of the course, during a discussion, one of the female participants from Argentina, recalling this habit of Mike, said: 'the course in hydroinformatics has opened so many windows for all of us'. This is one of the principles that we continue to follow in this course: to open the windows, to present various technologies and approaches, to show creative ways of using various tools for solving different problems, and to show to young people a range of wonderful opportunities in their professional life.

REFERENCES

Abbott M. B. (1991a). Hydroinformatics: Information Technology and the Aquatic Environment. Ashgate (later Avebury), Aldershot, UK; Brookfield, VT, USA.

Abbott M. B. (1991b). A central issue of teaching within the hydroinformatics paradigm. *La Houille Blanche*, **3–4**(Juin 1991), 257–261, https://doi.org/10.1051/lhb/1991024

Abbott M. B. (1993a). The electronic encapsulation of knowledge in hydraulics, hydrology and water resources. *Advances in Water Resources*, **16**, 21–39, https://doi.org/10.1016/0309-1708(93)90027-D

Abbott M. B. (1993b). The impact of the knowledge revolution in the water industries. Proc. Watercomp 1993. Aust. Inst of Engineers, Melbourne, pp. 123–131.

Abbott M. B. (1996). The sociotechnical dimension of hydroinformatics. Proceedings of the 2nd International Conference on Hydroinformatics, 9–13 September 1996, Zürich, Switzerland, pp. 3–18.

Abbott M. B. (2008). Some future prospects in hydroinformatics. In: Practical Hydroinformatics, R. J. Abrahart, L. M. See and D. P. Solomatine (eds), Water Science and Technology Library. Springer, Berlin, Germany, **68**, pp. 3–16.

Abbott M. B., Babovic V. M. and Cunge J. A. (2001). Towards the hydraulics of the hydroinformatics era. *Journal of Hydraulic Research*, **39**(4), 339–349, https://doi.org/10.1080/00221680109499839

Abbott M. B., Solomatine D., Minns A. W., Verwey A. and Van Nievelt W. (1994). Education and training in hydroinformatics. *Journal of Hydraulic Research*, **32**, 203–214, https://doi.org/10.1080/00221689409498812

Craven J., Angarita H., Corzo Perez G. A. and Vasquez D. (2017). Development and testing of a river basin management simulation game for integrated management of the Magdalena-Cauca river basin. *Environmental Modelling & Software*, **90**, 78–88, https://doi.org/10.1016/j.envsoft.2017.01.002

Goodwin P. (2000). A view of hydroinformatics in the United States. *Journal of Hydroinformatics*, **02**(3), 149–153, IWA Publishing, https://doi.org/10.2166/hydro.2000.0013

Gourbesville P. (2021). EuroAquae + HydroEurope: 20 years of Hydroinformatics education, presentation in 1st IAHR Online Forum: Hydro-Environmental Challenges, Solutions and Trends for Water Security (https://www.iahr.org/video/clip?id=690 [last accessed 15 July 2022].

Gourbesville P. and Cunge J. (2010). EuroAquae and its links to RTD (Research & Technological Development) & Management. In: Water System Science and Policy Interfacing, P. Quevauviller (ed.), RSC Publishing, Cambridge, UK, pp. 347–369, https://doi.org/10.1039/9781847556622-00347

Gourbesville P. and Molkenthin F. (2020). Added value of collaborative engineering approach for hydroinformatics training. WaterEurope Project, 6th IAHR Europe Congress, June 30th–July 2nd, 2020, Warsaw, Poland, IAHR Abstract book, 2020, pp. 834–835.

ICT4Water. (2018). Digital Single Market for Water Services Action Plan Luxembourg, Publications Office of the European Union. ISBN 978-92-79-79351-6.

Jonoski A. and Popescu I. (2012). Distance learning in support of water resources management: an online course on decision support systems in river basin management. *Water Resources Management*, **26**, 1287–1305, https://doi.org/10.1007/s11269-011-9959-y

Kumar P., Folk M., Markus M. and Alameda J. C. (2005). Hydroinformatics: Data Integrative Approaches in Computation, Analysis, and Modelling. CRC Press, Boca Raton, FL, USA

Loucks D. P. (2000). Modeling the biophysical and social dynamics of a 'River of Grass': a challenge for hydroinformatics. *Journal of Hydroinformatics*, **02**(3), 207–217, IWA Publishing, https://doi.org/10.2166/hydro.2000.0018

Makropoulos C. and Savic D. (2019). Urban hydroinformatics: past, present and future. *Water*, **11**(10), 1959, https://doi.org/10.3390/w11101959

Popescu I., Jonoski A. and Bhattacharya B. (2012). Experiences from online and classroom education in hydroinformatics. *Hydrology and Earth System Sciences*, **16**, 3935–3944, https://doi.org/10.5194/hess-16-3935-2012

Price R. K., Ahmad K. and Holz P. (1998). Hydroinformatics concepts. In: Hydroinformatics Tools for Planning, Design, Operation and Rehabilitation of Sewer Systems. NATO ASI Series (Series 2: Environment), J. Marsalek, C. Maksimovic, E. Zeman and R. Price (eds), Springer, Dordrecht, **44**, pp. 47–76, https://doi.org/10.1007/978-94-017-1818-9_3

Price R. K., Bhattacharya B., Popescu I. and Jonoski A. (2007). Flood modelling for management: UNESCO-IHE's online course in hydrology. *Bulletin of World Meteorological Organisation*, **56**(2), 102–106.

Roser M. and Ritchie H. (2022). Technological change. https://ourworldindata.org/technological-progress# [last accessed 15 July 2022].

Sarni W., White C., Webb R., Cross K. and Glotzbach R. (2019). Digital Water: Industry Leaders Chart the Transformation Journey. International Water Association and Xylem Inc.

Wagener T., Savic D., Butler D., Ahmadian R., Arnot T., Dawes J., Djordjevic S., Falconer R., Farmani R., Ford D., Hofman J., Kapelan Z., Pan S. and Woods R. (2021). Hydroinformatics education – the Water Informatics in Science and Engineering (WISE) Centre for Doctoral Training. *Hydrology and Earth System Sciences*, **25**, 2721–2738, https://doi.org/10.5194/hess-25-2721-2021

The following article has been reprinted from *Journal of Hydraulic Research*, Vol. 5, No. 2, pp. 97–117, with permission from the copyright holders, IAHR International Association for Hydro-Environment Engineering and Research.

ON THE NUMERICAL COMPUTATION OF NEARLY HORIZONTAL FLOWS

SUR LE CALCUL NUMERIQUE DES ECOULEMENTS QUASI HORIZONTAUX

M. B. Abbott
Reader for Hydraulics, International Courses,
Delft Technological University, The Netherlands

F. Ionescu
Institutul de Studii și Cercetări Hidrotechnice, Bucharest, Rumania

Summary

A method is described for solving the quasi-linear equations of nearly-horizontal flow (e.g. one-dimensional long-wave equations, flood routing equations) under very general conditions. The method is unconditionally stable, so that is it especially useful when computing with non-uniform distance intervals and non-uniform time steps. The effects of bed slope, change of section, inflow, resistance terms and hydraulic jump "pseudoviscosities" can be easily introduced.

The method consists of an implicit numerical operator, so arranged that it can be solved by a pair of simultaneous tri-diagonal algorithms. The programming is simple, even for non-equidistant distance and time steps. For equidistant steps the running time is found to be no greater than existing explicit operators while with non-equidistant steps it is considerably shorter. The present scheme also has a generally higher order accuracy than existing explicit schemes.

Attention is directed to the danger, when using such unconditionally stable schemes, of introducing excessive errors by taking excessive time steps. Criteria for this and a means of circumventing it are described.

Résumé

On décrit une méthode très générale pour la résolution des équations quasi linéaires des écoulements quasi horizontaux comme, par exemple, les équations d'onde à grande amplitude unidimensionnelles ou les équations de propagation de crues. La méthode n'est jamais instable quelles que soient les conditions, d'où sa grande utilité dans les calculs fondés sur des intervalles de distance ou de temps non uniformes. Les influences des termes associés à la pente du lit, à un changement de section, au débit entrant et à la résistance, ainsi que des „pseudoviscosités" de ressaut hydraulique, peuvent aisément y être incorporées.

La méthode est essentiellement constituée par un opérateur numérique implicite, disposé de telle sorte que sa résolution s'obtient moyennant deux algorithmes simultanés tri-diagonaux. La programmation en est simple, même dans le cas où les intervalles de temps ou de distance ne sont pas constants. Pour des intervalles équidistants, le temps de calcul ne dépasse pas celui des opérateurs explicites existants, tandis que pour des intervalles non équidistants, il est sensiblement plus faible. En outre, la précision de la présente méthode est, de facon générale, plus élevée que celle des procédés explicites existants.

On attire l'attention sur le fait que, quand on utilise des schémas inconditionellement stables, on court le risque d'introduire des erreurs excessives en adoptant des intervalles de temps trop importants. Les critères à respecter pour éviter cet écueil sont précisés.

Paper received 7th November 1966

L'article reçu le 7 novembre 1966

The formal numerical properties of the scheme are given briefly in an appendix.

Les propriétés numériques du schéma sont indiquées brièvement en annexe.

Introduction

It often happens in practice that levels and velocities are available initially, and required subsequently, at unequal intervals, while at least one set of boundary data may also be given at unequal times. The process of interpolating this data back and forth from the physical space and time points to equidistant points may then be so undesirable that a computer programme capable of working as well on the physically defined grid as on a uniform grid may be required. The present note describes such a programme, suitable for solving quasi-linear flow equations with or without energy diffusing terms.

In order to hold stability with arbitrary distance and time intervals it is necessary – and can usually be made sufficient – that the domain of the complete numerical operator entirely encompass the domain of the solution. This condition is attained here, in the usual way, by using an implicit difference system, but the original aspect is the introduction of a pair of simultaneous tri-diagonal algorithms for solving the quasi-linear implicit system in an economical manner. In this way the existing tri-diagonal algorithm, already applied to linear systems, is generalised to quasi-linear wave systems (e.g. CUNGE and WEGNER, 1964; GARY, 1964).

Introduction

Dans les problèmes pratiques, il arrive fréquemment que les valeurs des niveaux et des vitesses, aussi bien celles qui sont données au départ que celles que l'on recherche, soient à intervalles inégaux. En outre, il se peut que l'un au moins des groupes de conditions aux limites soit indiqué pour des temps inégaux. S'il en est ainsi, le travail d'interpolation consistant à „faire la navette" entre les points réels de l'espace et du temps et des points uniformément espacés peut devenir si fastidieux qu'un programme de calcul capable de fonctionner aussi bien avec un quadrillage défini par le réel qu'avec un quadrillage uniforme peut se montrer nécessaire. La présente note décrit un tel programme, qui convient à la résolution d'équations d'écoulement quasi linéaires comportant ou non des termes de diffusion d'énergie.

Pour maintenir la stabilité avec des intervalles arbitraires de temps et de distance, il est nécessaire (et l'on peut en général faire en sorte qu'il soit suffisant) que le domaine de l'opérateur numérique complet renferme intégralement le domaine de la solution. On satisfait ici à cette condition, de la façon habituelle, par l'emploi d'un schema implicite de différences. L'aspect original du présent procédé réside, cependant, dans l'introduction de deux algorithmes simultanés tri-diagonaux pour résoudre le système implicite quasi linéaire de façon économique. De la sorte, l'algorithme tri-diagonal existant, déjà appliqué aux systèmes linéaires, se laisse généraliser et étendre aux systèmes d'onde quasi linéaires (par exemple CUNGE et WEGNER, 1964; GARY, 1964).

Equation systems

Solutions are required to the equations

Systèmes d'équations

On recherche des solutions aux équations:

$$\frac{\partial u}{\partial t} + u\frac{\partial u}{\partial x} + g\frac{\partial h}{\partial x} = 0$$

$$\frac{\partial h}{\partial t} + u\frac{\partial h}{\partial x} + h\frac{\partial u}{\partial x} = 0 \qquad (1)$$

In normal form, the system (1) becomes

Sous une forme normale, le système (1) devient:

$$\left\{\frac{\partial}{\partial t}+(u+c)\frac{\partial}{\partial x}\right\}J_{+}=0$$
$$\left\{\frac{\partial}{\partial t}+(u-c)\frac{\partial}{\partial x}\right\}J_{-}=0 \tag{2}$$

Where $J_{+}=(u+2c)$ and $J_{-}=(u-2c)$. The normal form provides characteristic directions and Riemann invariants or quasi-invariants for checking the difference procedure.

Equations (1) can be rearranged in the form

où $J_{+}=(u+2c)$ et $J_{-}=(u-2c)$. La forme normale fournit des directions caractéristiques ainsi que des invariants ou quasi invariants Riemann-iens pour la vérification du processus par différence.

Les équations (1) peuvent s'écrire sous la forme:

$$h\frac{\partial u}{\partial t}-u\frac{\partial h}{\partial t}-(u^2-gh)\frac{\partial h}{\partial x}=0$$
$$g\frac{\partial h}{\partial t}-u\frac{\partial u}{\partial t}-(u^2-gh)\frac{\partial u}{\partial x}=0 \tag{3}$$

They may then be approximated by the system

On peut alors les remplacer approximativement par le système:

$$h\frac{(u_j^{n+1}-u_j^{n})}{\Delta t}-\frac{u}{2}\left\{\frac{h_{j+1}^{n+1}-h_{j+1}^{n}}{\Delta t}+\frac{h_{j-1}^{n+1}-h_{j-1}^{n}}{\Delta t}\right\}-\frac{(u^2-gh)}{4}\left\{\frac{h_{j+1}^{n+1}-h_{j-1}^{n+1}}{\Delta x}+\frac{h_{j+1}^{n}-h_{j-1}^{n}}{\Delta x}\right\}=0$$
$$g\frac{(h_j^{n+1}-h_j^{n})}{\Delta t}-\frac{u}{2}\left\{\frac{u_{j+1}^{n+1}-u_{j+1}^{n}}{\Delta t}+\frac{u_{j-1}^{n+1}-u_{j-1}^{n}}{\Delta t}\right\}-\frac{(u^2-gh)}{4}\left\{\frac{u_{j+1}^{n+1}-u_{j-1}^{n+1}}{\Delta x}+\frac{u_{j+1}^{n}-u_{j-1}^{n}}{\Delta x}\right\}=0 \tag{4}$$

(This system, given above for Δx constant for all j, can easily be extended for variable Δx. In any event, its truncation error can be reduced to $0\,(\Delta x^2, \Delta t^2)$, while it remains unconditionally stable. These properties of stability and truncation error are discussed further in an appendix.

(Ce système, qui figure ci-dessus pour le cas de Δx constant pour tous j, peut être facilement étendu au cas de Δx variable. Quoi qu'il en soit, l'erreur de troncation peut être réduite jusqu'à $0(\Delta x^2, \Delta t^2)$ pendant que le système demeure inconditionnellement stable. Ces propriétés de stabilité et d'erreur de troncation sont considérées en annexe.

(We remark certain details and generalisations in parenthesis. Equations (1) to (3) constitute especially easily checked cases of the more general equations

(Avant de poursuivre, il convient de signaler quelques points de détail et quelques généralisations. Les équations (1), (2) et (3) constituent des exemples, particulièrement faciles à vérifier, des équations plus générales:

$$\frac{\partial A}{\partial t}+\frac{\partial Q}{\partial x}=q$$
$$\frac{\partial Q}{\partial t}+\left[gD-\left(\frac{Q}{A}\right)^2\right]\frac{\partial A}{\partial x}-2\frac{Q}{A}\frac{\partial A}{\partial t}=S \tag{3a}$$

where où

$$S = -Q/A(2.5q + 0.5|q|) + gA(I - I_f).$$

In this more general form,

Q = flow quantity
A = area
q = inflow per unit run of channel, outflow taken as negative. (The q term was introduced by JUUL HANSEN, 1965).
D = effective depth
I = bed slope
I_f = specific energy slope

Dans cette forme plus générale,

Q = débit
A = surface
q = débit entrant par unité de longueur du canal, le débit sortant étant négatif. (Ce terme fut introduit par JUUL HANSEN, 1965).
D = profondeur efficace
I = pente du lit
I_f = pente de la ligne de charge

Because of its practical importance, this form will be described in more detail later. Equations (3) are modified for an uncoupled stratified fluid layer only to the extent that gh is replaced by $g(1-\lambda)h$, where λ is the density of the fluid relative to the ambient fluid (ABBOTT, 1966). The more general coupled case is at present under investigation.

Equations (3) can be written as conservations laws, i.e. in the form

Nous reviendrons plus loin sur cette forme générale, en raison de son importance pratique. Lorsqu'il s'agit d'une couche de fluide stratifié non couplée, la seule modification des équations (3) est le remplacement de gh par $g\ (1-\lambda)h$, λ étant la densité du fluide par rapport à celle du fluide ambiant (ABOTT, 1966). Le cas couplé plus général est actuellement examiné.

Les équations (3) peuvent s'écrire sous la forme de lois de conservation

$$\frac{\partial f}{\partial t} + \frac{\partial g}{\partial x} = 0$$

as follows:

comme suit:

$$\left.\begin{aligned} &\frac{\partial h}{\partial t} + \frac{\partial}{\partial x}(hu) = 0 \\ &\frac{\partial}{\partial t}(hu) + \frac{\partial}{\partial x}\left(hu^2 + \frac{gh^2}{2}\right) = 0 \end{aligned}\right\} \quad (3b)$$

In the present case, where only mechanical coordinates are present, the mass and momentum equations are equivalent to the energy conservation equation

Dans le cas présent, où seules interviennent des coordonnées mécaniques, les équations de masse et de quantité de mouvement sont équivalentes à celle de la conservation d'énergie:

$$\frac{\partial}{\partial t}\left(\frac{gh^2}{2} + \frac{hu^2}{2}\right) + \frac{\partial}{\partial x}\left(u\left(gh^2 + \frac{hu^2}{2}\right)\right) = 0 \quad (3c)$$

As this equation contains no other energy terms than those of nearly-horizontal flow, it cannot account for an hydraulic jump.)

Comme, dans cette équation, les seuls termes énergétiques sont ceux de l'écoulement quasi horizontal, aucun compte ne peut être tenu du ressaut hydraulique.)

The double tri-diagonal algorithm

The implicit system (4), or its generalisations to variable section and slope, or variable Δx, can be solved by using two simultaneous tri-diagonal algorithms.

L'algorithme tri-diagonal double

Le système implicite (4), ou ses généralisations tenant compte de variations de section, de pente ou de Δx, peut être résolu moyennant l'emploi de deux algorithmes tri-diagonaux simultanés.

The procedure is as follows.
Equations (4) can be written as

Le processus est le suivant:
Les équations (4) peuvent s'écrire:

$$\left\{-\frac{u}{2\Delta t}-\frac{(u^2-gh)}{4\Delta x}\right\}h_{j+1}^{n+1}+\left\{\frac{h}{\Delta t}\right\}u_j^{n+1}+\left\{-\frac{u}{2\Delta t}+\frac{(u^2-gh)}{4\Delta x}\right\}h_{j-1}^{n+1}=$$
$$=\left\{-\frac{u}{2\Delta t}+\frac{(u^2-gh)}{4\Delta x}\right\}h_{j+1}^{n}+\left\{\frac{h}{\Delta t}\right\}u_j^{n}+\left\{-\frac{u}{2\Delta t}-\frac{(u^2-gh)}{4\Delta x}\right\}h_{j-1}^{n} \tag{5}$$

$$\left\{-\frac{u}{2\Delta t}-\frac{(u^2-gh)}{4\Delta x}\right\}u_{j+1}^{n+1}+\left\{\frac{g}{\Delta t}\right\}h_j^{n+1}+\left\{-\frac{u}{2\Delta t}+\frac{(u^2-gh)}{4\Delta x}\right\}u_{j-1}^{n+1}=$$
$$=\left\{-\frac{u}{2\Delta t}+\frac{(u^2-gh)}{4\Delta x}\right\}u_{j+1}^{n}+\left\{\frac{g}{\Delta t}\right\}h_j^{n}+\left\{-\frac{u}{2\Delta t}-\frac{(u^2-gh)}{4\Delta x}\right\}u_{j-1}^{n} \tag{6}$$

We write these equations as

Ces équations peuvent être mises sous la forme:

$$A_j h_{j+1}^{n+1} + B_j u_j^{n+1} + C_j h_{j-1}^{n+1} = D_j \tag{7}$$
$$A_j u_{j+1}^{n+1} + B_j^* h_j^{n+1} + C_j u_{j-1}^{n+1} = D_j^* \tag{8}$$

We suppose that u_j^{n+1} and h_j^{n+1} can be related such that

Nous supposons maintenant que l'on peut relier u_j^{n+1} et h_j^{n+1} de telle sorte que:

$$h_{j+1}^{n+1} = E_j u_j^{n+1} + F_j \tag{9}$$
$$u_{j+1}^{n+1} = E_j^* h_j^{n+1} + F_j^* \tag{10}$$

Then (7) and (9) provide

Nous tirons alors de (7) et (9):

$$A_j E_j u_j^{n+1} + B_j u_j^{n+1} + C_j h_{j-1}^{n+1} = D_j - A_j F_j$$

or soit:

$$u_j^{n+1} = \frac{-C_j}{A_j E_j + B_j} h_{j-1}^{n+1} + \frac{D_j - A_j F_j}{A_j E_j + B_j} \tag{11}$$

while (8) and (10) provide

tandis que (8) et (10) conduisent à

$$A_j E_j^* h_j^{n+1} + B_j^* h_j^{n+1} + C_j u_{j-1}^{n+1} = D_j^* - A_j F_j^*$$

or soit:

$$h_j^{n+1} = \frac{-C_j}{A_j E_j^* + B_j^*} u_{j-1}^{n+1} + \frac{D_j^* - A_j F_j^*}{A_j E_j^* + B_j^*} \tag{12}$$

Comparison of equations (9), (10), (11) and (12) then suggests the identities

En comparant les équations (9), (10), (11) et (12), on pense alors aux identités:

$$E_{j-1} = \frac{-C_j}{A_j E_j^* + B_j^*} \quad , \quad F_{j-1} = \frac{D_j^* - A_j F_j^*}{A_j E_j^* + B_j^*} \tag{13}$$

$$E^*_{j-1} = \frac{-C_j}{A_j E_j + B_j} \quad , \quad F^*_{j-1} = \frac{D_j - A_j F_j}{A_j E_j + B_j} \tag{14}$$

The algorithm is initiated from one or the other of the boundary conditions. Thus, for example, if the flow is everywhere subcritical and the velocity u_J is given at one end, say at $x = l$ ($= J\Delta x$ for constant Δx), then the ultimate equation of the system (7), (8) provides

On commence l'algorithme à partir de l'une ou de l'autre des conditions aux limites. De la sorte, si, par exemple, l'écoulement est partout sous-critique et que la vitesse soit donnée à l'une des extrémités, disons à $x = l$ ($= J\Delta x$ pour Δx constant), la dernière équation du système (7,) (8) fournit:

$$A_{J-1} u_J^{n+1} + B^*_{J-1} h_{J-1}^{n+1} + C_{J-1} u_{J-2}^{n+1} = D^*_{J-1}$$

Thence d'où

$$h_{J-1}^{n+1} = -\frac{C_{J-1}}{B^*_{J-1}} u_{J-2}^{n+1} + \frac{D^*_{J-1} - A_{J-1} u_J^{n+1}}{B^*_{J-1}} \tag{15}$$

Comparing (15) with (9) then suggests the identities

La comparaison de (15) et de (9) suggère alors des identités:

$$E_{J-2} = -\frac{C_{J-1}}{B^*_{J-1}} \quad , \quad F_{J-2} = \frac{D^*_{J-1} - A_{J-1} u_J^{n+1}}{B^*_{J-1}} \tag{16}$$

Comparing (16) with (13) indicates that

Et en confrontant (16) et (13) on observe que:

$$E^*_{J-1} = 0 \quad , \quad F^*_{J-1} = u_J^{n+1} \tag{17}$$

Similarly, if h_J^{n+1} is given as a boundary condition at $j = J$ we find

De façon analogue, si h_J^{n+1} est donné comme une condition à la limite $j = J$, nous trouverons que:

$$E_{J-1} = 0 \quad , \quad F_{J-1} = h_J^{n+1} \tag{18}$$

Keeping to the case that u_J^{n+1} is given, we see that, from condition (17), E_{J-2} and F_{J-2} can be computed using equations (13) and A_{J-1}, B_{J-1}, C_{J-1} values at time $n\Delta t$. With E_{J-2} and F_{J-2} determined, E^*_{J-3} and F^*_{J-3} are found from (14), and so successively until arriving at either E_0, F_0 or $E_0^*\ F_0^*$, according to the nature of the other end conditions. Thus if h is given at the other end, the grid is chosen such that E_0^* and F_0^* are last determined. Then u_1^{n+1} is determined from (10), h_2^{n+1} from (9) and so successively back to u_J. The grid used in this case is schematised in Fig. 1.

Revenant au cas de u_J^{n+1} donné, nous voyons, d'après la condition (17), que l'on peut calculer E_{J-2} et F_{J-2} à l'aide des équations (13) et des valeurs de A_{J-1}, B_{J-1} et de C_{J-1} au temps $n\Delta t$. Une fois que E_{J-2} et F_{J-2} seront déterminés, on obtient E^*_{J-3} et F^*_{J-3} moyennant (14) et l'on poursuit successivement de la sorte jusqu'à ce que, selon la nature des autres conditions aux extrémités, on parvienne soit à E_0, F_0, soit à E_0^*, F_0^*.

De la sorte, si h est donné à l'autre extrémité, le quadrillage est choisi de manière que E_0^* et F_0^* soient déterminés en dernier. On obtient

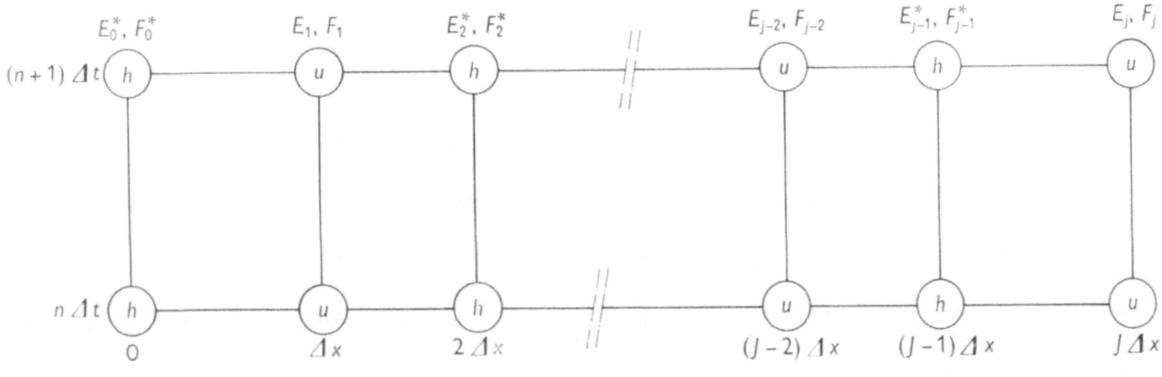

Fig. 1.

alors u_1^{n+1} à partir de (10), h_2^{n+1} à partir de (9), et l'on remonte de cette façon jusqu'à u_J. Le quadrillage utilisé dans ce cas est représenté schématiquement sur la figure 1.

As developed above the scheme has time and space symmetry in first derivatives, and thus has truncation error 0 $(\Delta x^2, \Delta t^2)$ in these parts, but uses coefficients A_j, B_j, and C_j taken entirely at $n\Delta t$, and thus introducing truncation errors of $0(\Delta t)$. This effect can be much reduced, to give an error of approximately $0(\Delta x^2, \Delta t^2)$

Le procédé tel que nous venons de le voir possède une symétrie spatiale et temporelle dans les dérivées premières et, par conséquent, l'erreur d'approximation y est de 0 $(\Delta x^2, \Delta t^2)$. Il emploie cependant les coefficients A_j, B_j et C_j, pris entièrement à $n\Delta t$, ce qui introduit des erreurs d'approximation de 0 (Δt). Cette influence peut être très sensiblement réduite pour donner une erreur d'environ 0 $(\Delta x^2, \Delta t^2)$.

In the event that the extended equations (3a) are employed, the algorithm coefficients in the expressions

Si l'on emploie les équations plus générales (3a), les coefficients d'algorithme des expressions:

$$\alpha_j A_{j-1}^{n+1} + \beta_j Q_j^{n+1} + \gamma_j A_{j-1}^{n+1} = \delta_j.$$

$$\alpha_j^* Q_{j+1}^{n+1} + \beta_j^* A_j^{n+1} + \gamma_j^* Q_{j-1}^{n+1} = \delta_j^*$$

take the following forms:

se mettent alors sous les formes suivantes:

$$\alpha_j = + \frac{1}{4\Delta x}\left[gD - \left(\frac{Q}{A}\right)^2\right] - \frac{Q}{A\Delta t}; \quad \beta_j = \frac{1}{\Delta t}; \quad \gamma_j = -\frac{1}{4\Delta x}\left[gD - \left(\frac{Q}{A}\right)^2\right] - \frac{Q}{A\Delta t};$$

$$\delta_j = \left\{-\frac{1}{4\Delta x}\left[gD - \left(\frac{Q}{A}\right)^2\right] - \frac{Q}{A\Delta t}\right\} A_{j+1}^n + \frac{1}{\Delta t} Q_j^n + \left\{\frac{1}{4\Delta x}\left[gD - \left(\frac{Q}{A}\right)^2\right] - \frac{Q}{A\Delta t}\right\} A_{j-1}^n + \tfrac{1}{2}\left(S_j^n + S_j^{n+1}\right);$$

$$\alpha_j^* = \frac{1}{\Delta x}; \quad \beta_j^* = \frac{4}{\Delta t}; \quad \gamma_j^* = -\frac{1}{\Delta x}$$

$$\delta_j^* = -\frac{1}{\Delta x} Q_{j-1}^n + \frac{4}{\Delta t} A_j^n + \frac{1}{\Delta x} Q_{j+1}^n + 2q_{j+1}^n + 2q_j^{n+1}$$

by repeating the above process using averages of u_j^n, h_j^n and the newly acquired u_j^{n+1} and h_j^{n+1}. For all reasonably smooth data and representative Δx_j, one such repetition appears to be quite

en répétant le processus décrit ci-dessus et en utilisant les moyennes de u_j^n, h_j^n et les valeurs nouvellement obtenues de u_j^{n+1} et h_j^{n+1}. Pour des données relativement cohérentes et

sufficient. An experimental investigation of the error is described later.

In the event that unequal distances are required, the terms in brackets { } in (4) must be replaced by

pour des valeurs représentatives de Δx_j, une seule répétition paraît tout à fait suffisante. On décrira plus loin une étude expérimentale de l'erreur.

Si l'on a besoin de distances inégales, les termes en parenthèses { } en (4) doivent être remplacés par:

$$\left\{\frac{2}{\Delta t}\left[(h_{j+1}^{n+1}-h_{j+1}^{n})q_j+(h_{j-1}^{n+1}-h_{j-1}^{n})(1-q_j)\right]\right\}\ldots\left\{\frac{2}{\Delta x_j+\Delta x_{j+1}}\left[(h_{j+1}^{n+1}-h_{j-1}^{n+1})+(h_{j+1}^{n}-h_{j-1}^{n})\right]\right\}$$

and et

$$\left\{\frac{2}{\Delta t}\left[(u_{j+1}^{n+1}-u_{j+1}^{n})q_j+(u_{j-1}^{n+1}-u_{j-1}^{n})(1-q_j)\right]\right\}\ldots\left\{\frac{2}{\Delta x_j+\Delta x_{j+1}}\left[(u_{j+1}^{n+1}-u_{j-1}^{n+1})+(u_{j+1}^{n}-u_{j-1}^{n})\right]\right\}$$

where: où:

$$\Delta x_j = x_j - x_{j-1}$$

$$q_j = \frac{\Delta x_j}{\Delta x_j + \Delta x_{j+1}}$$

in order that the truncation error should remain $\approx 0\,(\Delta x^2)$.* We remark that the weighting terms are constant for all n in any given problem.

pour que l'erreur d'approximation reste égale à $\approx 0\,(\Delta x^2)$.* Il faut noter que les termes de pondération sont constants pour tous n pour un problème donné.

Programming of coefficients

The coefficients in the above algorithm can be programmed as follows. A computational grid is defined such that if both end-conditions ($j = 0$ and $j = J$) are of the same type, then J is set as an even number, and otherwise J is set as an odd number. In the following, it will be supposed that the boundary conditions are given as:

La programmation des coefficients

Les coefficients de l'algorithme ci-dessus peuvent être programmés comme suit. On définit un quadrillage de calcul tel que J soit un nombre pair si les deux conditions aux extrémités ($j = 0, j = J$) sont de même nature, J étant un nombre impair dans le cas contraire.

Dans ce qui suit, on supposera que les conditions aux limites sont:

$$h_0^n = h_0^n(t) \quad \text{at/à} \quad j = 0$$

$$u_J^n = u_J^n(t) \quad \text{at/à} \quad j = J$$

so that J will be odd. (In the authors' computations, J was taken as 25 for accurate results, and as 11 for certain extreme examples. Then for all odd j's we have:

de sorte que J sera impair. (Dans ses calculs, l'auteur a pris $J = 25$ pour obtenir des résultats précis et $J = 11$ pour quelques cas extrêmes). Nous avons alors, pour tous j impairs.

* There remains a small term

$$\frac{2(\Delta x_{j+1}-\Delta x_j)}{(\Delta x_{j+1}+\Delta x_j)^2}\,(h_j^{n+1}+h_j^n) \quad \text{or/ou} \quad \frac{2(\Delta x_{j+1}-\Delta x_j)}{(\Delta x_{j+1}+\Delta x_j)^2}\,(u_j^{n+1}+u_j^n)$$

that can only be introduced with the first iteration.

* Il reste un petit terme seulement possible d'être introduit en employant la première itération.

$$A_j = -\left(\frac{u_j^{n+1}+u_j^n}{2}\right) q_j - \frac{\alpha_j}{2}\left\{\left(\frac{u_j^{n+1}+u_j^n}{2}\right)^2 - g\left[\left(\frac{h_{j+1}^{n+1}+h_{j+1}^n}{2}\right) q_j + \left(\frac{h_{j-1}^{n+1}+h_{j-1}^n}{2}\right)(1-q_j)\right]\right\}$$

$$B_j = \left(\frac{h_{j+1}^{n+1}+h_{j+1}^n}{2}\right) q_j + \left(\frac{h_{j-1}^{n+1}+h_{j-1}^n}{2}\right)(1-q_j)$$

$$C_j = -\left(\frac{u_j^{n+1}+u_j^n}{2}\right)(1-q_j) + \frac{\alpha_j}{2}\left\{\left(\frac{u_j^{n+1}+u_j^n}{2}\right)^2 - g\left[\left(\frac{h_{j+1}^{n+1}+h_{j+1}^n}{2}\right) q_j + \left(\frac{h_{j-1}^{n+1}+h_{j-1}^n}{2}\right)(1-q_j)\right]\right\}$$

$$D_j = \left\{C_j + \left(\frac{u_j^{n+1}+u_j^n}{2}\right)(1-2q_j)\right\} h_{j+1}^n + B_j u_j^n + \left\{A_j - \frac{(u_j^{n+1}+u_j^n)}{2}(1-2q_j)\right\} h_{j-1}^n$$

where où

$$\alpha_j = \frac{\Delta t}{\Delta x_j + \Delta x_{j+1}}, \qquad q_j = \frac{\Delta x_j}{\Delta x_j - \Delta x_{j+1}},$$

with avec

$$\Delta x_j = x_j - x_{j-1}$$

Similarly for all even j's we have

De façon analogue, nous avons pour tous j pairs:

$$A_j = \left\{-\left(\frac{u_{j+1}^{n+1}+u_{j+1}^n}{2}\right) q_j + \left(\frac{u_{j-1}^{n+1}+u_{j-1}^n}{2}\right)(1-q_j)\right\}\cdot q_j - \frac{\alpha_j}{2}\left\{\left[\left(\frac{u_{j+1}^{n+1}+u_{j+1}^n}{2}\right) q_j + \left(\frac{u_{j-1}^{n+1}+u_{j-1}^n}{2}\right)(1-q_j)\right]^2 - g\left(\frac{h_j^{n+1}+h_j^n}{2}\right)\right\}$$

$$B_j^* = g$$

$$C_j = \left\{-\left(\frac{u_{j+1}^{n+1}+u_{j+1}^n}{2}\right) q_j + \left(\frac{u_{j-1}^{n+1}+u_{j-1}^n}{2}\right)(1-q_j)\right\}\cdot(1-q_j) + \frac{\alpha_j}{2}\left\{\left[\left(\frac{u_{j+1}^{n+1}+u_{j+1}^n}{2}\right) q_j + \left(\frac{u_{j-1}^{n+1}+u_{j-1}^n}{2}\right)(1-q_j)\right]^2 - g\left(\frac{h_j^{n+1}+h_j^n}{2}\right)\right\}$$

$$D_j^* = \left\{C_j + \left[\left(\frac{u_{j+1}^{n+1}+u_{j-1}^{n+1}}{2}\right) q_j + \left(\frac{u_{j+1}^n+u_{j-1}^n}{2}\right)(1-q_j)\right](1-2q_j)\right\} u_{j+1}^n + B_j u_j^n + \left\{A_j - \left[\left(\frac{u_{j+1}^{n+1}+u_{j-1}^{n+1}}{2}\right) q_j + \left(\frac{u_{j+1}^n+u_{j-1}^n}{2}\right)(1-q_j)\right](1-2q_j)\right\} u_{j-1}^n$$

where, again, où, encore,

$$q_j = \frac{\Delta x_j}{\Delta x_1 + \Delta x_{j+1}} \text{ with/avec } \Delta x_j = x_j - x_{j-1}$$

For the initial application of the algorithm u_j^{n+1} and h_j^{n+1} are set equal to u_j^n and h_j^n. In the event that the lower order truncation error is acceptable, the iteration process on equation coefficients becomes unnessary, with obvious simplifications in the above computational coefficients.

A flow diagram of the authors' programme is given in Fig. 2 (see page 114).

Lors de la première application de l'algorithme, on pose u_j^{n+1} égal à u_j^n et h_j^{n+1} à h_j^n. Si l'erreur d'approximation d'ordre inférieur est acceptable, le processus d'itération pour déterminer les coefficients des équations n'est plus nécessaire d'où d'évidentes simplifications dans les coefficients ci-dessus.

Un organigramme du programme de l'auteur est indiqué dans la figure 2 (voir page 115).

Applications

This scheme has been used to solve a number of problems with satisfactory results. The more general formulation with shear stresses has been investigated for the case of a quasi-steady de Chézy resistance, and found to introduce no difficulties. The introduction of a pseudo-viscosity according to von NEUMANN and RICHTMEYER (1950), or LAX (1954, 1957), whereby a hydraulic jump can be accomodated by the scheme, has been investigated and appeared to introduce no difficulties.

(If this is carried out using the approach of von NEUMANN and RICHTMEYER (1950) a viscous term q of the form

Applications

Cette méthode a été employée de façon satisfaisante dans la résolution d'un certain nombre de problèmes. La formulation générale faisant intervenir les contraintes de cisaillement a été étudiée avec succès dans le cas d'une résistance de Chézy quasi permanente. L'introduction d'une pseudo-viscosité selon von NEUMANN et RICHTMEYER (1950) ou LAX (1954–1957) qui permet d'inclure un ressaut hydraulique, a été examinée sans qu'aucune difficulté particulière n'apparaisse.

(Si l'on aborde ce dernier problème à la manière de VON NEUMANN et RICHTMEYER (1950) un terme visqueux q de la forme:

$$q = a^2(\Delta x)^2 h \left(\frac{\partial u}{\partial x}\right)^2 \quad , \quad \left(\frac{\partial u}{\partial x}\right) < 0$$

$$q = 0 \quad , \quad \left(\frac{\partial u}{\partial x}\right) \geq 0$$

must be added to the potential energy terms in (3b) and (3c) (i.e. only once to the right side of (3c)). In the present case it would appear possible to further determine a value of q consistent with equations (3) and the running hydraulic jump solutions, where

doit être ajouté aux termes représentant l'énergie potentielle dans les équations (3b) et (3c) (c'est-à-dire une fois seulement au deuxième membre de (3c)). Dans le cas présent, il semblerait possible de poursuivre en déterminant une valeur de q qui satisfasse aux équations (3) et aux solutions du ressaut hydraulique mobile, où:

$$\frac{\partial}{\partial t} = -c\frac{\partial}{\partial x},$$

with c the jump velocity, given approximately as the average of celerities of incident characteristics of the same sense. This case has not, however, been incorporated into the present programme, so that its effect on stability remains uncertain. (See, for example, RICHTMEYER, 1957, p. 218; PREISSMANN and CUNGE, 1961;

c, la vitesse du saut, étant représentée approximativement par la moyenne des célérités des caractéristiques incidentes de même sens. Cette possibilité n'a cependant pas été incorporée dans le présent programme: son influence sur la stabilité demeure donc incertaine (voir par exemple, RICHTMEYER, 1957, p. 218; PREISS-

VASILIEV, GLADYSHEV and PRITVITS, 1965).

The pseudo viscosity introduced by LAX (1954) is generated essentially from the truncation error of the second operator in Fig. 5, when applied to the equations (3b). The later development of LAX (1957) uses the third operator illustrated, while further developments again are described by LAX and WENDROFF (1960, 1964). However, in the gas dynamic cases so discussed, the form of irreversible energy transfer is quite different from that obtaining in most practical hydraulic jumps (ratio of depths < 2), in that the one involves only second order derivatives while the other involves only third order derivatives (e.g. DAUBERT and MARVAUD, 1965). This entire matter, however, will best form the subject of a separate communication.)

Tests have also been carried out to determine the behaviour of errors as the time step increases for fixed distance step. Roughly speaking, it appears that the scheme attains its maximum accuracy as the computational time step Δt passes through the characteristic time step Δt_c, which is taken as the arithmetic mean of the individual characteristic times, as defined by the Δx_j and the characteristic directions $(u_j \pm \sqrt{gh_j})$. By taking the arithmetic mean, Δt_c may always be computed as though the computational points were equidistant. As Δt decreases with $\Delta t < \Delta t_c$ the error at fixed $T = J\Delta t$ changes only slightly, but as Δt increases with $\Delta t > \Delta t_c$ the error at fixed $T = J\Delta t$ increases more rapidly. This is to be expected, not only from a TAYLOR series analysis, but also because, as Δt departs more and more from Δt_c, so the response properties (energy transfer properties) of the computed system depart further and further from those of the physical system. Thus, whereas for very fine computational nets good results could be obtained with values of $\Delta t/\Delta t_c$ of 10, and even more, for very coarse grids the error became exceedingly sensitive to $\Delta t/\Delta t_c$ variations. An extreme example of this behaviour is illustrated in Fig. 3, where only 5 values of h and 6 values of u have been employed. The curves of Fig. 3 have all been generated with (4) from the same initial and boundary data. (See also RICHTMEYER, 1957, p. 200).

MANN CUNGE, 1961; VASILIEV, GLADYSHEV et PRITVITS, 1965).

La pseudo-viscosité introduite par LAX (1954) est engendrée essentiellement par l'erreur de troncation du second opérateur de la figure 5 lors de son application aux équations (3b). Le développement ultérieur de LAX (1957) emploie le troisième opérateur figuré; d'autres développements ont été élaborés par LAX et WENDROFF (1960, 1964). Dans les problèmes aérodynamiques envisagés par ces auteurs, la forme du transfert énergétique irréversible est cependant tout à fait différente de celle correspondant à la plupart des ressauts hydrauliques (rapport de profondeur < 2): la première ne comporte que des dérivés de second ordre, alors que dans la seconde n'interviennent que des dérivés de troisième ordre (p.e., DAUBERT et MARVAUD, 1965). Nous n'entrerons pas plus avant dans ces considérations qui feront l'objet d'une autre communication.)

Des essais ont également été effectués pour déterminer le comportement des erreurs lorsque le pas de temps augmente à un pas de distance constant.

Il semble, de façon assez grossière, que le processus atteigne un maximum de précision quand le pas de temps du calcul, Δt, s'identifie au pas de temps caractéristique, Δt_c, la valeur adoptée pour ce dernier étant la moyenne arithmétique des temps caractéristiques individuels (définis par les valeurs de Δx_j et par les directions caractéristiques $u_j \pm \sqrt{gh_j}$). En prenant la moyenne arithmétique, la détermination de Δt_c s'effectue comme si les points du calcul étaient équidistants:
Quand $\Delta t (< \Delta t_c)$ décroît, l'erreur à $T = J\Delta t$ constant est peu modifiée, mais quand Δt $(> \Delta t_c)$ augmente, l'erreur à $T = J\Delta t$ s'élève plus rapidement; un tel comportement peut d'ailleurs être pressenti, non seulement moyennant une analyse en série de TAYLOR, mais aussi parce que, à mesure Δt s'écarte de Δt_c, les propriétés de la réponse du système de calcul (propriétés de transfert énergétiques) s'éloignent de plus en plus de celles du système réel. Ainsi, tandis que, pour des réseaux de calcul très serrés, de bons résultats ont pu être obtenus avec $\Delta t/\Delta t_c = 10$ ou même davantage, l'erreur intervenant avec des réseaux très grossiers est extrêmement sensible à des variations de $\Delta t/\Delta t_c$. Un cas extrême de ce comportement est représenté par

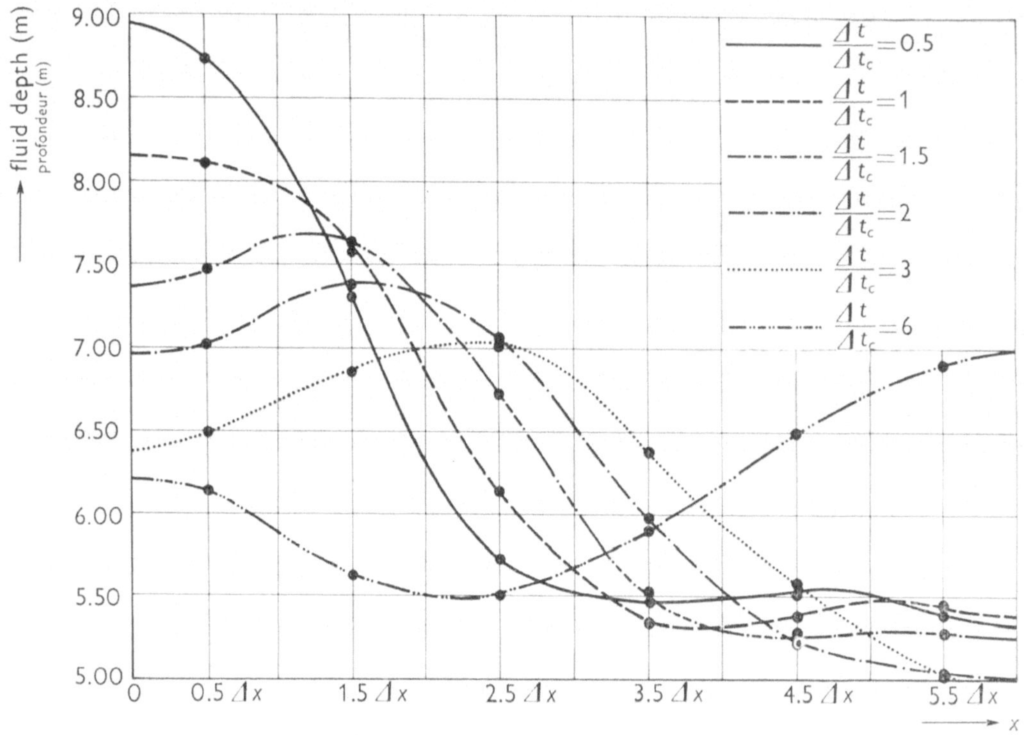

Fig. 3

Fig. 3 shows how solutions of the difference equations (4) rapidly depart from the true solutions as $\Delta t/\Delta t_c$ increases, even though these solutions remain stable.

Although the computational grid used in the example was much coarser than should ever be employed in practice, it illustrates the magnitude of errors that can arise without the usual warnings of instability, when this unconditionally stable scheme is used. (With the same grid, an explicit scheme would, of course, give either equally poor or unstable results). The corresponding total energy plots are given in Fig. 4.* In this connection it is therefore recommended that when the Δt steps are excessive in relation to the Δt_c – as can easily be checked with a

* In the event that the system (3b) had been employed, this plot would have given no indication of stability or accuracy, for the present scheme would then always assure conservation of mass and energy.

la figure 3, où n'ont été employées que 5 valeurs de h et 6 valeurs de u. Les courbes de la figure 3 ont été établies à partir de (4) avec les mêmes conditions initiales et les mêmes conditions aux limites (voir également Richtmeyer, 1957, p. 200).

Ces courbes montrent comment, à $\Delta t/\Delta t_c$ croissant, les solutions des équations de différence s'écartent rapidement des solutions exactes et ceci bien que les premières demeurent stables.

Quoique le quadrillage employé dans cet exemple fût plus grossier que celui que l'on devrait utiliser en pratique, la figure 3 illustre l'importance des erreurs qui peuvent intervenir quand l'instabilité, et l'alerte qu'elle donne, font défaut – comme c'est le cas avec la présente méthode. (Avec le même quadrillage, une méthode explicite donnerait, bien entendu, des résultats soit aussi mauvais, soit aussi instables). Les courbes correspondantes représentant l'énergie totale sont indiquées dans la figure 4.*

Il est par conséquent recommandé d'intro-

* Si l'on avait employé le système (3b), ces courbes n'auraient fourni aucune indication de stabilité ni de précision, car la présente méthode satisfait toujours alors à la conservation de masse et d'énergie.

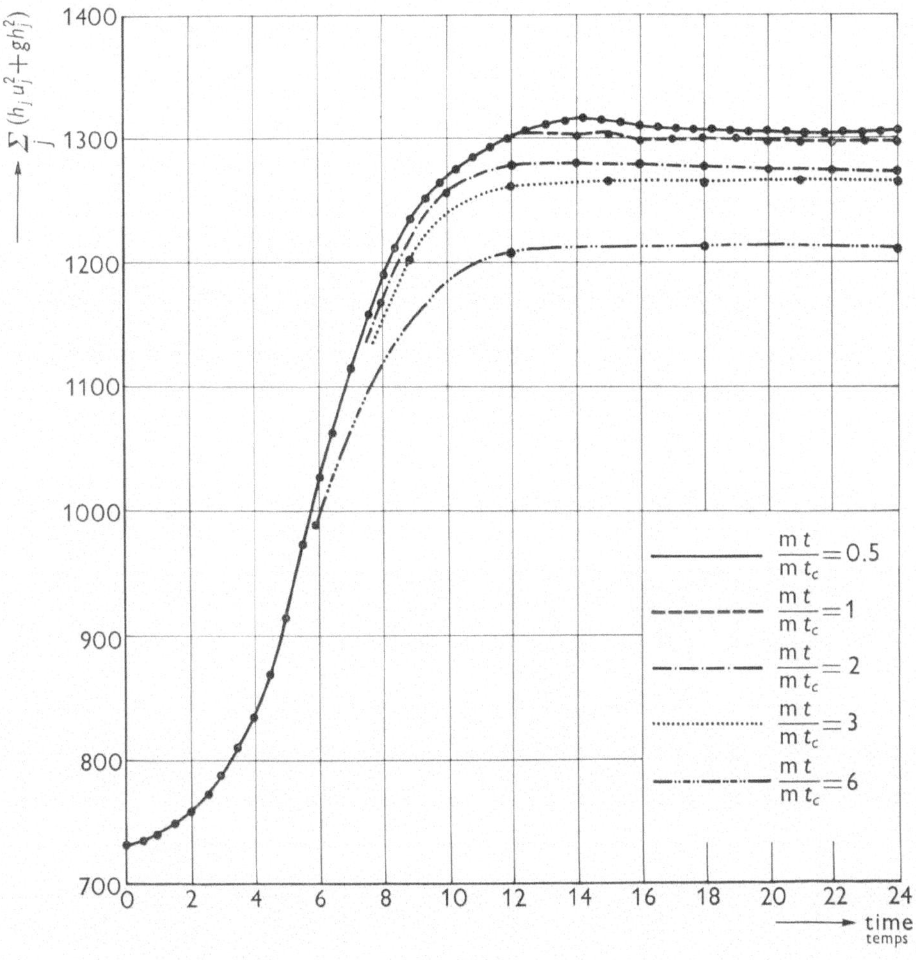

Fig. 4

scanning sub-programme – intermediate time steps should be introduced. When these are introduced at aliquot parts of the given steps they need introduce only very simple interpolations.

The running time of the present scheme, programmed in Algol 60, appears to be similar to that used for the corresponding explicit operator, as schematised in Fig. 5 (1), with Δx and Δt constant within the explicit stable range (RICHTMEYER 1957). Programming time appears also to be similar for constant time steps, while the present scheme is much easier to programme for variable time steps. Compared with the usual explicit scheme, as illustrated in Fig. 5 (1), the present scheme has an increasingly superior accuracy with increasing Froude number. Although this accuracy might be improved even further to compare with the explicit scheme proposed by LAX and WENDROFF (1960), such an improvement was found to introduce such

duire des intervalles de temps intermédiaires chaque fois que les pas Δt sont trop grands par rapport à Δt_c, ce que l'on peut aisément vérifier à l'aide d'un sous-programme de balayage. Si les nouveaux intervalles sont des fractions simples des premiers, les interpolations nécessaires ne compliquent guère les calculs.

Le temps de calcul de la présente méthode, programmée en Algol 60, paraît semblable à celui exigé par l'opérateur explicite correspondant tel que représenté figure 5 (1) avec Δx et Δt constants à l'intérieur du domaine explicite stable (RICHTMEYER, 1957). Pour des intervalles de temps constants, les temps de programmation semblent voisins également. Par contre, la méthode exposée ici est beaucoup plus facile à programmer lorsqu'il s'agit d'intervalles de temps variables. La précision de notre méthode l'emporte d'autant plus sur celle du procédé explicite habituel (figure 5, (1)) que le nombre

programming difficulties that it could not be attempted.

We finally remark that one-point boundary conditions may always be entered into the present scheme in their usual form (u or h), whereas in the conservation formulations only (uh) can be entered directly.

de Froude est plus grand. Bien qu'il soit en principe possible d'augmenter encore cette précision de façon à soutenir la comparaison avec le procédé explicite proposé par LAX et WENDROFF (1960), nous n'avons pu tenter une telle amélioration, en raison des difficultés de programmation.

Enfin, nous signalons que des conditions aux limites à point unique peuvent toujours figurer dans la présente méthode sous leur forme habituelle (u ou h), tandis que seulement (uh) peut être introduit directement dans les formulations de conservation.

Fig. 5

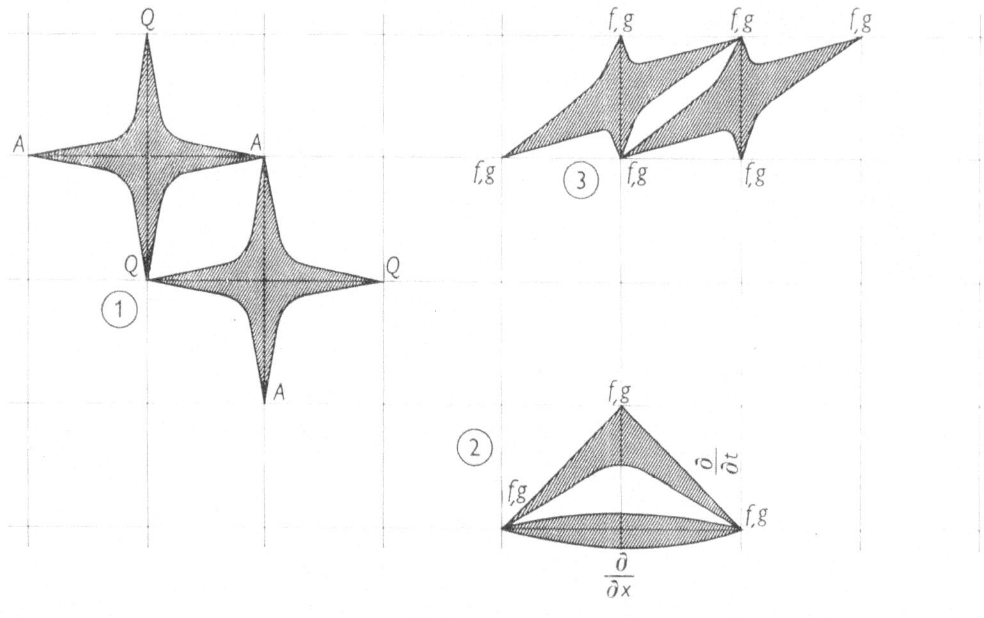

Conclusions

A numerical method has been developed and tested that provides accurate solutions to the quasi-linear equations of nearly-horizontal flow under very general conditions. The method is as easily programmed and is as economical in machine time as existing explicit schemes, but unlike the explicit schemes the present method is easily extended to arbitrarily distributed computational points and is unconditionally stable. The present method is also usually more accurate. The introduction of bed slope, change of section, inflow and hydraulic jump resistance terms introduces no new difficulties, while further generalisation of the method to computations in stratified fluids appear to be possible.

Conclusions

Une méthode numérique a été mise au point et expérimentée, qui fournit des solutions précises aux équations quasi linéaires des écoulements quasi horizontaux et ceci dans des conditions très générales. Par rapport aux procédés explicites, la méthode est aussi facile à programmer et aussi économique au point de vue temps de machine. Par contre, notre méthode s'en distingue par la facilité avec laquelle elle s'accommode de points de calculs répartis de façon arbitraire, et par sa stabilité ,,inconditionnelle''. En outre, elle est en général plus précise. Sans introduire de nouvelles difficultés, on peut y incorporer des termes tenant compte de la pente du lit, de changements de section, du débit entrant et de la résistance associée au ressaut

Although the implicit operator with tri-diagonal algorithm provides a simple and exceptionally flexible method for solving quasi-linear problems of nearly-horizontal flow, there is the danger of introducing considerable errors if it is applied indiscriminately. For, whereas explicit operators become unstable before excessive time steps can introduce large truncation errors, the present scheme, being unconditionally stable, can introduce large truncation errors without any indication from stability, or even from energy norm control. A measure of how excessive the time steps Δt may be is provided by their ratio to the characteristic time step Δt_c defined by the distance intervals and the characteristic directions. For fine computational nets, values of $\Delta t/\Delta t_c$ of 10, or even more, can be employed, but when coarse nets are used $\Delta t/\Delta t_c$ should not be taken as more than 2. These criteria may often be attained quite easily by introducing additional time levels at aliquot parts of those specified.

hydraulique et il semble possible de l'étendre également au cas des fluides stratifiés.

L'opérateur implicite et l'algorithme tri-diagonal fournissent une méthode simple et exceptionellement souple pour la résolution des problèmes quasi linéaires relatifs aux écoulements presque horizontaux.

Toutefois, si l'on s'en sert aveuglément, on court le risque d'introduire des erreurs considérables car, tandis que les opérateurs explicites deviennent instables avant que les intervalles de temps excessifs aient pu introduire des erreurs de troncation importantes, la présente méthode, étant parfaitement stable, peut engendrer de grandes erreurs sans que la stabilité, ni même le contrôle des normes énergétiques, nous en avertisse. Le rapport des intervalles de temps Δt à l'intervalle caractéristique Δt_c (défini par les intervalles de distance et par les les directions caractéristiques) constitue un critère permettant d'apprécier la validité d'un intervalle de temps que l'on se propose d'employer. Pour des réseaux très sérrés, $\Delta t/\Delta t_c$ peut être de 10 ou même davantage. Cependant, pour des réseaux grossiers, $\Delta t/\Delta t_c$ ne devrait pas dépasser 2. On peut souvent satisfaire aisément à ces critères en ajoutant des niveaux de temps supplémentaires, définis en fractionnant les premiers intervalles.

Acknowledgement

This work was carried out during Eng. Ionescu's participation in the International Courses in Hydraulic Engineering at Delft. The programme testing and the preparation of the flow diagram was carried out by Eng. E. Pardo Castro, of the International Courses, Delft.

Remerciements

Ce travail a été exécuté pendant le séjour à Delft de l'Ingénieur Ionescu qui suivait les Cours Internationaux d'Engineering Hydraulique. Les auteurs remercient l'Ing. E. Pardo Castro qui a effectué les essais de programme et qui a établi l'organigramme.

Appendix: Stability and truncation error

A well known necessary condition for the stability of difference schemes for hyperbolic equations is that the domain of every point of the difference operator should entirely encompass the domain of dependence of the corresponding differential equation (e.g. Courant, Isaacson and Rees, 1949). However, when sufficient weight is placed upon the forward time values, the domain of the operator constitutes the domain of the entire solution, so

Annexe: La stabilité et les erreurs de troncation

Une condition nécessaire, bien connue, de la stabilité des procédés par différences pour les équations hyperboliques est que le domaine de chaque point de l'opérateur de différence doit renfermer intégralement le domaine de dépendance de l'équation différentielle correspondante (par ex. Courant, Isaacson et Rees, 1949). Cependant, quand on place un poids suffisant sur les valeurs de temps futures, le domaine de l'opérateur constitue le domaine de la solution

that this necessary condition may always be satisfied. The stability problem is then to find the set of necessary weight distributions and to select from this set those elements that are also sufficient. The parallel truncation error problem is to select from these, again, those distributions that make the truncation error a minimum. It is usually found in implicit schemes, that when the truncation error is a minimum, the necessary and sufficient conditions coincide. In the present case, programming simplicity and truncation error requirements, as explained later, indicate a distribution of half the weight on forward times and half on backward times, giving the equation system (4). It is then only required to confirm that the system(4) simultaneously satisfies necessary and sufficient conditions for stability. We use the von NEUMANN condition, whereby necessary and sufficient conditions coincide for normal operators, and we follow von NEUMANN's conjecture that linear operators with variable coefficients are stable if and only if all their localised operators, in which coefficients are taken as constant, are stable. We accordingly transform (4), in the form of (5) and (6) to homogeneous coordinates $p_j{}^n = \sqrt{g}h_j{}^n$, $q_j{}^n = \sqrt{h}u_j{}^n$ (which are essentially kinetic and potential energy components in $R_{\bar{j}}^2$) to obtain the normal forms:

entière, de sorte qu'il est toujours possible de satisfaire à cette condition nécessaire. Le problème de la stabilité est de rechercher un groupe de distributions de pondération nécessaires et de retenir parmi ce groupe, les éléments qui sont également suffisants. Le problème parallèle de l'erreur de troncation consiste alors à sélectionner, à partir de ces éléments, ceux qui rendent l'erreur minimale. Pour les procédés implicites, on trouve généralement que, quand l'erreur de troncation est minimale, les conditions nécessaires et suffisantes coïncident. Dans le cas présent, les exigences de simplicité de programmation et d'erreur de troncation que l'on verra plus loin, conduisent à une distribution qui partage le poids également entre les temps „futurs" et les temps „précédents", ce qui donne le système d'équations (4). Il suffit alors de confirmer que le système (4) satisfait simultanément aux conditions nécessaires et aux conditions suffisantes de stabilité. Nous utilisons la condition de VON NEUMANN, selon laquelle les conditions nécessaires et suffisantes sont identiques pour des opérateurs normaux et nous adoptons son hypothèse suivant laquelle les opérateurs linéaires à coefficients variables ne sont stables que si tous leurs opérateurs localisés, dans lesquels les coefficients sont supposés constants, sont stables. Par conséquent, le mettant sous la même forme que (5) et (6), nous transformons (4) en coordonnées homogènes $p_j{}^n = \sqrt{g}h_j{}^n$, $q_j{}^n = \sqrt{h}u_j{}^n$ (qui sont essentiellement des composantes en $R_{\bar{j}}^2$ de l'énergie cinétique et potentielle) et nous obtenons les formes normales:

$$\left\{-\frac{u}{2\Delta t}-\frac{(u^2-gh)}{4\Delta x}\right\}p_{j+1}^{n+1}+\left\{\frac{\sqrt{gh}}{\Delta t}\right\}q_j^{n+1}+\left\{-\frac{u}{2\Delta t}+\frac{(u^2-gh)}{4\Delta x}\right\}p_{j-1}^{n+1}=$$

$$=\left\{-\frac{u}{2\Delta t}+\frac{(u^2-gh)}{4\Delta x}\right\}p_{j+1}^{n}+\left\{\frac{\sqrt{gh}}{\Delta t}\right\}q_j^{n}+\left\{-\frac{u}{2\Delta t}-\frac{(u^2-gh)}{4\Delta x}\right\}p_{j-1}^{n}$$

$$\left\{-\frac{u}{2\Delta t}-\frac{(u^2-gh)}{4\Delta x}\right\}q_{j+1}^{n+1}+\left\{\frac{\sqrt{gh}}{\Delta t}\right\}p_j^{n+1}+\left\{-\frac{u}{2\Delta t}+\frac{(u^2-gh)}{4\Delta x}\right\}q_{j-1}^{n}=$$

$$=\left\{-\frac{u}{2\Delta t}+\frac{(u^2-gh)}{4\Delta x}\right\}q_{j+1}^{n}+\left\{\frac{\sqrt{gh}}{\Delta t}\right\}p_j^{n}+\left\{-\frac{u}{2\Delta t}-\frac{(u^2-gh)}{4\Delta x}\right\}q_{j-1}^{n}$$

Taking the Fourier transform and simplifying in the usual way (RICHTMEYER, 1957) provides the relation:

En prenant la transformée de Fourier et en simplifiant de la façon habituelle (RICHTMEYER, 1957), il vient:

$$\begin{bmatrix} a+ib & c \\ c & a+ib \end{bmatrix} \begin{bmatrix} \xi_{1m}^{n+1} \\ \xi_{2m}^{n+1} \end{bmatrix} = \begin{bmatrix} a-ib & c \\ c & a-ib \end{bmatrix} \begin{bmatrix} \xi_{1m}^{n} \\ \xi_{2m}^{n} \end{bmatrix} \quad \text{(A1)}$$

between the *mth* Fourier coefficients ξ_{1m}^n and ξ_{2m}^n of p_j^n and q_j^n respectively. In (A1) the coefficients are seen to be

$$a = -\frac{u}{\Delta t} \cdot \cos m\Delta x, \quad b = -\frac{(u^2-gh)}{2\Delta x} \cdot \sin m\Delta x, \quad c = \frac{\sqrt{gh}}{\Delta t}$$

so that the coefficient b becomes zero when a is a maximum, and vice versa. As in all such locally linear wave equations, there exists an equipartition of energy between the kinetic and the potential, over the periods af all Fourier components, so that the system (Al) is invariant under interchanges of potential and kinetic terms, i.e. under interchanges of the corresponding Fourier coefficients ξ_{1m} and ξ_{2m}.* Moreover, the left matrix operator of (A1) appears as the complex conjugate of the right matrix operator, so that the modulus of each vector $\{\xi_{1m}^{n+1}, \xi_{2m}^{n+1}\}$ is amplified to the same extent as is the vector $\{\xi_{1m}^n, \xi_{2m}^n\}$.

It follows at once that the corresponding amplification matrix must have a unit spectral radius. In this case the Parseval equality, which generally expresses the isomorphism of the L_2 space of square integrable functions and the l_2 space of their corresponding Fourier coefficients, can be regarded simply as a statement of the law of additivity of energies. Since the energy of each Fourier component vector $\{\xi_{1m}^n, \xi_{2m}^n\}$ remains constant under every mapping (A1), it then follows that the energy of every vector function $\{p, q\}$ remains constant under every mapping (4), and the system (4) is stable. The extension to include energy diffusing terms is covered by the usual generalisation of the von Neumann necessary condition, which states that the spectral radius should vary from unity only by terms $0(\Delta t)$. The above argument appears to have the advantage (over the usual direct investigation of spectral radii, in this case of the form $(a^2-(ib\pm c)^2)/((a+ib)^2-c^2))$ that it generalises at once to unevenly spaced grid points.

* This implies, of course, that h_j^n is behaving as a variation of h.

relation liant les m^{es} coefficients ξ_{1m}^n et ξ_{2m}^n de p_j^n et q_j^n respectivement. Dans l'équation (A1), on voit que:

de sorte que le coefficient b s'identifie à zéro, quand a est maximal et inversement. Comme c'est le cas dans toute équation d'onde présentant une linéarité locale, il existe, pour toutes les composantes de Fourier, une équirépartition d'énergie parmi l'énergie cinétique et l'énergie potentielle. Il en résulte que le système (A1) est invariant lors que les termes cinétique et potentiel sont interchangés, c'est-à-dire lorsque l'on intervertit les coefficients de Fourier correspondants: ξ_{1m} et ξ_{2m}.*

D'ailleurs, l'opérateur matriciel du premier membre de (A1) est le conjugué complexe de celui du second membre, de sorte que le module de chaque vecteur $\{\xi_{1m}^{n+1}, \xi_{2m}^{n+1}\}$ est amplifié dans le même rapport que celui du vecteur $\{\xi_{1m}^n, \xi_{2m}^n\}$.

Il s'ensuit immédiatement que la matrice d'amplification correspondante doit avoir un rayon spectral unitaire. Dans ce cas, l'égalité de Parseval, qui exprime généralement l'isomorphisme de l'espace L_2 des fonctions carrées intégrables et l'espace l_2 des coefficients de Fourier correspondants, peut tout simplement être considérée comme étant un énoncé de la loi d'additivité des énergies. Puisque l'énergie de chaque composante de vecteur de Fourier $\{\xi_{1m}^n, \xi_{2m}^n\}$ demeure constante pour chaque transformation (A1), il s'ensuit que l'énergie de chaque fonction vectorielle $\{p, q\}$ demeure constante pour chaque transformation (4) et que le système (4) est stable. Pour inclure des termes de diffusion énergétique, on fait appel à la généralisation habituelle de la condition nécessaire de von Neumann: le rayon spectral ne devrait s'écarter de l'unité que par des termes $0\ (\Delta t)$. Le raison-

* Ceci veut dire, bien entendu, que h_j^n se comporte comme une variation de h.

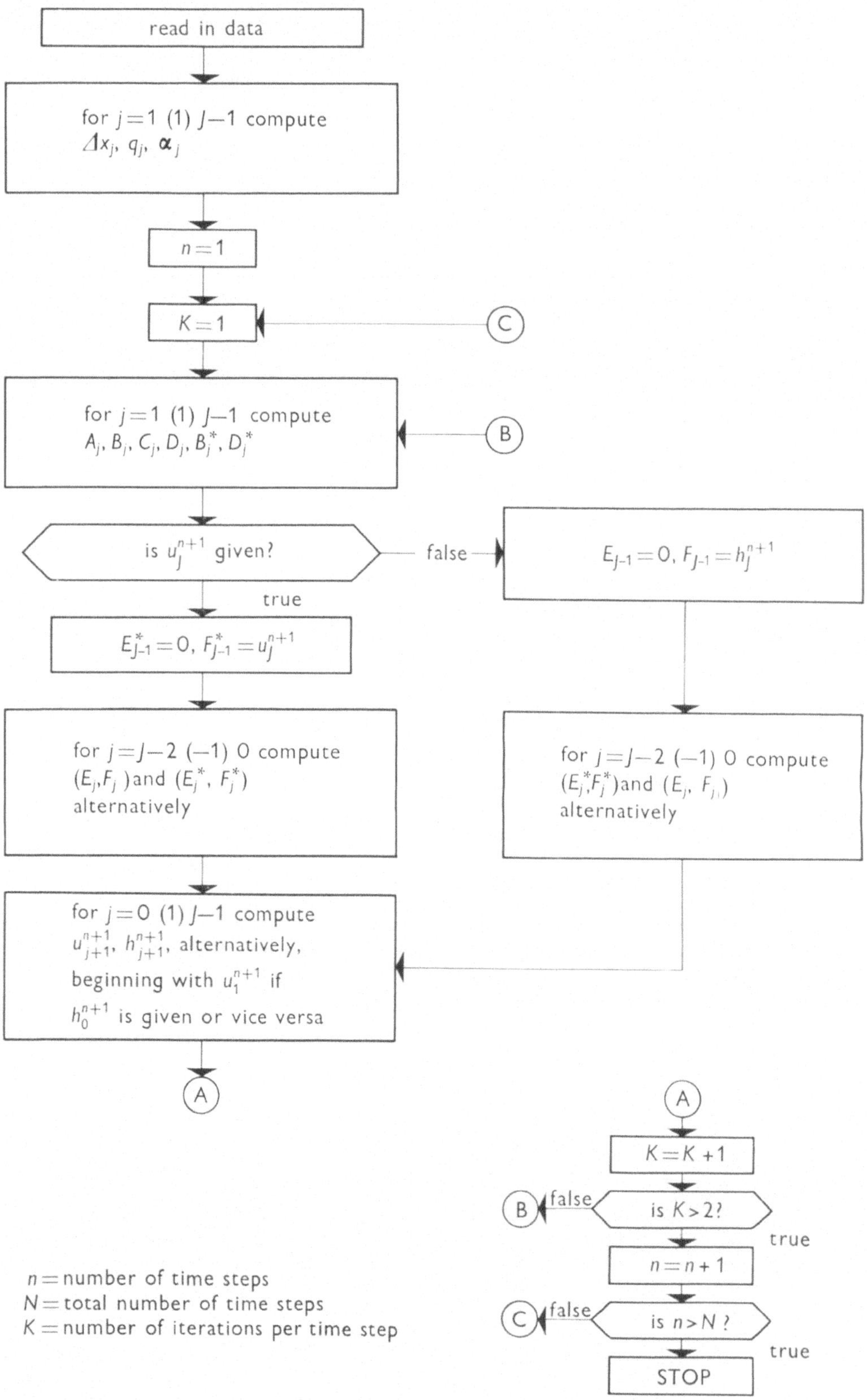
read in data
for $j=1\ (1)\ J-1$ compute $\Delta x_j,\ q_j,\ \alpha_j$
$n=1$
$K=1$
C
for $j=1\ (1)\ J-1$ compute $A_j, B_j, C_j, D_j, B_j^*, D_j^*$
B
is u_J^{n+1} given?
false
$E_{J-1}=0,\ F_{J-1}=h_J^{n+1}$
true
$E_{J-1}^*=0,\ F_{J-1}^*=u_J^{n+1}$
for $j=J-2\ (-1)\ 0$ compute (E_j,F_j) and (E_j^*,F_j^*) alternatively
for $j=J-2\ (-1)\ 0$ compute (E_j^*,F_j^*) and (E_j,F_j) alternatively
for $j=0\ (1)\ J-1$ compute $u_{j+1}^{n+1}, h_{j+1}^{n+1}$, alternatively, beginning with u_1^{n+1} if h_0^{n+1} is given or vice versa
A
A
$K=K+1$
B
false
is $K>2$?
true
$n=n+1$
C
false
is $n>N$?
true
STOP
n = number of time steps
N = total number of time steps
K = number of iterations per time step

Fig. 2

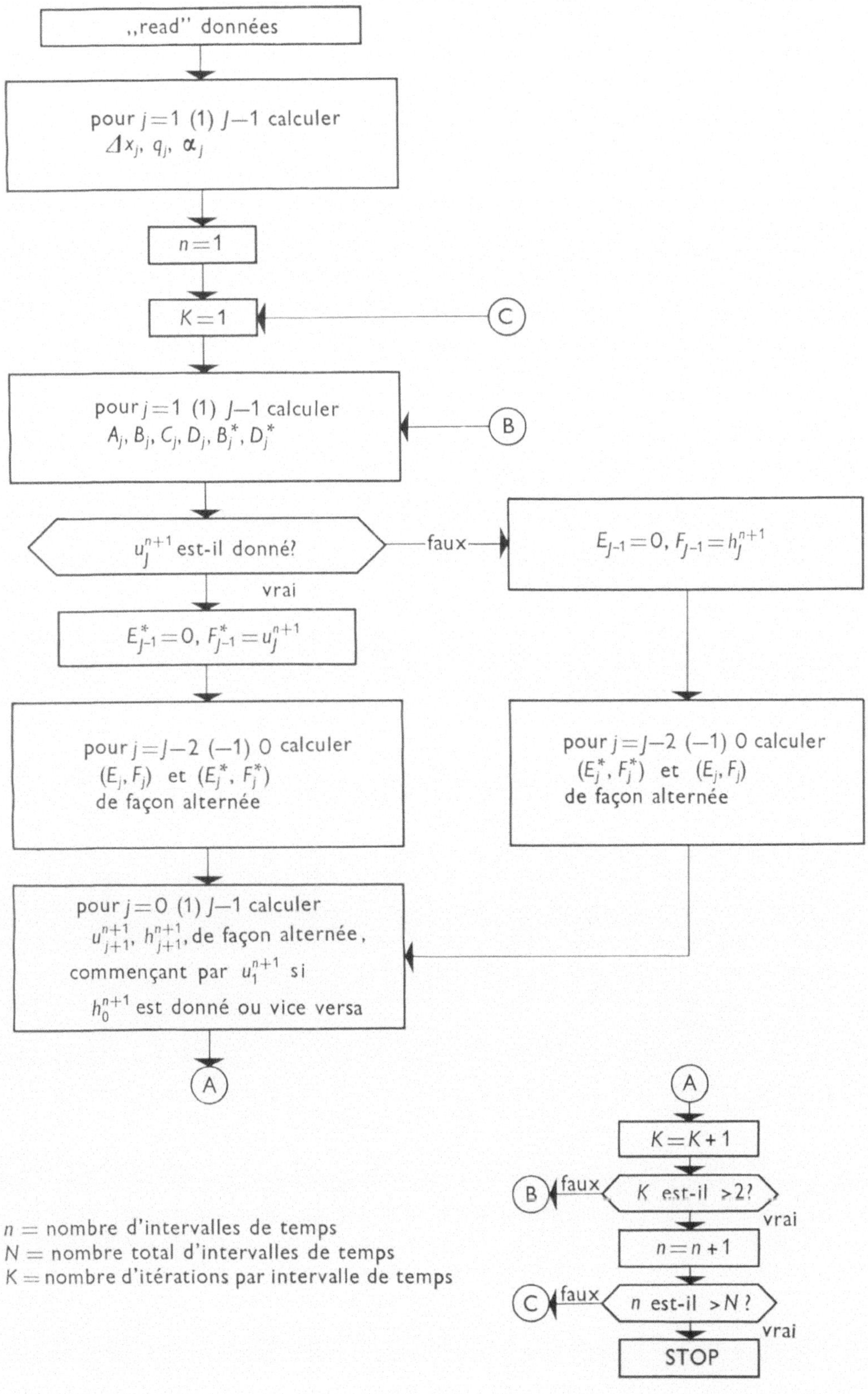
,,read'' données
pour $j=1$ (1) $J-1$ calculer
Δx_j, q_j, α_j
$n=1$
$K=1$
C
pour $j=1$ (1) $J-1$ calculer
$A_j, B_j, C_j, D_j, B_j^*, D_j^*$
B
u_J^{n+1} est-il donné?
faux
$E_{J-1}=0$, $F_{J-1}=h_J^{n+1}$
vrai
$E_{J-1}^*=0$, $F_{J-1}^*=u_J^{n+1}$
pour $j=J-2$ (−1) 0 calculer
(E_j, F_j) et (E_j^*, F_j^*)
de façon alternée
pour $j=J-2$ (−1) 0 calculer
(E_j^*, F_j^*) et (E_j, F_j)
de façon alternée
pour $j=0$ (1) $J-1$ calculer
u_{j+1}^{n+1}, h_{j+1}^{n+1}, de façon alternée,
commençant par u_1^{n+1} si
h_0^{n+1} est donné ou vice versa
A
A
$K=K+1$
B
faux
K est-il >2?
vrai
$n=n+1$
C
faux
n est-il $>N$?
vrai
STOP
n = nombre d'intervalles de temps
N = nombre total d'intervalles de temps
K = nombre d'itérations par intervalle de temps

Fig. 2

That the truncation error is $0(\Delta t^2)$ follows at once from the centering of the operator with weighting $\frac{1}{2} : \frac{1}{2}$, which is as illustrated in Fig. A1.

nement ci-dessus semble avoir l'avantage (par rapport à l'examen direct usuel des rayons spectraux qui sont, dans ce cas de la forme $(a^2 - (ib \pm c)^2)/((a + ib)^2 - c^2))$ de pouvoir s'appliquer immédiatement à un quadrillage d'espacements irréguliers. Que l'erreur d'approximation soit de $0(\Delta t^2)$ résulte immédiatement du centrage de l'opérateur avec la pondération $\frac{1}{2} : \frac{1}{2}$, ce qui est représenté par la figure A1. On voit que, dans le développement en série de TAYLOR, tous les termes d'ordre impair au point 1 seront annulés par les termes d'ordre

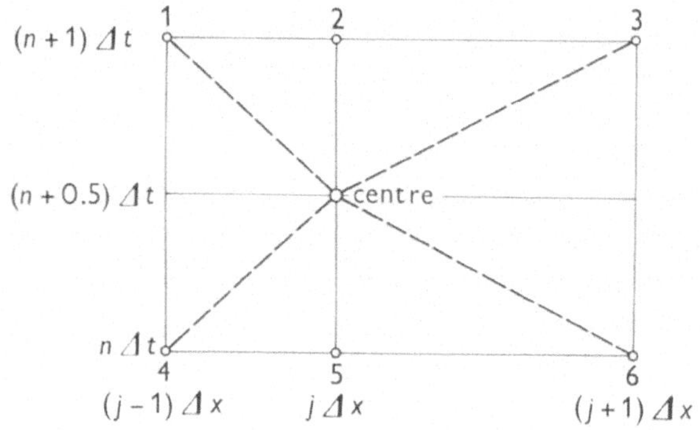

Fig. A1

It is seen that, in the TAYLER series expansion, all odd order terms at point 1 will be cancelled by equel and opposite odd order terms at point 4, and similarly for the pairs 2, 5 and 3, 6. Other forms of centering, although allowing more simple programming, do not allow such a uniform reduction of error.

Although the present scheme possesses unused degrees of freedom in weighting, so that higher order truncation errors would appear to be attainable, the associated algebraic complications precluded their introduction.

impair de valeur égale et de signe contraire au point 4. Il en va de même pour les paires de points 2, 5 et 3, 6. D'autres formes de centrage quoique permettant des programmations plus simples, ne conduisent pas à une réduction d'erreur aussi uniforme.

Bien que, dans la pondération, la présente méthode possède des degrés de liberté inutilisés et que par conséquent on pourrait espérer atteindre des erreurs de troncation d'un ordre supérieur, les complexités algébriques qui entrent en jeu empêchent leur emploi.

Notations

Fluid velocity in x-direction	u	Vitesse du fluide le long de l'axe des x
Fluid depth	h	Profondeur du fluide
Fluid density	ϱ	Densité du fluide
Shear stress, due to wind, bed slope or bed resistance	τ	Taux de cisaillement, engendré soit par le vent, soit par la pente ou par la résistance du lit
Characteristic celerity: $c = +\sqrt{gh}$	c	Célérité caractéristique: $c = +\sqrt{gh}$
Riemann invariants; i.e. the value of $J_{\pm}$ remains constant along a characteristic $(\pm)$	$J_{\pm}$	Invariants Riemann-iens (c'est-à-dire que la valeur de $J_{\pm}$ demeure constante le long d'une caractéristique $(\pm)$)
Coefficients, each replacing a set of terms in u and h	A_j, B_j, C_j, D_j	Coefficients remplaçant un groupe de termes en u et h
Weighting factor	q_j	Facteur de pondération

References / Bibliographie

1. Abbott, M. B., An introduction to the method of characteristics. London, Thames and Hudson, 1966. Chapter 4.
2. Courant, R., E. Isaacson and M. Rees, On the solution of non-linear hyperbolic equations by difference methods. Comm. Pure and App. Math. *2* (1949) p. 255.
3. Cunge, J. A. and M. Wegner, Intégration numérique des équations d'écoulement de Barré de Saint-Venant par un schéma implicite de différences finies. Grenoble: La Houille Blanche, 1, 1964.
4. Daubert et Marvaud, Etude de la formation d'ondes secondaires par une méthode d'intégration numérique. Leningrad: XIth Congress I.A.H.R. paper 3.12, 1965.
5. Gary, J., On certain finite difference schemes for hyperbolic systems. Math. of Comp. Jan. 1964.
6. Juul Hansen, A., Numerical analysis of unsteady channel flow. Hydraulics Laboratory, Technical University of Denmark. Bulletin 10, (1966).
7. Lax, P. D., Weak solutions of non-linear hyperbolic equations and their numerical computations. Comm. Pure and Appl. Math. *7* (1954) p. 159.
8. Lax, P. D., Hyperbolic systems of conservation laws. Comm. Pure and Appl. Math. 10 (1957) p. 537
9. Lax, P. D. and B. Wendroff, Systems of conservation laws. Comm. Pure and Appl. Math. *13* (1960) p. 217.
10. Neumann, J. von and R. D. Richtmeyer, A method for the numerical computation of hydrodynamic shocks. J. Appl. Phys. *21* (1950) p. 232.
11. Preissmann, A. et J. A. Cunge, Calcul du mascaret sur machine électronic. Grenoble: La Houille Blanche, 5, 1964.
12. Richtmeyer, R. D., Difference methods for initial value problems. New York, Interscience, 1957.
13. Vasiliev, O. F., M. T. Gladyshev and N. A. Pritvits, Numerical methods for the calculation of shock waves in open channels. Leningrad: XIth Congress I.A.H.R. paper 3.44, 1965.
14. Wendroff, B., Difference methods for hyperbolic equations with high order of accuracy. Comm. Pure and Appl. Math. 17 (1964) p. 381.

The following article has been reprinted from *Journal of Hydraulic Research*, Vol. 16, pp. 173–204, with permission from the copyright holders, IAHR International Association for Hydro-Environment Engineering and Research.

ON THE NUMERICAL MODELLING OF SHORT WAVES IN SHALLOW WATER

REPRÉSENTATION, SUR MODÈLE NUMÉRIQUE, DES ONDES DE FAIBLE LONGUEUR SE PROPAGEANT EN EAU PEU PROFONDE

by

M. B. ABBOTT

Reader, International Institute for Hydraulic and Environmental Engineering, Delft, Netherlands, Head, Computational Hydraulic Centre, Danish Hydraulic Institute, Hørsholm, Denmark

and

H. M. PETERSEN and O. SKOVGAARD

Engineers, Computational Hydraulic Centre, Danish Hydraulic Institute, Hørsholm, Denmark*

Summary A modelling system is described that generates and runs models of short waves of any form (periodic or irregular), with any desired physically realistic current field over any given bathymetry. As this system constitutes the eighth version of the general System 21, "Jupiter", described in an earlier contribution, it is called the "Mark 8". The system-generated models are based upon Boussinesq equations, in which *the vertical velocity is supposed to increase linearly from zero at the bed to a maximum magnitude at* the surface, in two independent (horizontal) space variables and time. The Boussinesq equations are formulated as mass and momentum conservation laws while, by virtue of the high order of accuracy of the difference approximations, there is very little numerical energy falsification. This formulation also appears to provide genuine weak solutions, for correctly simulating breaking waves, and thus assures the correct simulation of wave thrusts, or radiation stresses, and associated longshore currents.

The System has been tested against analytical results in one and two dimensions and also against physical model tests, for all its main capabilities. In all cases, the agreement of the System model results with the analytical and physical results are satisfactory. The System is already being applied in engineering practice. A discussion is presented of future applications of the System to ship motion simulations, to sediment transport computations and also to more efficient nearly-horizontal flow computations.

Résumé Description d'un système de représentation sur modèle permettant la création et l'exploitation de modèles d'ondes de faible longueur de n'importe quelle forme (périodique ou irrégulière), compte tenu de toute répartition physiquement réaliste des courants sur une répartition bathymétrique donnée quelconque. Du fait que ce Système correspond à la 8ème version du système général 21 "Jupiter" (exposé dans un rapport de publication antérieure), il lui a été affecté la désignation "Mark 8". Les modèles découlant du Système sont établis sur la base des équations de Boussinesq, qui admettent la linéarité de l'accroissement de la vitesse verticale, de la valeur nulle au fond à la valeur maximale au niveau de la surface, en fonction de deux variables spatiales indépendantes (horizontales) et du temps.

Les équations de Boussinesq sont formulées sous la forme de lois de conservation de la masse et de la quantité de mouvement: du fait de l'ordre de précision élevé des approximations de différence, les valeurs numériques de l'energie sont très peu faussées. D'autre part, cette formulation paraît livrer de véritables solutions "faibles" pour la représentation correcte des ondes déferlantes, et par conséquent, pour assurer

* O. Skovgaard now at Lab. Appl. Math. Phys., Tech. Univ. Denmark, Lyngby, Denmark.

Received December 27, 1977 and revised March 11, 1978

la simulation précise des poussées, ou des contraintes de rayonnement et des courants littoraux correspondants, qu'engendrent les ondes considérées.
Les possibilités essentielles du Système présenté ont été vérifiées par confrontation de celui-ci, d'une part à des résultats analytiques uni et bi-dimensionnels, et d'autre part, à des valeurs déterminées sur des modèles réduits. Les résultats déterminés au moyen du modèle du Système s'accordent bien avec les résultats analytiques et du modèle réduit dans tous les cas. Le Système est déja utilisé pour résoudre des problèmes d'hydraulique d'ordre pratique.
En conclusion, les auteurs exposes les différentes possibilités d'extension du Système à des fins telles que la simulation des mouvements des navires, le calcul des débits solides, et d'autre part, en vue d'un meilleur calcul des écoulements quasi-horizontaux.

Introduction

Engineering practice has long been concerned with the behaviour of short-waves in shallow water. For the design of harbours, for example, a detailed knowledge is required of the directions of propagation and the magnitudes of short waves. These waves attack moles, training works and other structures, they infiltrate through harbour entrances to disturb the waters within the harbour area, both directly and through accumulated, seiching actions, they are instrumental in bringing sediments into suspension while they often induce the currents that transport these sediments to quieter regions of deposition, and they may also adversely influence navigation directly. For the control of coastal erosion, as well, a knowledge of short wave motions and the currents that these induce is of fundamental importance, while there are numerous related applications to the design of terminals and other offshore constructions, and to the planning and control of dredging operations.

As a result of the practical interest in these waves, a considerable effort has been made to predict their behaviour along coasts and in and around harbours, terminals and other engineering works. By far the greatest part of this effort has been directed towards developing techniques for physical modelling and a veritable arsenal has been accumulated for this purpose: e.g. field and laboratory instrumentation, wave generators with their associated data-processing, control and evaluation equipment and a wide range of analytical procedures. A particularly significant development has been the introduction of irregular wave generators, whereby a local wave climate can be established that is far more realistic for most applications than that obtained with periodic waves [e.g. SORENSEN, 1973].

The effort expended in numerical modelling, although considerably less than that used in physical modelling, has still produced several useful techniques. Most of these techniques are based upon analogies between periodic wave propagation and the propagation of light waves, whereby advanced methods of geometric optics can be carried over to the study of the behaviour of periodic water waves. However, the range of water waves that can be so treated is very restricted, being essentially limited to small-amplitude periodic waves, described by linearised wave theories, as outlined in more detail below. Although the behaviour of these waves may often provide some guide to real wave behaviour, the experience of physical modelling suggests that the restrictions of small amplitude and periodicity, so the restrictions essential to the linearised theory, give unacceptable errors in many practical situations. Thus, if numerical simulation is to meet the needs of practice, even if only going so far as is possible with physical simulations, so as to interact with these simulations, numerical methods have to be introduced that are not subject to the linearising restrictions. The present paper describes such methods and shows how they are capable of providing comparable information to that obtainable from physical modelling, so as to be compatible with physical modelling techniques, thus improving the overall reliability of the engineering investigation and design process.

Review of background material

In order to relate the present work and the methods that it proposes to earlier works and their methods, and to provide a basis for some comparisons, a rather extensive review of theories and methodologies is necessary. This review encompasses the four main streams of development that have influenced the present work, namely (1) the analysis of alternative differential formulations, (2) analyses of numerical formulations, (3) recent work on the orders of approximation implicit in differential formulations and the consequences of this work for the numerical formulations and (4) the development and refinement of modelling systems for nearly-horizontal flows. These are considered in order, as follows:

(1) The terminology of "short" waves is that of common engineering practice. In classical hydrodynamics, however, what are here called "short" shallow water waves are viewed as waves whose length is still large compared with the depth of the water in which they propagate and they are, correspondingly, usually subsumed under "long waves". The various theories constructed to describe the behaviour of these waves are generally characterized by an URSELL number:

$$U_* = \frac{\zeta_* L_*^2}{h_*^3}$$

where ζ_* is a measure of the wave amplitude, L_* is a characteristic horizontal length of the surface profile and h_* is a measure of the water depth [URSELL, 1953]. The earliest relevant theory, of

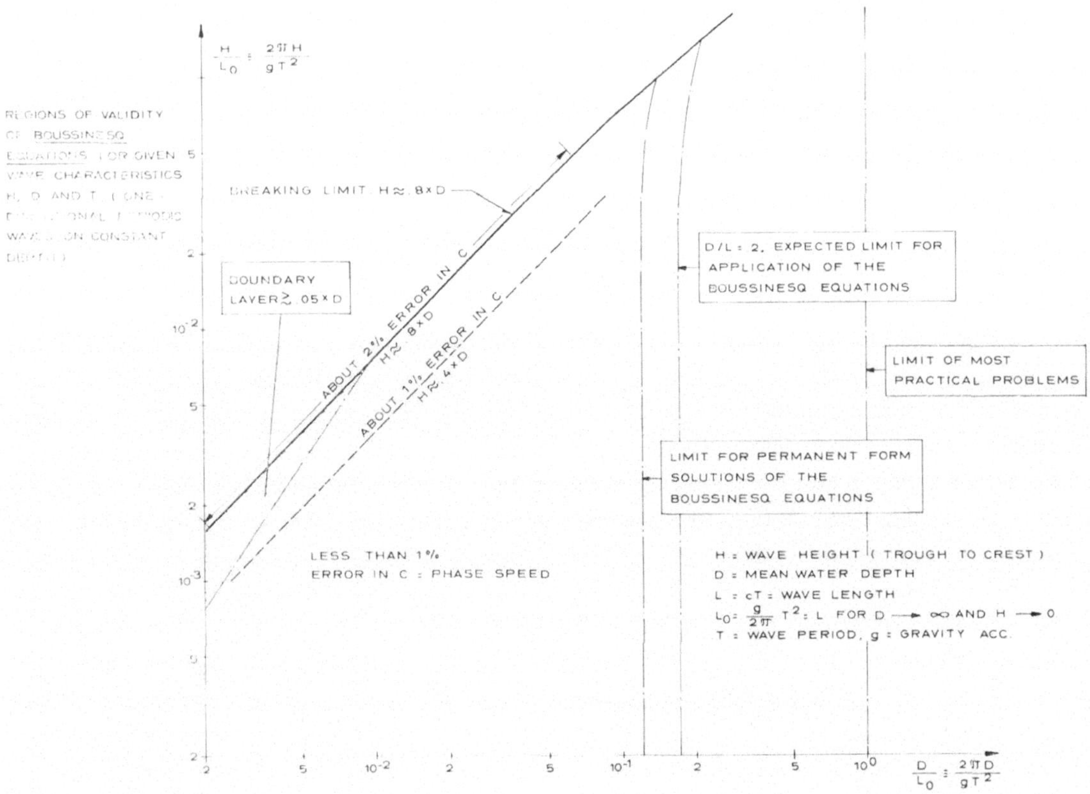

a. Range of application of the mass and BOUSSINESQ equation system (4, 5, 6).

Fig. 1.

a. Domaine de validité du système des équations de masse et de BOUSSINESQ (4, 5, 6).

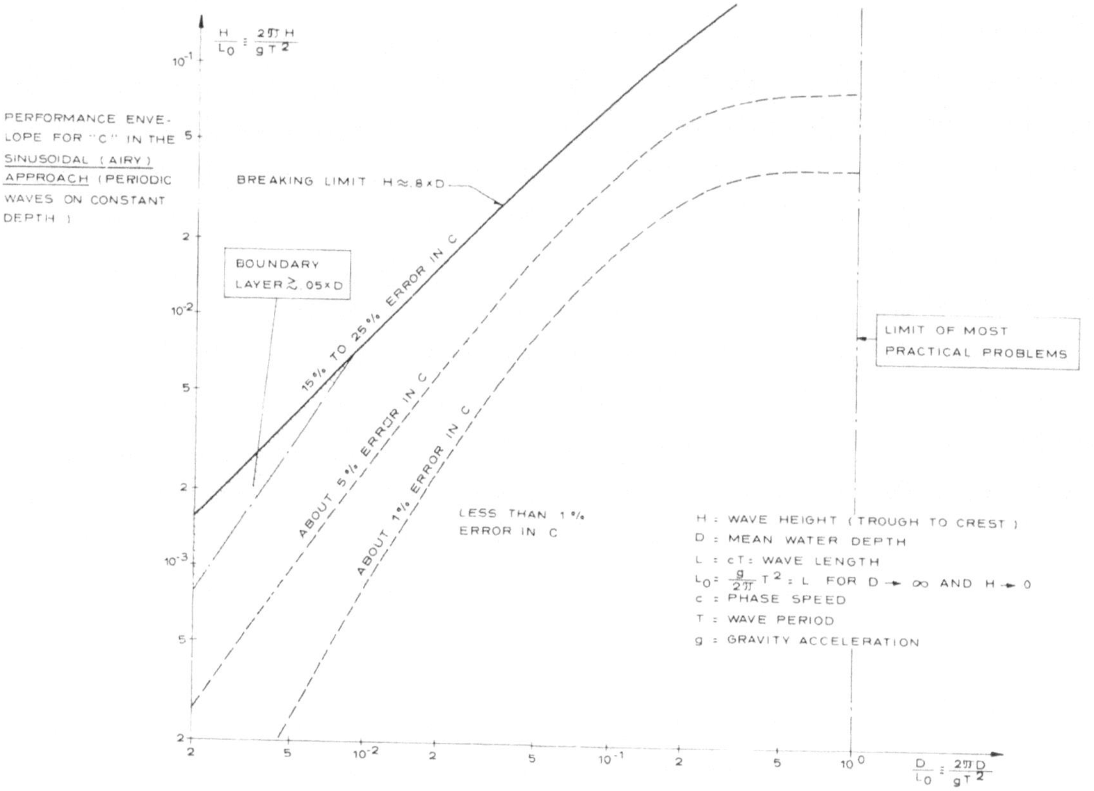

b. The corresponding range of application of the sinusoidal wave approximation.

Fig. 1.

b. Domaine de validité correspondant de l'approximation d'onde sinusoïdale.

AIRY [1845], makes the assumption that the pressure distribution in the vertical is hydrostatic. The resulting waves are non-dispersive, while waves of finite amplitude cannot propagate without change of shape. AIRY's theory corresponds to $U_* \gg 0(1)$. The symmetric case of $U_* \ll 0(1)$ is covered by the linearised shallow water theory of JEFFREYS and JEFFREYS [1946], which also assumes a hydrostatic pressure distribution.

Between the theories of AIRY and JEFFREYS and JEFFREYS there is the theory of BOUSSINESQ [1872, 1877]. In this theory the curvature of streamlines in the vertical plane is described through a vertical velocity the magnitude of which increases linearly from zero at the bed to a maximum at the free surface. In this theory, therefore, the pressure distribution is no longer hydrostatic, but the vertical component of motion can be integrated out of the equations of motion to reduce the three dimensional description to a two-dimensional one. The original theory held for the irrotational motion of an incompressible, homogeneous, inviscid fluid over a horizontal bed, while only solutions for uni-directional wave propagation were considered. URSELL [1953] showed that the BOUSSINESQ theory included the AIRY and JEFFREYS and JEFFREYS theories as special cases. As shown later (e.g. Fig. 1), the BOUSSINESQ theory may in fact be regarded as the most uniformly valid basis for finite amplitude water waves so long as h_*/L_* remains small and breaking does not occur.

When $U_* = 0(1)$ the non-linear, governing equations for a horizontal bottom are of the KORTEWEG-DE VRIES [1895] type, that have unidirectional cnoidal waves as permanent solutions. LIN and CLARK [1959] extended solutions of the BOUSSINESQ theory to include diffracted and reflected plane waves over a horizontal bed while MEI and LE MÉHAUTÉ [1966] extended the theory to account for a slowly varying bathymetry [see also MADSEN and MEI, 1969]. In 1967, PEREGRINE derived the BOUSSINESQ-type equations governing the propagation of arbitrary, long-wave disturbances of small to moderate amplitude over a slowly-varying bathymetry. His equations read

$$\frac{\partial \zeta}{\partial t} + \frac{\partial}{\partial x}[(D+\zeta)u] + \frac{\partial}{\partial y}[(D+\zeta)v] = 0 \tag{1}$$

$$\frac{\partial u}{\partial t} + u\frac{\partial u}{\partial x} + v\frac{\partial u}{\partial y} + g\frac{\partial \zeta}{\partial x} = \tfrac{1}{2}D\left[\frac{\partial^3(Du)}{\partial x^2 \partial t} + \frac{\partial^3(Dv)}{\partial x \partial y \partial t}\right] - \tfrac{1}{6}D^2\left[\frac{\partial^3 u}{\partial x^2 \partial t} + \frac{\partial^3 v}{\partial x \partial y \partial t}\right] \tag{2}$$

$$\frac{\partial v}{\partial t} + u\frac{\partial v}{\partial x} + v\frac{\partial v}{\partial y} + g\frac{\partial \zeta}{\partial y} = \tfrac{1}{2}D\left[\frac{\partial^3(Dv)}{\partial y^2 \partial t} + \frac{\partial^3(Du)}{\partial x \partial y \partial t}\right] - \tfrac{1}{6}D^2\left[\frac{\partial^3 v}{\partial y^2 \partial t} + \frac{\partial^3 u}{\partial x \partial y \partial t}\right] \tag{3}$$

where $z = -D(x,y)$ corresponds to the bed elevation. In terms of depth-integrated velocities, which are alternatively volume flux densities or horizontal-momentum levels divided by the (specific mass) density, these governing equations become,

$$\frac{\partial \zeta}{\partial t} + \frac{\partial p}{\partial x} + \frac{\partial q}{\partial y} = 0 \tag{4}$$

$$\frac{\partial p}{\partial t} + \frac{\partial}{\partial x}\left(\frac{p^2}{h}\right) + \frac{\partial}{\partial y}\left(\frac{pq}{h}\right) + gh\frac{\partial \zeta}{\partial x} = \tfrac{1}{2}Dh\left[\frac{\partial^3}{\partial x^2 \partial t}\left(\frac{Dp}{h}\right) + \frac{\partial^3}{\partial x \partial y \partial t}\left(\frac{Dq}{h}\right)\right] - \tfrac{1}{6}D^2 h\left[\frac{\partial^3}{\partial x^2 \partial t}\left(\frac{p}{h}\right) + \frac{\partial^3}{\partial x \partial y \partial t}\left(\frac{q}{h}\right)\right] \tag{5}$$

$$\frac{\partial q}{\partial t} + \frac{\partial}{\partial y}\left(\frac{q^2}{h}\right) + \frac{\partial}{\partial x}\left(\frac{pq}{h}\right) + gh\frac{\partial \zeta}{\partial y} = \tfrac{1}{2}Dh\left[\frac{\partial^3}{\partial y^2 \partial t}\left(\frac{Dq}{h}\right) + \frac{\partial^3}{\partial x \partial y \partial t}\left(\frac{Dp}{h}\right)\right] - \tfrac{1}{6}D^2 h\left[\frac{\partial^3}{\partial y^2 \partial t}\left(\frac{q}{h}\right) + \frac{\partial^3}{\partial x \partial y \partial t}\left(\frac{p}{h}\right)\right] \tag{6}$$

where $h \equiv D+\zeta$, $p = uh$ and $q = vh$. It is equations (4, 5, 6), augmented with other terms, such as those that account for a reduced flow area (permeable breakwater) and corresponding resistance terms (equations 11, 12, 13), and also wave resistance terms, that are solved by the present System. As described in detail later, equations (4, 5, 6) do not have a unique form, but can be rewritten by using relations corresponding to equations of lower order. The range of applications of equations (4, 5, 6) is shown in fig. 1a. This figure is based upon a comparison between one-dimensional Cnoidal-wave solutions of (4, 5, 6) and the permanent wave solutions of DEAN (1974].

The theory constituted by equations (4, 5, 6) with their corresponding classes of solution functions evidently includes sub-theories corresponding to simpler linearised equations, such as those corresponding to the classical linearized long wave equations for periodic waves of small amplitude ($U_* \ll 0(1)$):

$$\frac{\partial}{\partial x}\left(D\frac{\partial \zeta^*}{\partial x}\right) + \frac{\partial}{\partial y}\left(D\frac{\partial \zeta^*}{\partial y}\right) + Dk^2\zeta^* = 0 \tag{7}$$

In (7) $\zeta^* = \zeta^*(x, y)$ is a complex wave amplitude so that $\zeta^* \exp(-i\omega t)$ is the instantaneous elevation of the water surface, ω is the angular frequency and k is the wave number. Equation (7) can also be regarded as a special non-dispersive version of BERKHOFF's [1973, 1976] wave equation for intermediate depths and mild slopes [see also JONSSON et al, 1976]:

$$\frac{\partial}{\partial x}\left(cc_g \frac{\partial \zeta^*}{\partial x}\right) + \frac{\partial}{\partial y}\left(cc_g \frac{\partial \zeta^*}{\partial y}\right) + cc_g k^2 \zeta^* = 0 \tag{8}$$

where c and c_g are the phase and group celerities respectively of classical small-amplitude sinusoidal wave theory. BERKHOFF's [1976] linear wave equation also assumes periodic waves with $U_* \ll 0(1)$ but it is not restricted to hydrodynamically "long" waves. By virtue of the linearisation, arbitrary input waves can also be built up from Fourier components in BERKHOFF's theory and their solutions obtained by superimposing the corresponding component solutions, but this leads to considerable complication over and above the restrictive conditions. The use of the linear long wave equation appears to be implicit in the solution method of CHEN and MEI [1974, 1976] while the use of this equation or BERKHOFF's linear equation appears to be implicit in the methods of ZIENKIEWICZ and BETTESS [1976] and BETTESS and BETTESS [1976].

BERKHOFF's wave equation describes combined diffraction-refraction, but if reflections can be ignored it is possible to reduce this equation so as to address depth refraction only, without diffraction, reflection or scattering. This equation then provides the well-known eiconal equation of geometric optics [e.g. BORN and WOLF, 1975. See also BERKHOFF, 1976].

The extension of the solution of the general equations (4, 5, 6) to allow for wave breaking, by the introduction of a suitable (non-linear) dissipative interface [ABBOTT, 1974] allows the simultaneous computation of wave-induced currents [LONGUET HIGGINS and STEWART, 1960, 1962; LUNDGREN, 1963], and thence a rational approach to sediment transport within the breaker zone [e.g. JONSSON et al, 1975]. This use of the present code involves so much other material, however, that it can best form the subject of a separate communication. Similarly, the computation of ship motions simultaneously with short-period wave motions, already realised with the present method, will be treated in a separate communication.

(2) The second point of departure for the present work was the numerical solution of Boussinesq-type equations by RODENHUIS [1966], described with a summary of earlier work by ABBOTT and RODENHUIS [1972]. This work showed the extreme sensitivity of the solutions of the equations to numerical error influences and the necessity, for practical applications, of a high-accuracy difference scheme. It also indicated, however, that an excellent agreement could be obtained between solutions of difference forms of the Boussinesq equations and real-world finite amplitude waves if sufficient numerical accuracy could be obtained.

(3) The third point of departure was the systematic study, in mathematical fluid mechanics, of the orders of approximation implicit in differential formulations of the present problem. The development of this study can be traced through the papers of FRIEDRICHS [1948], STOKER [1955], LONG [1964], RODENHUIS [1966], MEYER [1967], PEREGRINE [1967, 1972, 1974], BENJAMIN, BONA and MAHONY [1972] and BONA and SMITH [1976]. The most significant result of this development, from the present point of view, was the observation [originally, it seems, of LONG, 1964] that the higher-order terms of BOUSSINESQ-type equations could be rewritten by using the linearized wave equations

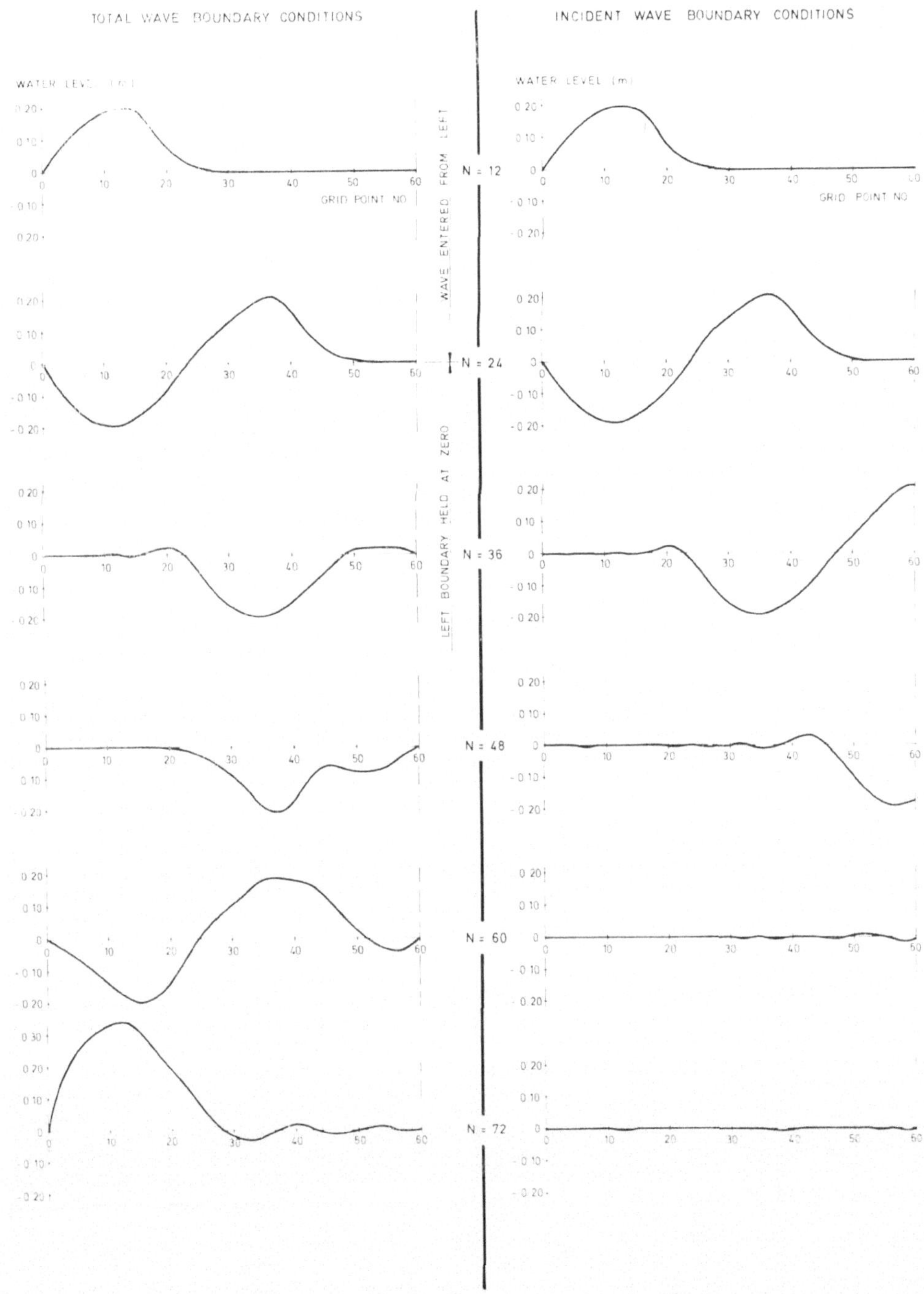

Fig. 2a.

Rig test of the wave-filtering boundary module, or SOMMERFELD boundary module, showing how:

a. As opposed to *allowing* a wave to be totally reflected (left) it allows it to be totally absorbed (right).

Essais en laboratoire du module filtre-d'onde pariétal (ou module pariétal de SOMMERFELD), mettant en évidence:

a. *Comment il permet* à une vague d'être completement amortie (droite) ou au contraire d'être completement réfléchie (gauche).

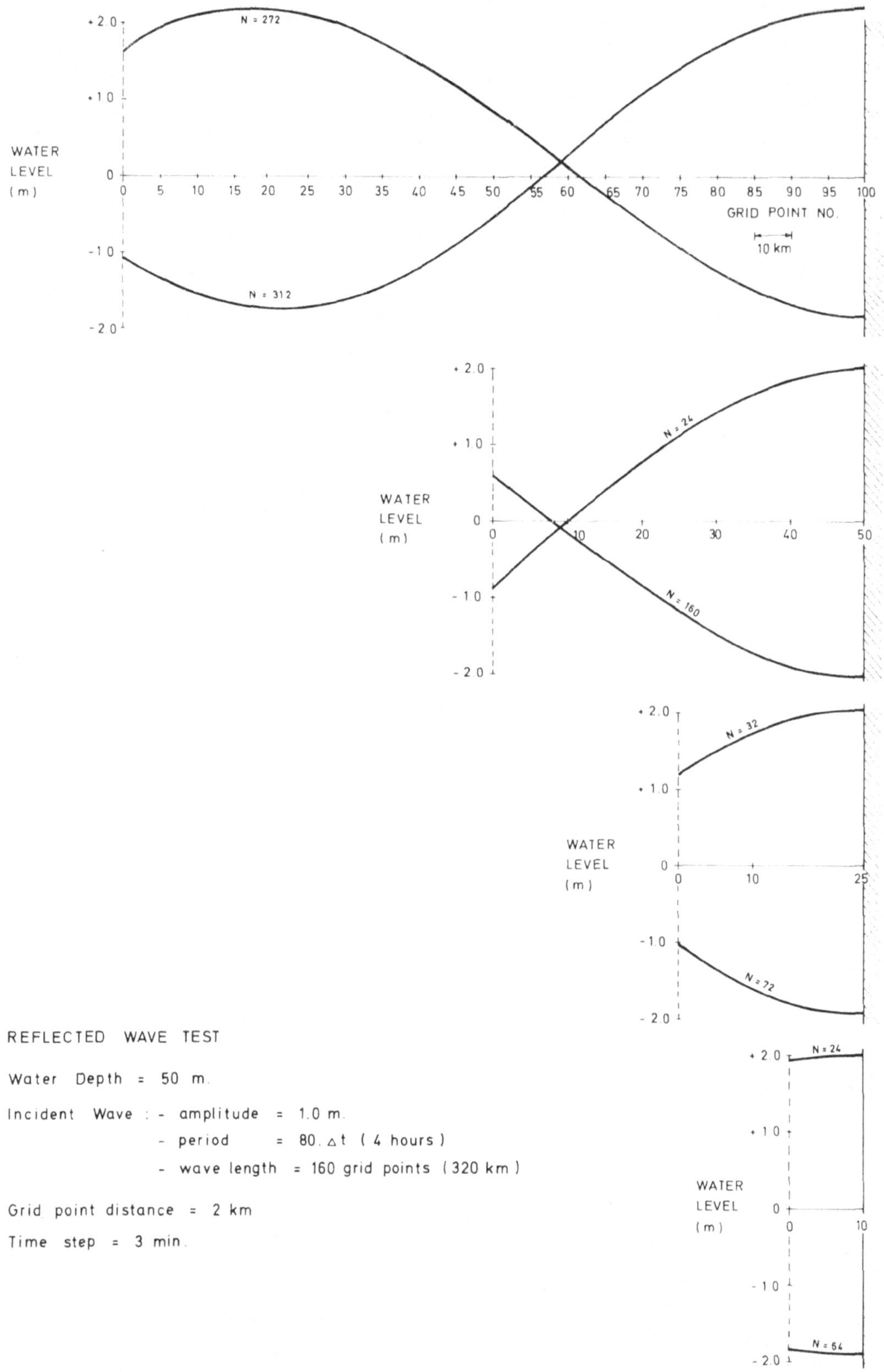

Fig. 2b.

b. It enables the model to be cut at any point and the solution to be generated on one side of the cut and

b. la possibilité de couper le modèle en tout point, et d'élaborer la solution d'un côté de la coupure,

$$\frac{\partial u}{\partial t} + g\frac{\partial \zeta}{\partial x} = 0 \tag{9}$$

$$\frac{\partial \zeta}{\partial t} + D\frac{\partial u}{\partial x} = 0 \tag{10}$$

without any loss of generality or accuracy. The parallel study of consistency of difference schemes then suggested that the same, much simplified substitution could be used at the level of the difference equations, so as to transform the high-accuracy but algorithmically intractable difference forms of the Boussinesq equations (4, 5, 6) into algorithmically efficient difference schemes, as exemplified in Appendix 1.

(4) The fourth development that was of decisive importance for the present work was the construction and refinement of the modelling system for nearly-horizontal flows, System 21, "Jupiter" [e.g. ABBOTT, DAMSGAARD and RODENHUIS, 1973; ABBOTT, 1976, 1, 2, 3]. For reasons quite

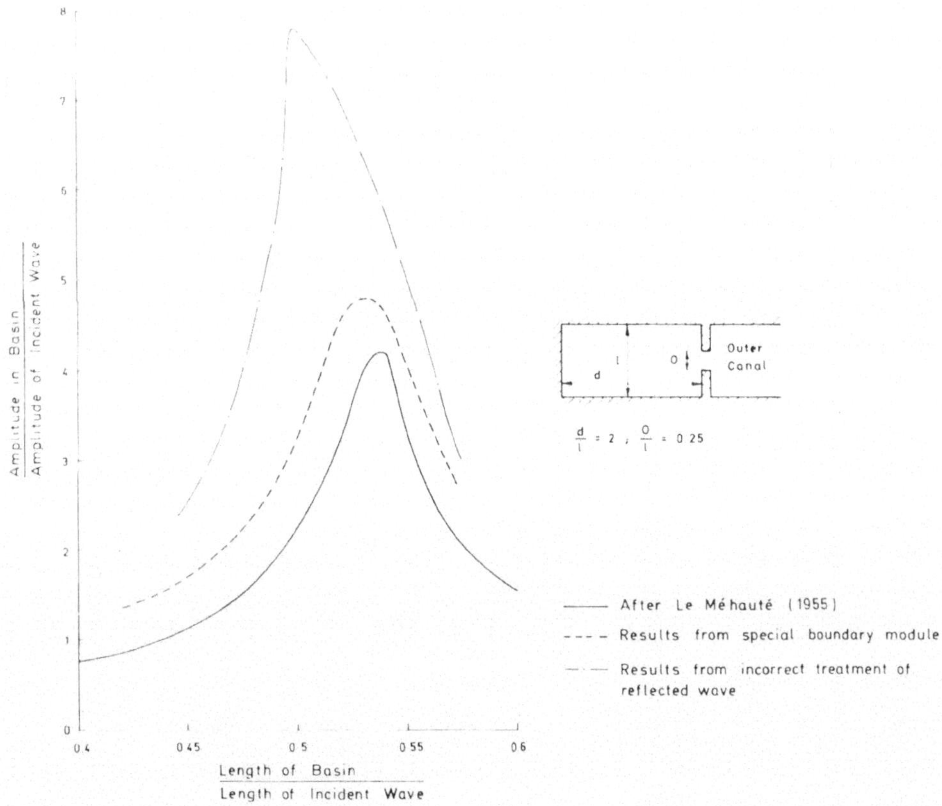

Fig 2c.

c. how the use of the module at the harbour entrance provides a better comparison with analysis and experiments in a harbour subject to nearly-horizontal flow seiching, as compared to the use of raw elevation data at the harbour entrance.

c. comment l'utilisation de module à l'entrée du port permet une meilleure confrontation des résultats à ceux déterminés par analyse ou sur modèle réduit, pour le cas d'un port affecté de balancements du type seiche de l'écoulement dans le sens quasi-horizontal, que si l'on ne tenait compte que des cotes d'altitude brutes à l'entrée du port.

unconnected with the present work, this had been brought to third-order accuracy in its convective components so that a relatively minor effort was required to bring its other mass-conservation and impulse-momentum components, both differentials and their coefficients, to the required level of accuracy, using the elimination techniques of the type described in Appendix 1. Moreover, the system speed and reliability had been increased to such an extent that it appeared that the new short-wave variant, even though necessarily working with a fine mesh, could still provide a useful instrument for engineering practice. In particular, the existing variant was both non-linearly and linearly stable for all physically possible boundary conditions [Daubert and Graffe, 1967], and these properties carried over to the new variant.

Among the many existing support and service programs that were carried over to the new variant, the most important was a filtering boundary module, that allowed the filtering-out of incoming wave trains from outgoing wave trains in any wave record, using a System-built model of the wave-recording situation. This then allowed models of water bodies containing post-measurement constructions to be cut out along any convenient line, with the incoming wave trains being maintained as measured and the new reflected wave trains being passed-out through the boundary. Two of the many test-rig trials of this filtering module are illustrated in Fig. 2. Among other applications, this module makes possible the maintenance of a steady realistic sea state within any region regardless of the objects (harbours, ships, terminals, platforms etc.) placed within that region. Since the first formulation of such a boundary condition was given by Sommerfeld (1964), this module is called the "Sommerfeld boundary module".

Further formulations

The differential equations solved in open water regions have already been given as (4, 5, 6) above. In the case that the flow area is restricted, as occurs especially at permeable breakwaters, these equations generalise to

$$n\frac{\partial \zeta}{\partial t}+\frac{\partial p}{\partial x}+\frac{\partial q}{\partial y}=0 \tag{11}$$

$$n\frac{\partial p}{\partial t}+\frac{\partial}{\partial x}\left(\frac{p^2}{h}\right)+\frac{\partial}{\partial y}\left(\frac{pq}{h}\right)+gh\frac{\partial \zeta}{\partial x}n^2+\frac{nhz}{3}\frac{\partial^3 p}{\partial x^2 \partial t}+\frac{nhz}{3}\frac{\partial^3 q}{\partial x \partial y \partial t}+$$

$$+(1-n)^3\frac{\alpha v}{d^2}p+\frac{(1-n)}{n}\frac{\beta}{d}\cdot p\sqrt{\left(\frac{p}{h}\right)^2+\left(\frac{q}{h}\right)^2}=0 \tag{12}$$

$$n\frac{\partial q}{\partial t}+\frac{\partial}{\partial y}\left(\frac{q^2}{h}\right)+\frac{\partial}{\partial x}\left(\frac{pq}{h}\right)+gh\frac{\partial \zeta}{\partial y}n^2+\frac{nhz}{3}\frac{\partial^3 q}{\partial y^2 \partial t}+\frac{nhz}{3}\frac{\partial^3 p}{\partial x \partial y \partial t}+$$

$$+(1-n)^3\frac{\alpha v}{d^2}q+\frac{(1-n)}{n}\frac{\beta}{d}\cdot q\sqrt{\left(\frac{p}{h}\right)^2+\left(\frac{q}{h}\right)^2}=0 \tag{13}$$

with n the pore volume. (Here z is held constant for simplicity of exposition.)

The laminar-flow resistance term is introduced in order to provide realistic simulations of reduced-scale, physical model tests. The equations are further augmented by wave resistance terms of the type developed by Jonsson [1967]. Conservations forms and distributional formulations are further necessary when treating breaking waves for wave-induced currents [WHITHAM, 1974; ABBOTT, 1974].

The construction of a high-accuracy (at least third-order accurate) difference approximation to (4, 5, 6) or its generalisation (11, 12, 13), necessitates the use of a four-stage difference scheme with memory components [ABBOTT, 1978]. The principle of the technique used is described in the Appendix for a one-dimensional descent based upon the implicit scheme of PREISSMANN [LIGGETT and CUNGE, 1975]. In this case, which is probably the simplest of all, it is seen how the zero-order equation system (9, 10) is used to provide a scheme that requires only one time level of memory more than is required by the nearly-horizontal flow system and which remains very fast in operation.

Performance envelope testing

A range of descriptive parameters within which a system provides usable results, together with error estimates over this range, consistutes a "performance envelope" for the system. The envelope for the differential basis of the Mark 8 has been described in Fig. 1a, while Fig. 1b provides an envelope for the differential basis of models restricted to the linear-approximation. The envelope for the difference scheme approximation is usually more restricted than is the envelope of the differential forms themselves, because of the errors inherent in numerical approximations, while correspondingly it contains additional numerical parameters over and above the physical parameters of the differential form. The objective here was to extend the envelope of the difference form to fill-out so far as possible the envelope of the differential form, schematized in Fig. 1a.

In view of the marked non-linearity of the Boussinesq equations, which must carry over to their numerical approximations, the definition of the performance envelope necessitated a very extensive program of testing.

The tests were conducted in terms of the dimensionless dependent variables:

$$P = \frac{\text{amplitude of waves generated by the difference equations}}{\text{amplitude of waves generated by the differential equations}}, \tag{14}$$

a measure of the amplification error, and

$$Q = \frac{\text{celerity of waves generated by the difference equations}}{\text{celerity of waves generated by the differential equations}}, \tag{15}$$

a measure of the phase error

where the waves generated by the differential equations are cnoidal waves in all performance envelope test cases. The dimensionless independent variables were:

$$R = \frac{\text{wave amplitude}}{\text{still water depth}} \tag{16}$$

$$S = \frac{\text{wave length}}{\text{still water depth}} \tag{17}$$

augmented by the numerical parameters

$$Cr = c_{\pm}\left(\frac{\Delta t}{\Delta x}\right) \tag{18}$$

with $c_{\pm}$ the dynamic wave celerities generated by the differential equations (4, 5, 6) or equivalently (11, 12, 13) and

$$N_x = \frac{2l}{k\Delta x}, \qquad N_t = \frac{N_x}{Cr} \tag{19}$$

the number of computational points per wave length in space and time respectively, at dimensionless wave number k.

Fig. 3a shows a test at a low ratio of wave amplitude to wave length, R, and low Courant number, Cr, in which a very small number of points per wave length ($N_x = 8$ to 12) are seen to be sufficient to provide $P \approx 1$ and $Q \approx 1$. Fig. 3b shows how, as the ratio of wave amplitude to wave length increases, so a significant phase error, measured by Q, appears, even for $N_x, \approx N_t, = 11$. Increasing $N_x, \approx N_t,$ to 22 removes the phase error in this case, as shown in Fig. 3c. This test illustrates the interaction between physical and numerical parameters in the performance envelope, corresponding to the marked non-linearity of the governing equations.

Two schematizations of the performance envelope for uni-directional propagation, based on some 120 test runs, are shown in Fig. 4. It is seen that, for $Cr \approx 1$, an acceptable accuracy is obtained with values of $N_x(\approx N_t)$ as low as 6.

Since much the greater part of the energy of irregular waves is usually distributed over wave

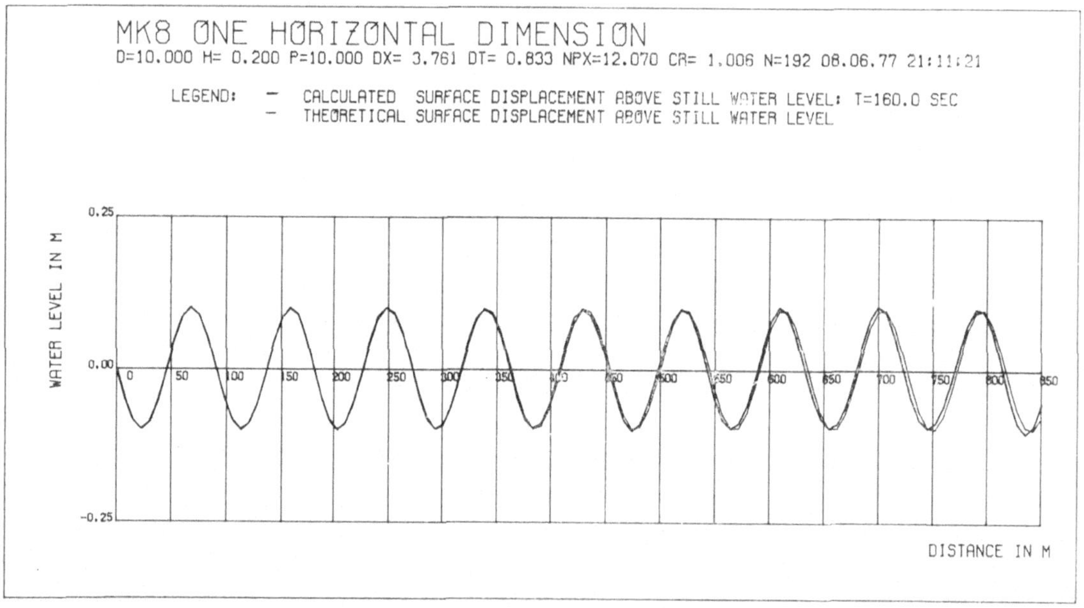

Fig. 3.

a. Comparison between a numerically computed wave train and the corresponding analytical (Cnoidal) wave train at $Fr \approx 1$ for a low ratio of wave amplitude to wave length but small N_x.

a. Confrontation d'un train d'ondes, déterminé par le calcul, au train analytique (cnoïdal) correspondant, avec $Fr \approx 1$ et une faible valeur du rapport de l'amplitude par la longeur de l'onde, mais aussi de N_x.

component periods greater than 4 seconds, at least for the applications envisaged for the present system, Fig. 4 indicates that time steps of about 1 second can be used in many practical situations, giving distance steps of the order of 10 m. These parameters are such as to assure the viability of the system in engineering practice.

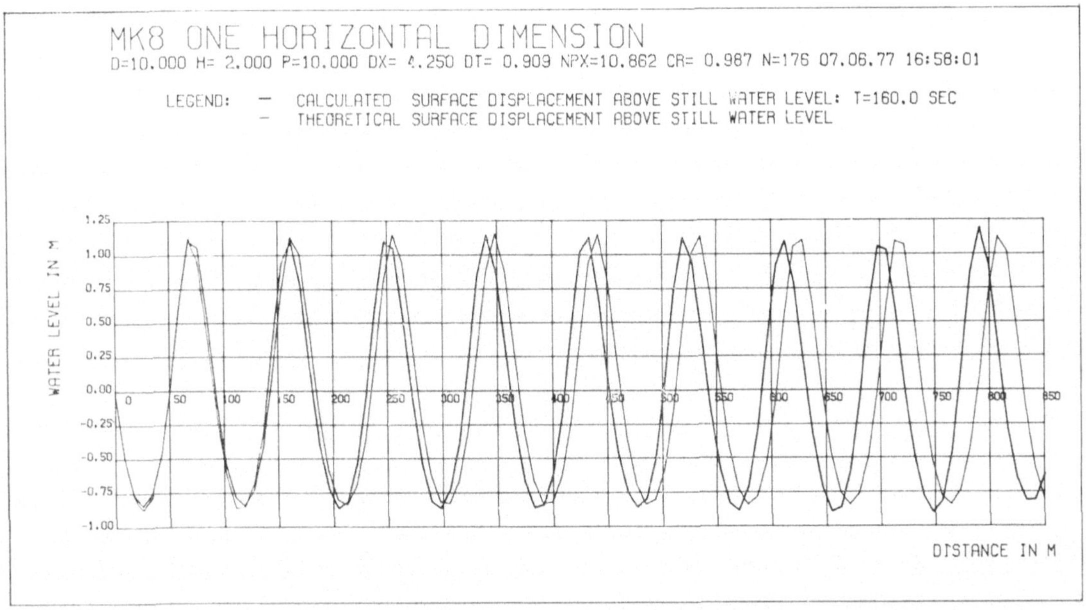

b. The comparison taken at a considerably higher ratio of wave amplitude to wave length, showing how a phase error appears even at $N_x(\approx N_t) = 11$.

b. Cette même confrontation, mais compte tenu d'un valeur très nettement plus élevée du rapport amplitude/longueur, mettant en évidence l'apparition d'un déphasage, même avec $N_x(\approx N_t) = 11$.

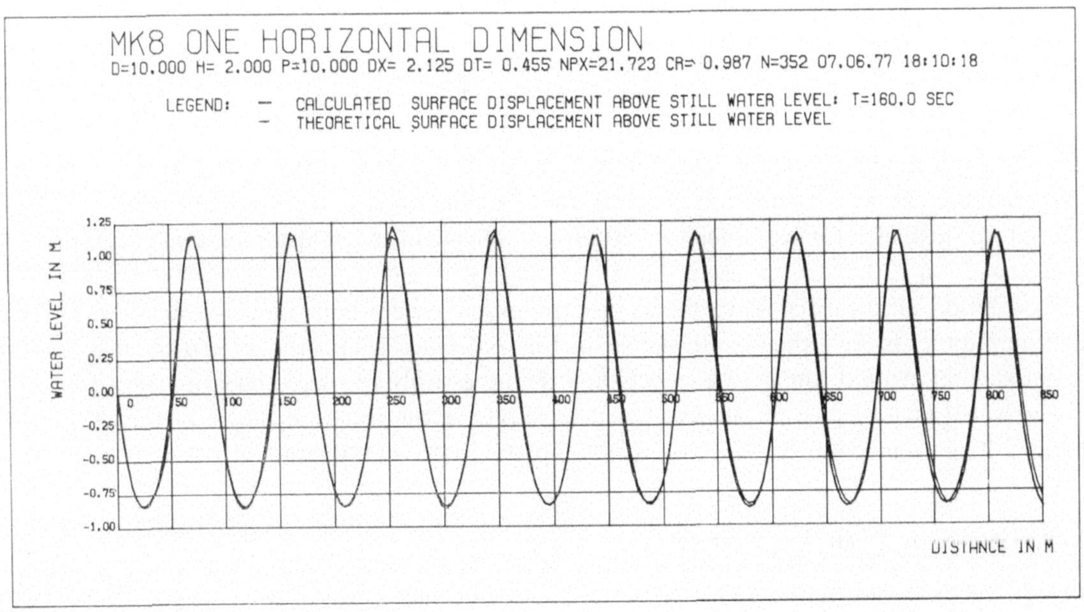

c. When $N_x(\approx N_t) = 22$, this phase error disappears again.

c. Ce déphasage disparaît de nouveau si $N_x(\approx N_t) = 22$.

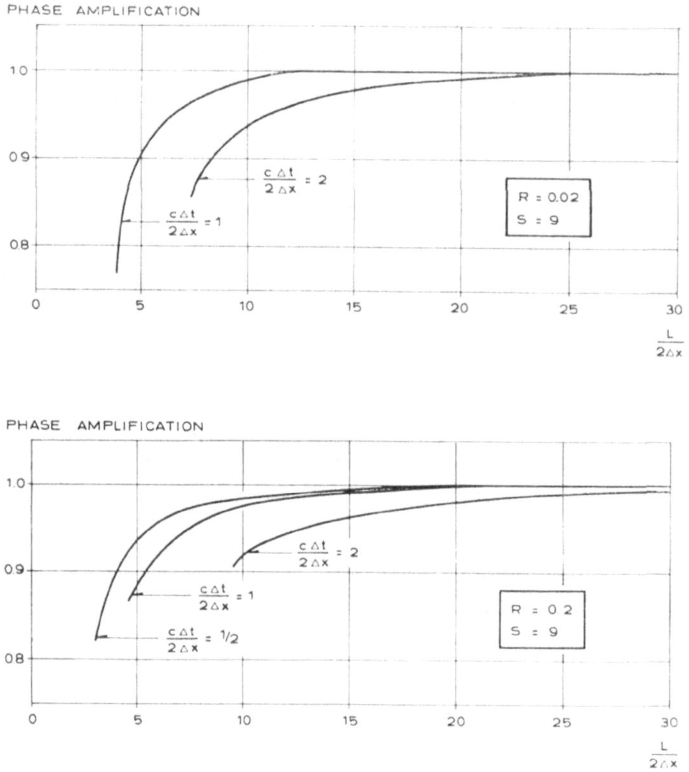

Fig. 4.

Experimental phase representation or"portrait" of the performance envelope, based upon test runs with the System in a onr-dimensional mode.

Représentation de phase expérimentale (ou „portrait") de l'enveloppe des caractéristiques, obtenue à partir des résultats déterminés en exploitant le Système en mode uni-dimensionnel.

One-dimensional tests were also made for partial reflections using a generalisation of the SOMMERFELD boundary module of the Mark 6. Some results are shown in Fig. 5. Although these were generally very successful, the field test described below indicated that the physical mechanism of reflection and transmission could be introduced into the system so accurately that the SOMMERFELD boundary module could be relegated to external boundaries of the model domain.

Whereas a great deal of reference material was available for one-dimensional testing, very little material was available for two dimensional testing, outside of the field testing that is described in the next section. Comparisons were, however, made with the classical SOMMERFELD diffraction (1896) solution of linearised wave theory, by running a system-built model with waves of very small amplitude, to give the results shown in Fig. 6a. Once again, the agreement between the present general method run so restrictively and the necessarily restricted linearised theory is seen to be very satisfactory. Fig. 6b shows a perspective plot of the situation shown in plan in Fig. 6a, whereby the partial standing waves, especially, appear in a visually familiar form.

Comparisons with physical experiments

After the method had been tested over the whole operational range of wave heights and lengths shown in Fig. 4, tests were made on shoaling waves comparing computed results with the experimental results reported by MADSEN and MEI [1969]. Some typical results are shown in Fig. 7.

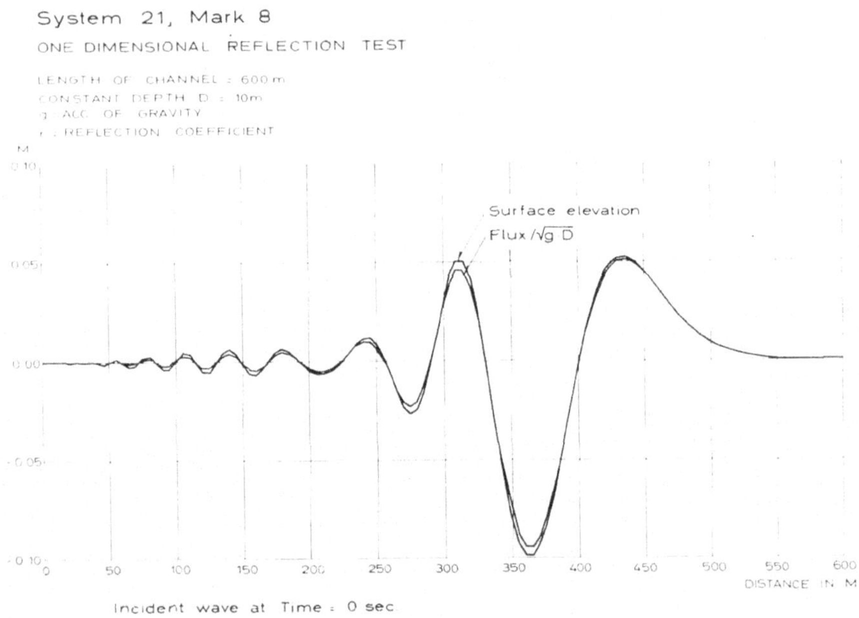

a.

Test of total, zero and partial wave energy reflections obtained using the SOMMERFELD boundary module.

Fig. 5.

Résultats d'une étude des réflexions totale, nulle et partielle de l'énergie de l'onde, déterminées au moyen du module pariétal de SOMMERFELD.

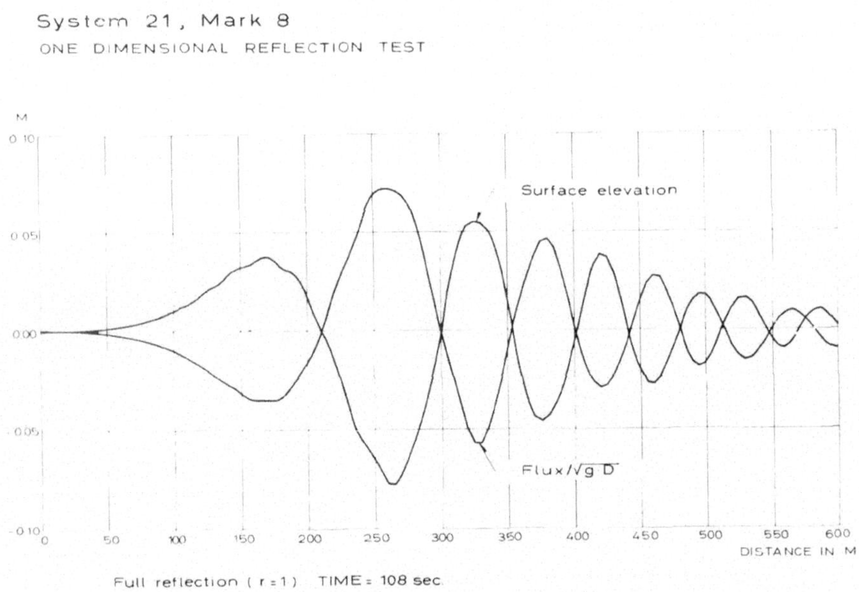

b.

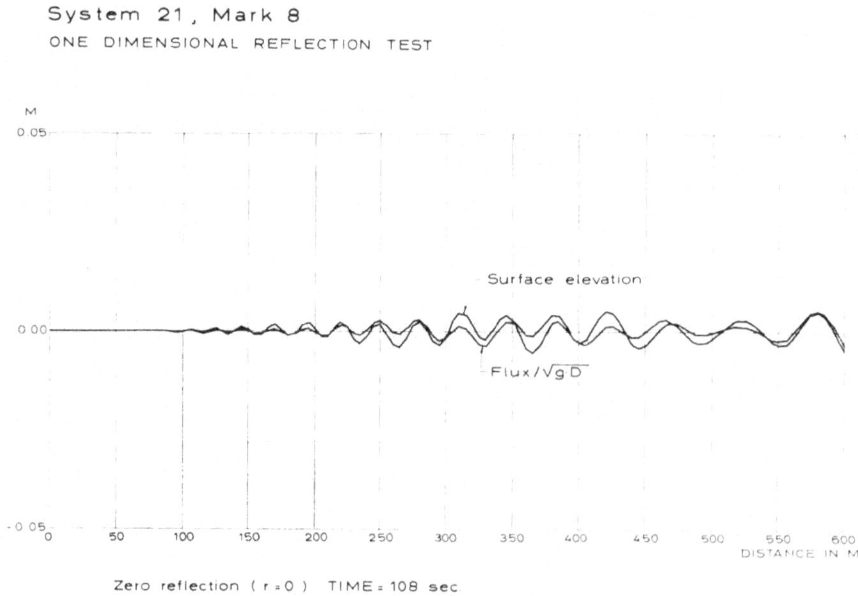
System 21, Mark 8
ONE DIMENSIONAL REFLECTION TEST
M
0.05
0.00
-0.05
0 50 100 150 200 250 300 350 400 450 500 550 600
DISTANCE IN M
Surface elevation
Flux/√gD
Zero reflection (r = 0) TIME = 108 sec

c.

Fig. 5.

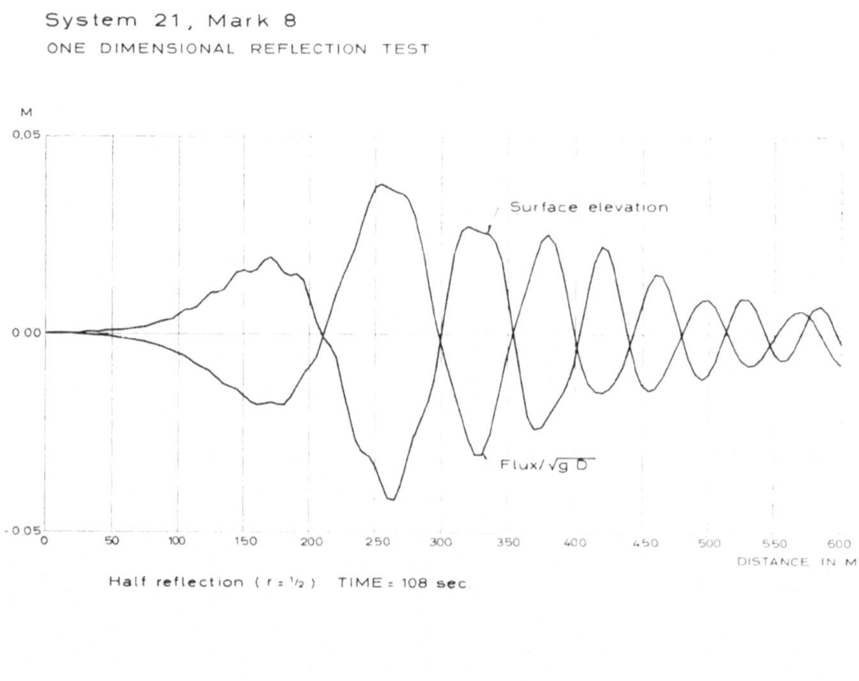
System 21, Mark 8
ONE DIMENSIONAL REFLECTION TEST
M
0.05
0.00
-0.05
0 50 100 150 200 250 300 350 400 450 500 550 600
DISTANCE IN M
Surface elevation
Flux/√gD
Half reflection (r = ½) TIME = 108 sec

d.

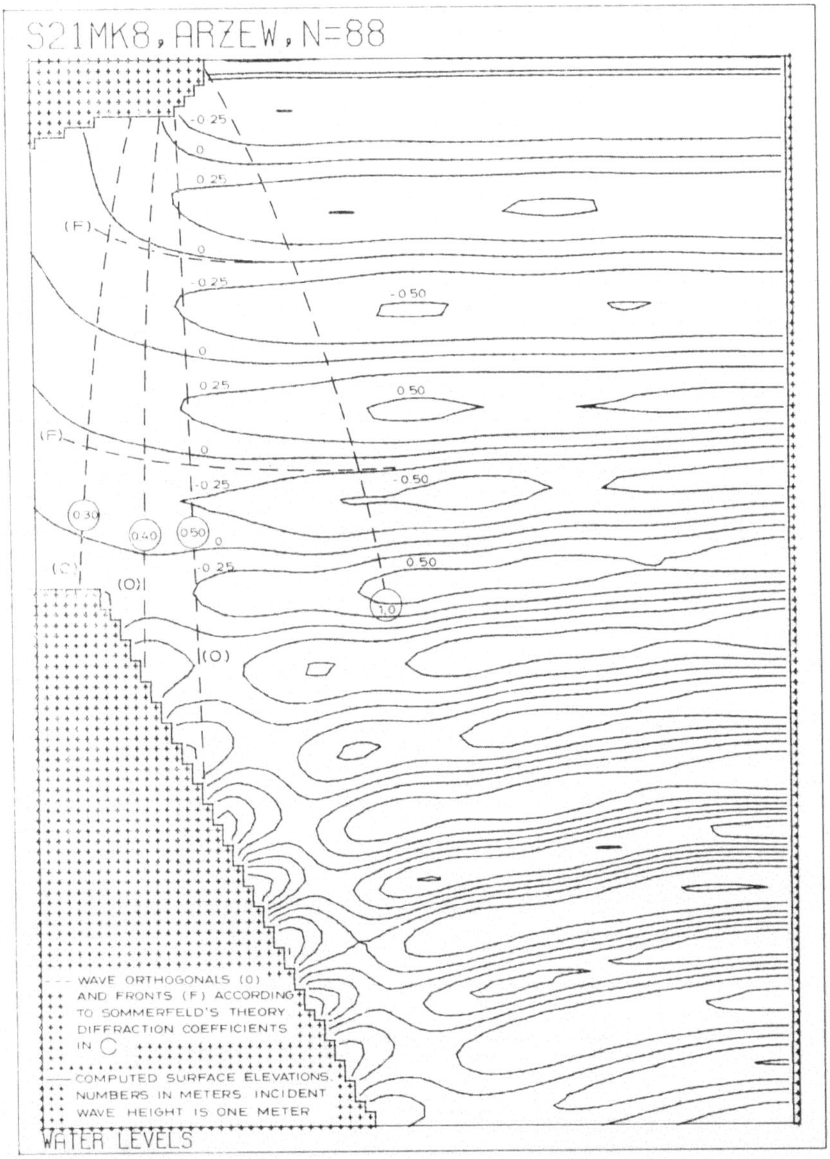

Fig. 6.

Comparison of two-dimensional computations of pure diffraction with the analytical results of SOMMERFELD.
a. Plan view of elevation contours;

Comparaison des résultats des calculs bi-dimensionnels de la diffraction pure, et des résultats analytiques déterminés par SOMMERFELD.
a. Représentation en plan des „courbes de niveau"

Agreement between computed and mean measured results were maintained to within 5% in elevation, over the whole test range, which represents a higher accuracy than that to be expected in the individual physical tests themselves.

Tests were next made of reflection and transmission of waves through permeable breakwaters, using a scheme generalised to be consistent with (11, 12, 13) throughout the domain of the computation, with porosity set to its physical value in the breakwater and set to unity in the open water, so that in open water this scheme degenerates to be consistent with (4, 5, 6). Simulations were

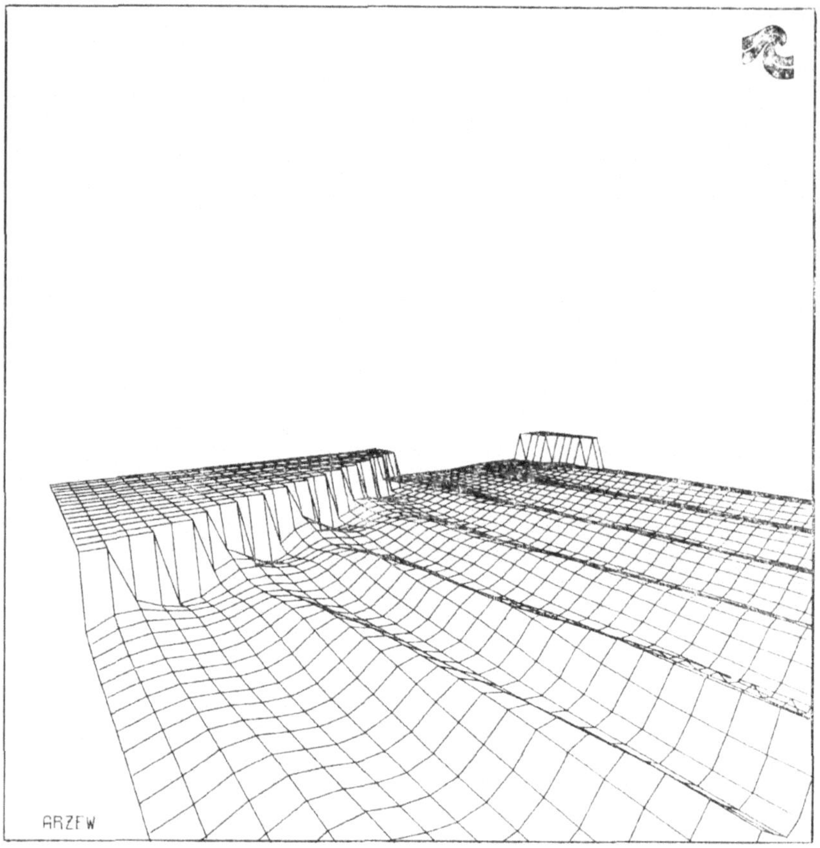

b. perspective plot of the situation shown in a.

b. Représentation en perspective de la configuration de la figure a.

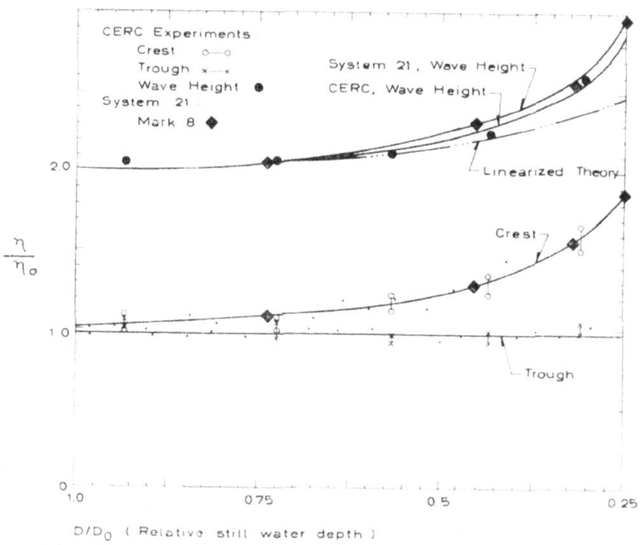

Comparison of numerical computations of shoaling waves obtained using the System in one-dimensional mode, as compared with the experimental results of MADSEN and MEI (1969).

Fig. 7.

Comparaison des résultats de calcul des ondes se propageant sur hauts-fonds, déterminés en exploitant le Système en mode uni-dimensionnel, et des résultats expérimentaux déterminés par MADSEN et MEI (1969).

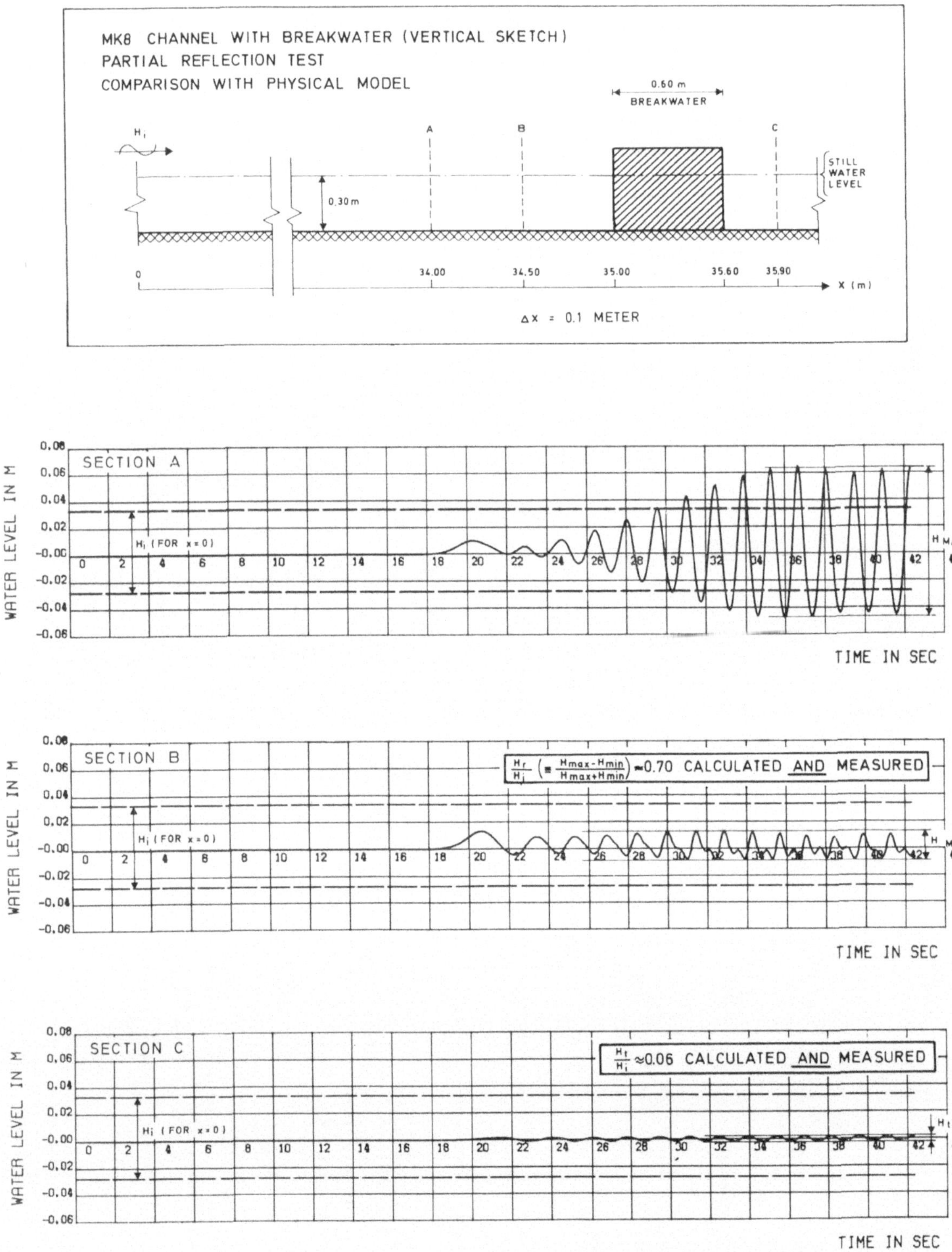

Water elevation for wave transmission through a permeable breakwater, with partial reflection, obtained using the System in one-dimensional mode. The experimental results obtained by the U.S. Army Engineer Waterways Experimental Station, Vicksburg, are used for the comparison.

Fig. 8.

Surélévation du plan d'eau, correspondant à la transmission de l'onde à travers une digue perméable en présence de réflexions partielles, déterminée en exploitant le Système en mode uni-dimensionnel.
Comparaison avec les résultats expérimentaux déterminés dans les laboratoires du U.S. Army Engineers Waterways Experimental Station, Vicksburg.

a. Aerial photograph of harbour at Hanstholm.

Fig. 9.

a. Photographie aérienne du port de Hanstholm.

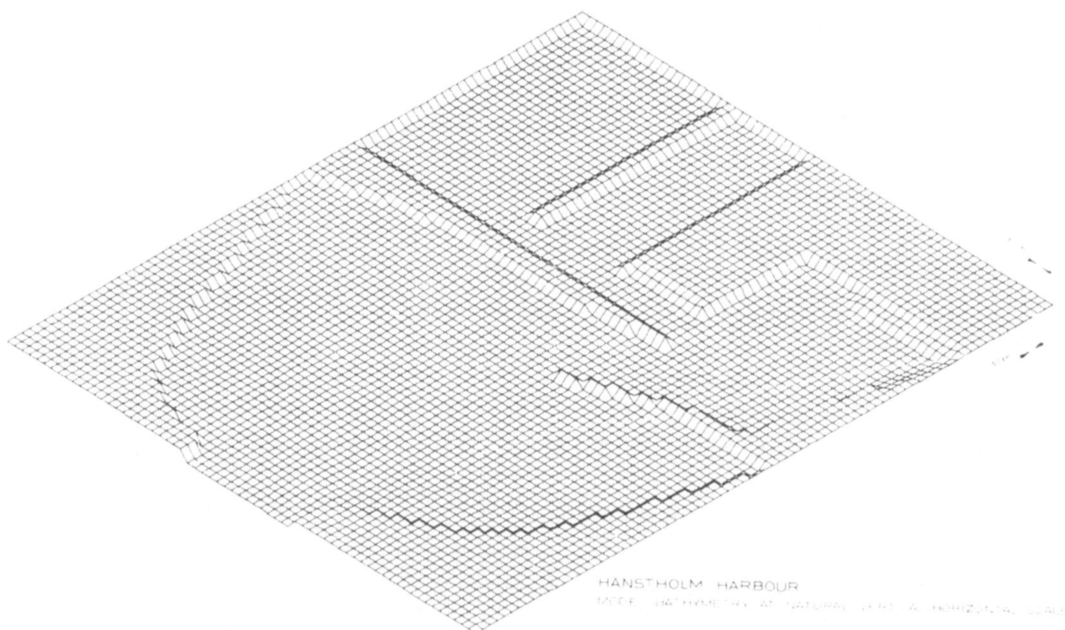

b. the numerical model bathymetry at natural vertical-horizontal scale.

b. Bathymétrie numérique, déterminée sur un modèle sans distorsion d'échelle.

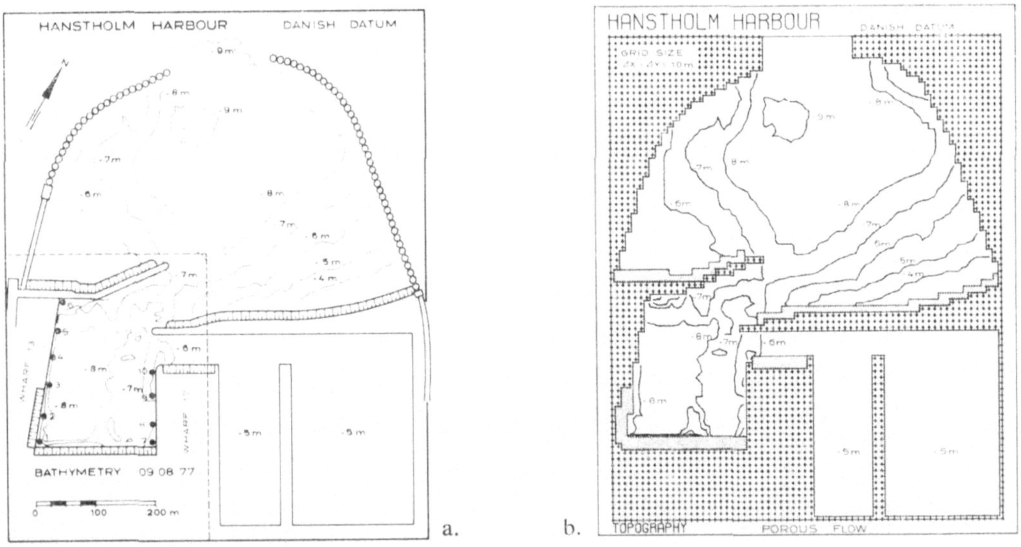

Fig. 10.

Contours of the Hanstholm harbour as
a. used in the physical model tests and
b. used in the numerical model tests. The porous areas are shown in both a and b.

Courbes de niveau bathymétriques du port de Hanstholm ,prises en compte:
a. pour les essais sur modèle réduit,
b. pour les essais sur modèle numérique.

made of physical tests reported by the U.S. Army Engineer Waterways Experimental Station, Vicksburg [Keulegan, 1973]. As shown in Fig. 8, the agreement between the computed and experimental results is within a few percent, and again within the range of the experimental error.

With the main capabilities of the System proven, comparisons could be made between results of the system running in its full two-dimensional mode and results obtained in a physical model of a real harbour. For the purpose of comparison, the Danish harbour of Hanstholm was selected, this having been exceptionally well-documented and physical-model studied. The layout of the harbour is shown in Fig. 9, while Fig. 10a and b show the physical model bathymetry and mathematical model bathymetry respectively.

Physical model tests had been made both with periodic waves, over the full range of observed wave periods, and also with irregular (field-measured) waves. As is now well-established [e.g. Sorensen, 1973], periodic wave tests usually give exceedingly unreliable results: small variations in incident wave period give considerable changes in amplifications of the incident wave at fixed measuring points while these amplifications can vary rapidly with the location of a measuring point. For this and other reasons, [Sorensen, 1973], it is now standard practice to run short wave models with irregular waves, generated using irregular wave generators that are themselves programmed from field-measured time series of water surface elevations.

Comparisons between the mathematical and physical models were first made for the case of periodic wave inputs. In view of the rapid variation of amplification with distance, each measuring point in the physical model was associated with its three most closely adjacent measuring points in the mathematical model. Comparisons were made for 12 seconds and 17 second periodic waves, with the results shown in Fig. 11. It is seen that the agreement is highly satisfactory, especially when account is taken of the uncertainties in the physical model testing. By way of providing a demonstration of these uncertainties, Fig. 12a shows one of the elevation time-series plots used to obtain Fig. 11, where the variation in amplification at three immediately adjacent points is easily seen. Similarly, Fig. 12b shows the effect of merely increasing the period from the 12 seconds of

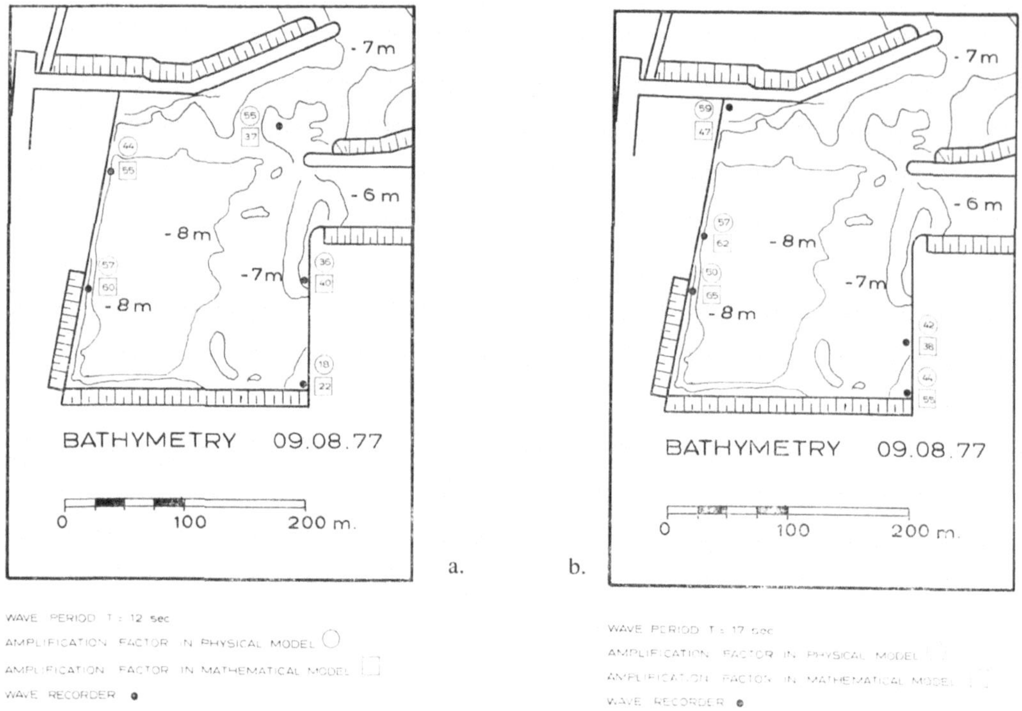

Fig. 11.

Comparisons between amplification factors obtained in the first inner harbour in the physical model with those computed using the System, for 12s and 17s periodic waves of 10 cm amplitude applied at the harbour entrance. With this small amplitude, the behaviour is almost linear.

Les zones poreuses sont réprésentées sur les deux figures.
Confrontation des coefficients d'amplification, déterminés dans le premier port intérieur sur le modèle réduit, à ceux calculés à l'aide du modèle du Système, pour des ondes incidentes de périodes 12s et 17s et d'amplitude 10 cm à l'entrée du port. Les caractéristiques correspondant à cette faible amplitude sont quasi-linéaires.

Fig. 12a to 12,5 seconds, demonstrating further the oversensivity of periodic wave testing with fixed measuring stations. Some improvements in periodic wave testing are obviously possible when using the mathematical model, as by averaging amplifications over areas of interest, but since the system – generated models run just as well with irregular waves as regular waves, there seems to be little point in refining investigation techniques along these lines. (Moreover, even these refinements would be of little help for combined mathematical and physical model testing, as envisaged for future investigations of combined ship-fendering-mooring systems).

The comparisons with irregular waves proceeded on the basis of surface elevation measurements made at the entrance to the complete harbour, so that here, as in the periodic wave tests as well, the Sommerfeld boundary module was not activated. A typical comparison between root mean square elevations in physical and mathematical models is shown in Fig. 13. In order to provide a more visually satisfactory presentation, comparable with that obtained with a physical model, Fig. 14 shows some contour plots, separated in time by one wave period, and a perspective plot from the Hanstholm computations, for periodic waves. When running on an IBM 370/165 installation in PL/1, the Hanstholm harbour computations shown used between 1 and 1,5 second of CPU time for each 1 second of physical simulation. This time can doubtless be improved considerably by program optimisation and the use of newer machines.

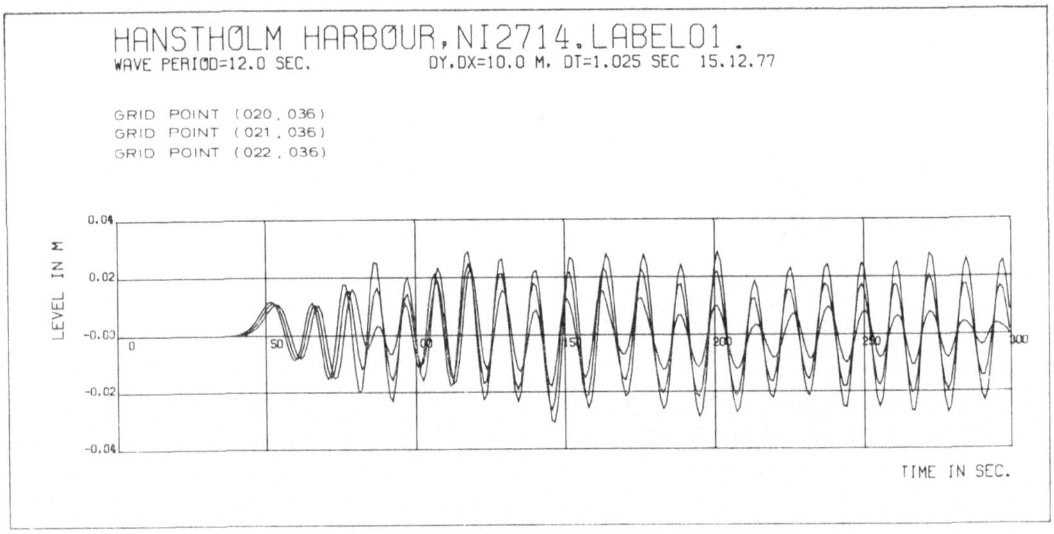

a. Time series of periodic water elevations taken at three adjacent points in the inner harbour of the numerical model, as used to generate Fig. 11, illustrating the rapid variations of elevations in space.

Fig. 12.

a. Cotes successives du plan d'eau dans le temps, relevées en trois points adjacents dans le port intérieur du modèle numérique, et utilisées pour établir la figure 11: la rapidité des variations du plan d'eau ressort très nettement de cette représentation.

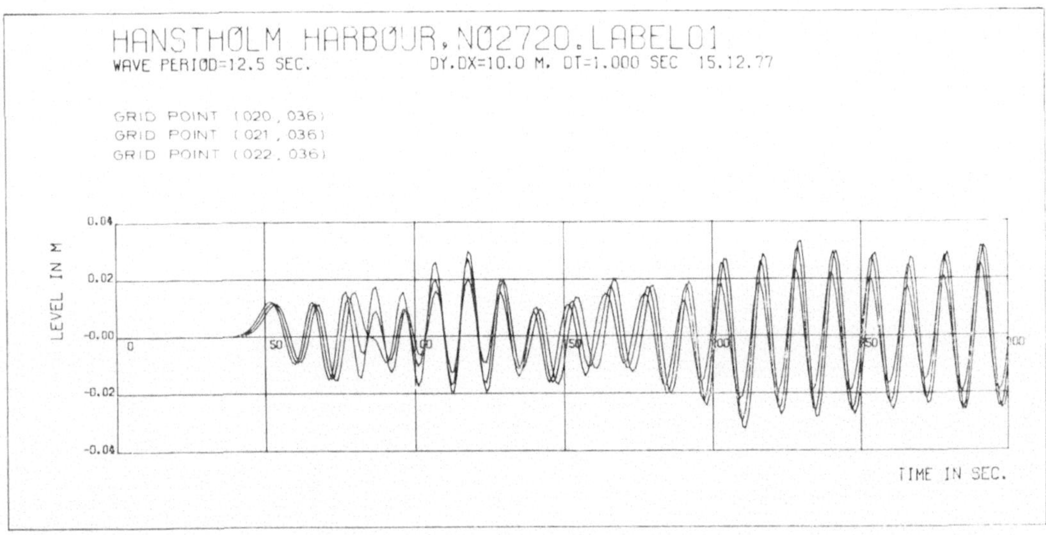

b. When the 12s period wave is replaced by a 12,5s period wave, the amplifications obtained at fixed points are seen to change considerably, a behaviour that is well-established in physical modelling.

b. Au passage de la période de 12s à 12,5s, on constate une très sensible variation des amplitudes relevées en des points fixes bien définis: il s'agit d'un phénomène bien connu sur les modèles réduits.

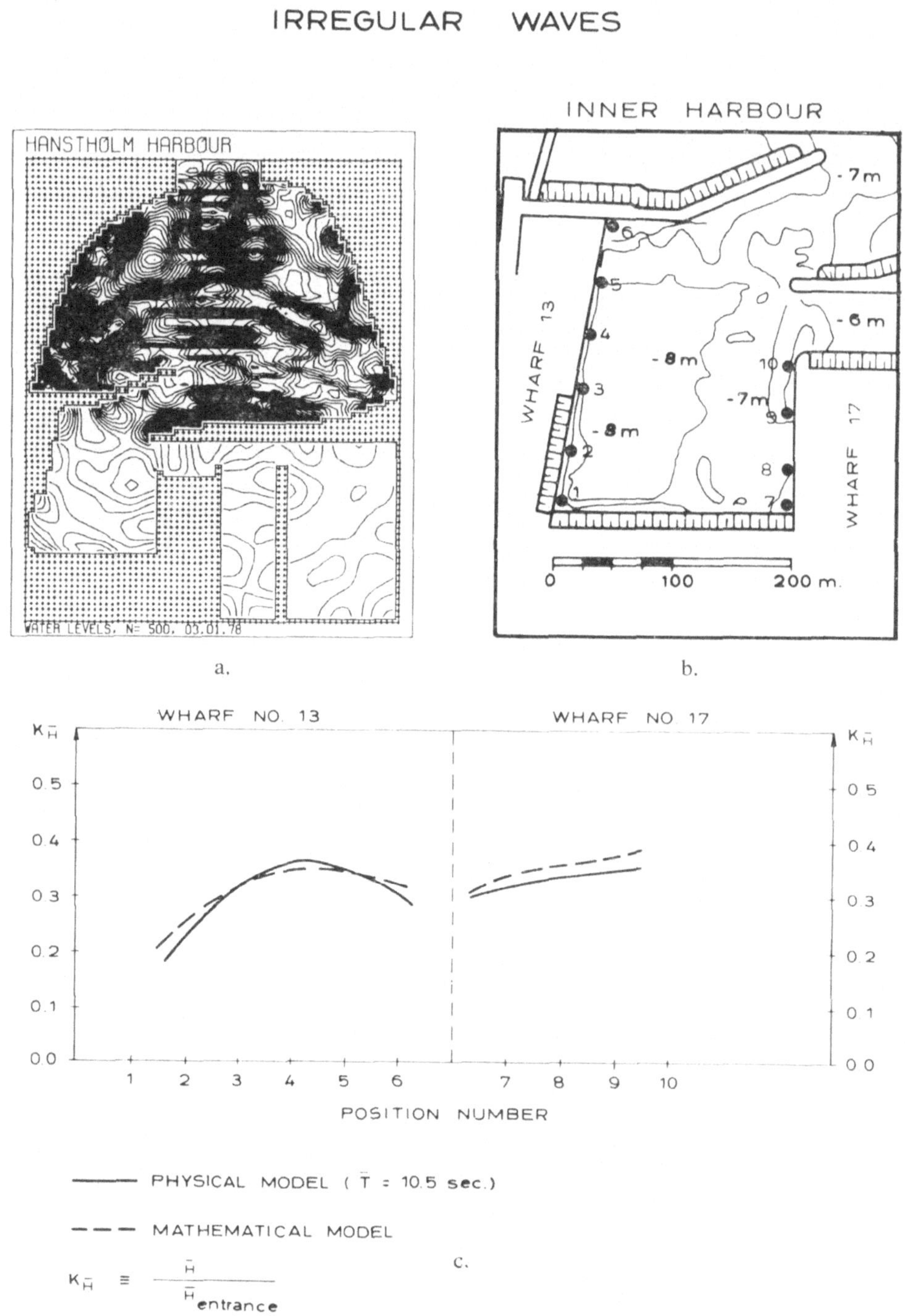

Fig. 13.

a. Surface elevation contours with irregular waves (Significant wave height ≈ 1.5 m).
b. Measuring stations in the physical model.
c. Comparison between amplification factors obtained in the first inner harbour in the physical model with those computed using the System model, for tests with irregular, field-measured waves, applied at the harbour entrance.

a. „Courbes de niveau" du plan d'eau, en présence d'ondes irrégulières (amplitude significative ≈ 1,5 m).
b. Emplacements des stations de mesure sur le modèle réduits.
c. Confrontation des coefficients d'amplification, déterminés dans le premier port intérieur sur le modèle réduit, à ceux calculés à l'aide du modèle du Système, correspondant à des essais effectués avec des ondes incidentes irrégulières à l'entrée du port (déterminées à partir des résultats de mesures effectuées dans la nature).

REGULAR WAVES

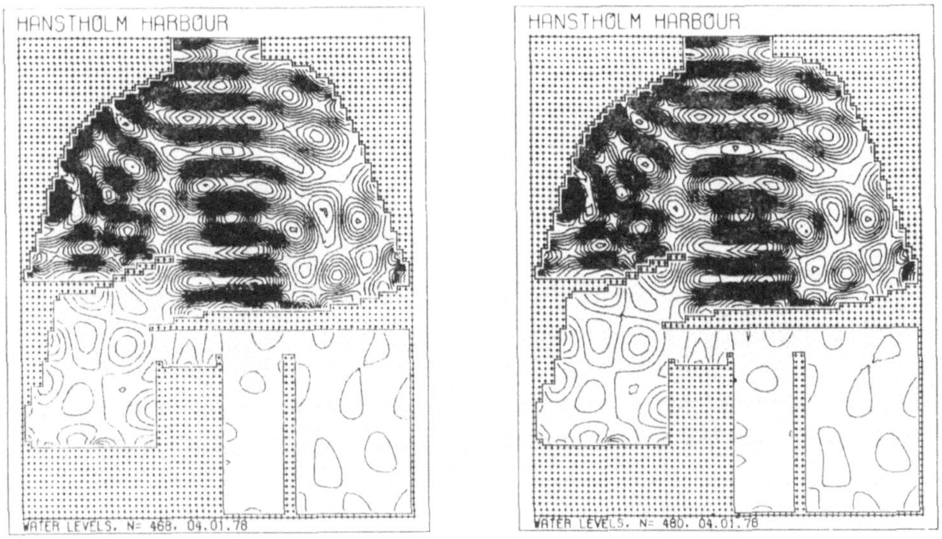

Fig. 14.

a. Surface elevation contours with regular waves compared for a one wave period time difference to illustrate the steady state attained.

a. „Courbes de niveau" du plan d'eau, en présence d'ondes régulières: comparaison pour une différence de temps correspondant à une période d'onde, destinée à mettre en évidence le régime permanent établi.

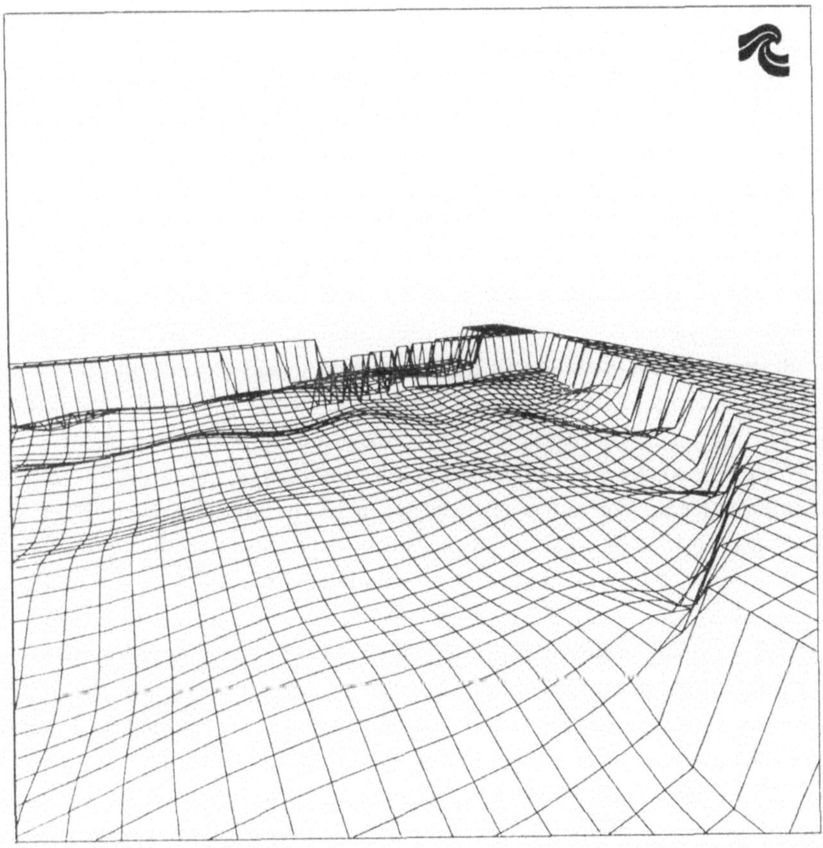

b. A perspective plot of waves in the outer harbour (view from the harbour entrance).

b. Représentation en perspective des ondes dans l'avant-port. (Vue du port de l'entrée).

Conclusions

On the basis of a development of the System 21, "Jupiter", orginally intended for simulations of flows with negligible vertical accelerations, or nearly-horizontal flows, it has been possible to develop a further version of that system for simulating a wide range of vertically accelerated flows. This range covers most short wave simulations required for engineering practice, while the flexibility and speed of the system are such that it can be employed economically. The extensive testing program, providing comparisons with independent analytical and physical model results, provides an acceptable level of confidence in the accuracy of the models generated by this new version.

Like the earlier versions, it generates models from a simple contour tracing or listing of topography and runs these models automatically, with flooding and drying and all other such operations controlled through its frame organisation. It is, of course ,simple to generate any physically realistic current field, steady or unsteady, across the domain of the computation and this will interact realistically with the short wave motions. Similarly, the nett, longer duration momentum fluxes that provide "wave thrusts" [LUNDGREN, 1963] or, what are essentially the same things, "radiation stresses" [LONGUET HIGGINS and STEWART, 1960, 1962], are generated automatically in the computations.

Additional Future Applications

The development described here already opens up a wide field of applications to numerical modelling, but it is obviously capable of extensive further development so as to suit it to further applications besides. The simplest of these extensions is to ship movements over a variable bathymetry in short waves and currents. The feasibility of this extension has already been demonstrated and a complete simulation system for a ship-fendering-mooring system in any region is currently being developed on this basis. It is already clear, however, that numerical modelling will quickly reach economic limitations in this area of applications, so that it will almost certainly be combined with physical-model testing for most applications. Physical-modelling technique is being developed accordingly. The projected combined numerical-physical modelling capability promises to increase further the design reliability of ship-fendering-mooring systems, especially for applications in very disturbed waters.

A further extension of the present system is to sediment transport under the influence of waves and currents. Since the system is based upon conservation laws of mass and momentum, while its difference formulations are such that these mass and momentum conserving properties are maintained with negligible energy falsification, it provides the energy loss to the main flow occasioned by resistances and other such "continuous" influences to a high order of accuracy [ABBOTT and RASMUSSEN, 1977]. This property makes possible an estimation of the rate of production of turbulent energy outside of the breaker zone at every point of the model at any time. Moreover, the conservation laws used admit of weak solutions in all cases, even in the case of the BOUSSINESQ equations [e.g. WHITHAM, 1974], so that, when provided with a suitable dissipative mechanism, the corresponding difference schemes can provide good approximations to cresting and breaking waves [ABBOTT, 1974]. Since, in nearly all practical situations, the energy loss to the main flow is accounted for by turbulent energy production, it is possible to compute the rate of turbulent energy production also in the breaker zone. It follows that the new modelling capability provides the most essential information needed in sediment-load calculations. The automatic appearance of radiation stresses in the models is naturally associated with the automatic simulation of longshore currents, "rip currents" and other flow features, so that a rational computation of sediment transport becomes possible at and inside the breaker zone.

Although the range of sediment transport computations can be extended in this way, the cost of computing over and inside the breaker zone is currently so great that it is doubtful whether these particular extensions are economically justified at the present time. The computations outside of the breaker zone, on the other hand, already appear to be justified by the needs of engineering practice.

The completion of the new system version raises other possibilities quite outside of the immediate applications to short wave simulations. This new version is consistently third order accurate so that, in principle at least, it provides a considerable better accuracy for a given amount of computation, or a smaller amount of computation for a given accuracy, as compared with existing nearly-horizontal flow models and system versions. Since the transport-dispersion stage of the present system is already of third-order accuracy [HINSTRUP, KEJ and KROSZYNSKI, 1977], the possibility arises of generating models that are consistently of the same accuracy, for nearly-horizontal flow computations as well. Although formidable problems would be presented when extending the present methods to account for the simultaneous computation of several grid scales [ABBOTT, 1976, 3], this is still an interesting development for many applications.

Acknowledgement

This work was carried out with the partial support of the Danish Technological Research Council.

APPENDIX

A method for approximating the Boussinesq equations by third order accurate finite differences

The method has been applied on two different schemes, viz a staged scheme as made by ABBOTT and IONESCU [1967] and PREISSMANN's scheme [see LIGGETT and CUNGE, 1975).

Although the staged scheme finally was chosen for the two-dimensional model described here, both schemes were found to be usable for the proposed approximations.

For simplicity of presentation the derivation is restricted to the plane motion of water on constant still water depth. The method used in the text covers also the general two-dimensional situation.

Governing equations

The basic differential equations are of the BOUSSINESQ type. They were derived by PEREGRINE [1967] for the three dimensional motion of water of variable depth.

For the plane situation with constant depth, the integrated equations can be written as:

$$\frac{\partial h}{\partial t} + \frac{\partial p}{\partial x} = 0 \tag{A.1}$$

$$\frac{\partial p}{\partial t} + \frac{2p}{h}\frac{\partial p}{\partial x} + \left(gh - \frac{p^2}{h^2}\right)\frac{\partial h}{\partial x} = \tfrac{1}{3}D\left(\frac{\partial^3 p}{\partial x^2 \partial t} - \frac{p}{h}\frac{\partial^3 h}{\partial x^2 \partial t}\right) \tag{A.1.2}$$

Several assumptions are required. The first of these is that the Taylor series expansions of dependent variables (h,p) are convergent, and indeed that

$$\frac{\partial^{(n+1)}}{\partial x^p \partial t^q} \ll \frac{\partial^n}{\partial x^r \partial t^s}; \qquad p+q=n+1, \qquad r+s=n$$

It is further assumed that all derivatives of third and higher order can be rewritten by means of first order relations:

$$\frac{\partial h}{\partial t}+\frac{\partial p}{\partial x}=0 \tag{A.1.3}$$

$$\frac{\partial p}{\partial t}+gD\frac{\partial h}{\partial x}=0 \tag{A.1.4}$$

while third order accuracy is still maintained [LONG, 1964].

Differentials are approximated by finite differences in a three level PREISSMANN scheme with centre point at $(n\cdot\Delta t,\ (j+\frac{1}{2})\cdot\Delta x)$.

The method is given by an example:

$$\left(\frac{\partial p}{\partial t}\right)^n_{j+\frac{1}{2}} = \text{1st order approx.} + \text{TRUNC3} + 0(\Delta x^4, \Delta t^4) \tag{A.1.5}$$

where

$$\text{1st order approx.} = \tfrac{1}{2}\left(\frac{p^{n+1}_{j+1}-p^{n-1}_{j+1}}{2\cdot\Delta t}+\frac{p^{n+1}_{j}-p^{n-1}_{j}}{2\cdot\Delta t}\right) \tag{A.1.6}$$

$$\text{TRUNC3} = -\tfrac{1}{6}\left[\tfrac{3}{4}\left(\frac{\partial^3 p}{\partial x^2 \partial t}\right)^n_{j+\frac{1}{2}}\cdot\Delta x^2+\left(\frac{\partial^3 p}{\partial t^3}\right)^n_{j+\frac{1}{2}}\cdot\Delta t^2\right] \tag{A.1.7}$$

It then appears that approximating by 1st order differences introduces an error (TRUNC 3) which is of the same type as terms in the basic differential equation (equation A.1.2).

A first order approximation can therefore not be expected to fit the solution of the differential equations satisfactorily.

In order to be able to express the various types of third order terms in PREISSMANN's scheme, the following manipulations are performed.

From equation (A.1.3):

$$\frac{\partial^3 p}{\partial x^2 \partial t} = \cdot - \frac{\partial^3 h}{\partial x^2 t^2} \tag{A.1.8}$$

and from equation (A.1.4):

$$\frac{\partial^3 p}{\partial t^3} = -gD\frac{\partial^3 h}{\partial x \partial t^2} \tag{A.1.9}$$

Substituting in equation (A.1.7) provides a correction term to subtract from (A.1.6):

$$\text{TRUNC3} = \left(\frac{\partial^3 h}{\partial x \partial t^2}\right)^n_{j+\frac{1}{2}} (\tfrac{1}{8}\cdot\Delta x^2 + \tfrac{1}{6}\cdot g\cdot D\cdot\Delta t^2) \approx$$

$$\approx \left(\frac{h^{n+1}_{j+1} - 2\cdot h^n_{j+1} + h^{n-1}_{j+1}}{\Delta x\cdot\Delta t^2} - \frac{h^{n+1}_j - 2\cdot h^n_j + h^{n-1}_j}{\Delta x\cdot\Delta t^2}\right)\cdot(\tfrac{1}{8}\cdot\Delta x^2 + \tfrac{1}{6}\cdot g\cdot D\cdot\Delta t^2)$$

Difference Equations

The result of discretizing the differential equations in the described way is a set of linear coupled differences equations:

$$A_1\cdot h^{n+1}_j + B_1\cdot p^{n+1}_j + C_1\cdot h^{n+1}_{j+1} + D_1\cdot p^{n+1}_{j+1} = E_1 \qquad \text{(A.1.1)}$$

$$A_2\cdot h^{n+1}_j + B_2\cdot p^{n+1}_j + C_2\cdot h^{n+1}_{j+1} + D_2\cdot p^{n+1}_{j+1} = E_2 \qquad \text{(A.1.11)}$$

where the coefficients and the right hand sides are functions of known parameters only. Equations (A.1.10) and (A.1.11) can be solved either by a method proposed by PREISSMANN [LIGGETT and CUNGE, 1975] or by elimination through row operations that allows the follow-up application of the tri-diagonal double sweep algorithm [ABBOTT, 1978].

Tests have shown that a third order accurate discretized formulation of the BOUSSINESQ equations is necessary in order not to invalidate the simulation through truncation error terms that are of the same order and of the same kind as the terms present in the differential equation that describe the vertical accelerations.

References **Bibliographie**

ABBOTT, M. B. (1974), Continuous Flows, Discontinuous Flows and Numerical Analysis, Journ. Hyd. Res., Vol. 12, No. 4, pp. 417–467.

ABBOTT, M. B. (1976, 1), Computational Hydraulics: A Short Pathology, Journ. Hyd. Res., Vol. 14, No. 4, pp.271–285.

ABBOTT, M. B., (1976, 2), Discussion of: Instability in Tidal Flow Computational Schemes, by T. J. WEARE, Journ. Hyd. Div. A.S.C.E., Vol. 102, No. HY12, pp. 1787–1790.

ABBOTT, M. B. (1976, 3), The Application of Design Systems to Problems of Unsteady Flow in Open Channels, Proc. Intern. Symp. on Unsteady Flow in Open Channels, Newcastle-upon-Tyne, April 1976, BHRA Fluid Engineering, Cranfield, Bedford, Paper No. E1. 1, 14 pp.

ABBOTT, M. B. (1978), Computational Hydraulics: Elements of the Theory of Free Surface Flows, Pitman, London.

ABBOTT, M. B. and F. IONESCU (1967), On the Numerical Computation of Nearly Horizontal Flows, Journ. Hyd. Res., Vol. 5, No. 2, pp. 97–117.

ABBOTT, M. B. and C. H. RASMUSSEN (1977), On the Numerical Modelling of Rapid Expansions and Contractions in Models that are Two-Dimensional in Plan, Proc. XVII IAHR Congress, Baden-Baden, Aug. 1977, Subject A.d, 8 pp.

ABBOTT, M. B. and G. S. RODENHUIS (1972), A Numerical Simulation of the Undular Hydraulic Jump, Journ. Hyd. Res., Vol. 10, No. 3, pp. 239–257.

ABBOTT, M. B., A. DAMSGAARD and G. S. RODENHUIS (1973), System 21, "Jupiter" (A Design System for Two-Dimensional Nearly Horizontal Flows), Journ. Hyd. Res., Vol. 11, No. 1, pp. 1–28.

AIRY, G. B. (1845), Tides and Waves, Art. 192 in Encyclopedia Metropolitana, London.

BENJAMIN, T. B. and M. J. LIGHTHILL (1954), On Cnoidal Waves and Bores, Proc. Roy. Soc. London, Series A, Vol. 224, pp. 448–460.

BENJAMIN, T. B., J. L. BONA and J. J. MAHONY (1972), Model Equations for Long Waves in Nonlinear Dispersive Systems, Proc. Roy. Soc. London, Series A, Vol. 272, pp. 47–78.

BERKHOFF, J. C. W. (1973), Computation of Combined Refraction-Diffraction, Proc. 13th Coastal Engrg. Conf., Vancouver, July 1972, Am. Soc. Civ. Engrs., New York, Vol. 1, Chapter 24, pp. 471–490.

BERKHOFF, J. C. W. (1976), Mathematical Models for Simple Harmonic Linear Water Waves, Wave Diffraction and Refraction, Publ. No. 163, Delft, Hydraulic Laboratory, 112 pp.

BETTESS, P. and J. A. BETTESS (1976), Wave, a Finite Element Program for Solving the Wave Equation, Dept., Civ. Engrg., Univ. Coll. Wales, Swansea, Computer Rep. No. 81.

BONA, J. L. and R. SMITH (1976), A Model for the Two-Way Propagation of Water Waves in a Channel, Math. Proc. Camb. Phil. Soc., Vol. 79, pp. 167–182.

BORN, M. and E. WOLF (1975), Principles of Optics, Electromagnetic Theory of Propagation, Interference and Diffraction of Light. Pergamon, Oxford, 5th ed.

BOUSSINESQ, J. (1872), Théorie des Ondes et des Remous qui se Propagent le Long d'un Canal Rectangulaire Horizontal, en Communiquant au Liquide Contenu dans ce Canal des Vitesses Sensiblement Pareilles de la Surface au Fond, Journ. Math. Pures et Appl. 2nd Series, Vol. 17, pp. 55–108.

BOUSSINESQ, J. (1877), Essai sur la Théorie des Eaux Courantes. Institut de France, Académie des Sciences, Memoires Présentés par Divers Savants, Vol. 23, No. 1, 680 pp.

CHEN, H. S. and C. C. MEI (1976), Oscillations and Wave Forces in a Man-Made Harbor in the Open Sea, Proc. 10th Naval Hydrodyn. Symp., Camb., Mass., June 1974, Eds. R. D. COOPER and S. W. DOROFF, ACR-204, Office of Naval Res., Dept. of the Navy, Arlington, Va,. pp. 573–594.

DAUBERT, A. and O. GRAFFE (1967), Quelques Aspects des Ecoulements Presque Horizontaux a Deux Dimensions en Plan et Non Permanents: Application aux Estuaires, La Houille Blanche, Vol. 22, No. 8, pp. 847–860.

DEAN, R. G. (1974), Evaluation and Development of Water Wave Theories for Engineering Application, Vols. I and II. Spec. Rep. No. 1, U.S. Army, Corps of Engineers, Coastal Engrg. Res. Center, Fort Belvoir, Virginia, U.S. Government Printing Office, Washington, D.C., (Vol. I - Presentation of Research Results, 127 pp. Vol. II - Tabulation of Dimensionless Stream Function Theory Variables, 534 pp.).

FRIEDRICHS, K. O. (1948), On the Derivation of the Shallow Water Theory. Appendix to J. J. STOKER, The Formation of Breakers and Bores, The Theory of Nonlinear Wave Propagation in Shallow Water and Open Channels, Comm. on Pure and Appl. Math., Vol. 1, No. 1, pp. 81–85.

HINSTRUP, P. I., A. KEY and U. KROSZYNSKY (1977), A High Accuracy Two-Dimensional Transport-Dispersion Model for Environmental Applications, Proc. XVII Congress I.A.H.R., Baden-Baden, 3, pp. 129–137.

JEFFREYS, H. and B. S. JEFFREYS (1946), Methods of Mathematical Physics, 1st ed. Cambridge Univ. Press, Cambridge.

JONSSON, I. G. (1967), Wave Boundary Layers and Friction Factor, Proc. 10th Conf. Coastal Engrg., Tokyo 1966, Vol. I, Chap. 10, ASCE, New York, pp. 127–148.

JONSSON, I. G., O. SKOVGAARD and O. BRINK-KJAER (1976), Diffraction and Refraction Calculations for Waves Incident on an Island, J. Mar. Res. 34, No. 3, pp. 469–496.

JONSSON, I. G., O. SKOVGAARD and T. S. JAKOBSEN (1975), Computation of Long-Shore Currents, Proc. 14th Coastal Engrg. Conf., Copenhagen, ASCE, New York, pp. 699–714.

KEULEGAN, G. H. (1973), Wave Transmission Through Rock Structures, Hydraulic Model Investigation, U.S. Army Engineer Waterways Experiment Station, Hydraulics Laboratory, Vicksburg, Mississippi, Res. Rep. H-73-1, 75 pp.

KEULEGAN, G. H. and G. W. PATTERSON (1940), Mathematical Theory of Irrotational Translation Waves, Res. Paper RP 1272, Journ. Res. Natl. Bureau Stand., Vol. 24, pp. 47–101.

KORTEWEG, D. J. and G. DE VRIES (1895), On the Change of Form of Long Waves Advancing in a Rectangular Canal, and on a New Type of Long-Stationary Waves. Phil. Mag., London, Dublin and Edinburgh, Ser. 5, Vol. 39, pp. 422–443.

LAX, P. D. (1954), Weak Solutions of Nonlinear Hyperbolic Equations and Their Numerical Computation, Comm. on Pure and Appl. Math., Vol. 7, pp. 159–193.

LE MÉHAUTÉ, B. (1955), Two-Dimensional Seiche in a Basin Subjected to Incident Waves. Proc. 5th Coastal Engrg. Conf., 1954, Am. Soc. Civ. Engrs., New York, pp. 119–149.

LIGGETT, J. A. and J. A. CUNGE (1975), Numerical Methods of Solution of the Unsteady Flow Equations, Ch. 4 of Unsteady Flow in Open Channels (Ed. MAHMOOD, K. and V. YEVJEVICH), Water Resources Publications, Fort Collins.

LIN, C. C. and CLARK, A. On the Theory of Shallow Water Waves, Tseng Hua Journal of Chinese Studies, Special Number 1, Dec. 1959, pp. 54–61.

LONG, R. R. (1964), The Initial-Value Problem for Long Waves of Finite Amplitude. Journ. Fluid Mech., Vol. 20, Part 1, pp. 161–170.

LONGUET-HIGGINS, M. S. and R. W. STEWART (1960), Changes in the Form of Short Gravity Waves on Long Waves and Tidal Currents, J. Fluid Mech., 8, 4, pp. 565–583.

LONGUET-HIGGINS, M. S. and R. W. STEWART (1962), Radiation Stress and Mass Transport in Gravity Waves, with Application to "Surf Beats", J. Fluid Mech. 13, pp. 481–504.

LUNDGREN, H. (1963), Wave Thrust and Wave Energy Level. Proc. 10th Congr. Int. Assoc. Hyd. Res. London, Vol. 1, paper 1.20, pp. 147–151.

MADSEN, O. S. and C. C. MEI (1969), The Transformation of a Solitary Wave Over an Uneven Bottom, Journ. Fluid Mech., Vol. 39, Part 4, pp. 781–797.

MEI, C. C. and B. LE MÉHAUTÉ (1969), Note on the Equations of Long Waves over an Uneven Bottom, Journ. Geophys. Res., Vol. 71, No. 2, pp. 393–400.

MEYER, R. E., (1967), Note on the Undular Jump, Journ. Fluid Mech., Vol. 28, Part 2, pp. 109–221.

PEREGRINE, G. H. (1967), Long Waves on a Beach, Journ. Fluid Mech., Vol. 27, Part 4, pp. 715–827.

PEREGRINE, D. H. (1972), Equations for Water Waves and the Approximation behind Them. In R. E. MEYER (ed.), Waves on Beaches and Resulting Sediment Tranports. Academic Press, New York, pp. 95–121.

PEREGRINE, D. H. (1974), Discussion of: A Numerical Simulation of the Undular Hydraulic Jump, by M. B. ABBOTT and G. S. RODENHUIS, Journ. Hyd. Res., Vol. 12, No. 1, pp. 141–157.

RODENHUIS, G. S. (1966), The Undular Hydraulic Jump. Rep. Series No. 1, Intern. Courses in Hydraulic and Sanitary Engineering, Delft, 43 pp.

RODENHUIS, G. S. (1973), Difference Method for Higher-Order Equations of Flow. Journ. Hyd. Div. ASCE, Vol. 99, N. HY3, pp. 471–483.

SKOVGAARD, O. and J. AA. BERTELSEN (1974), Refraction Computations for Practical Applications, Proc. Int. Symp. Ocean Wave Measurement and Analysis, ASCE, New Orleans, La, pp. 761–773.

SKOVGAARD, O. and I. G. JONSSON (1976), Connection Between the Kinematics of Current-Depth Refraction and the Conservation Equations, Progr. Rep. No. 38, Inst. Hydrodyn. and Hydraulic Engrg (ISVA), Tech. Univ. Denmark, pp. 29–36.

SOMMERFELD, A. (1896), Mathematische Theorie der Diffraktion, Mathematische Annalen, Vol. 47, p. 317.

SOMMERFELD, A. (1964), Partial Differential Equations in Physics (Trans. E. G. STRAUS), Academic Press, New York, 335 pp.

SORENSEN, T. (1973), Model Testing with Irregular Waves, Dock and Harbour, Authority 5.

STOKER, J. J. (1957), Water Waves. Interscience., Publ., New York, 567 pp.

STREET, R. L., R. K. C. CHAN and J. E. FROMM (1970), Two Methods for the Computation of the Motion of Long Water Waves – A Review and Applications. In Proc. 8th Naval Hydrodynamics Symp., ACR-179 Editor, M. S. PLESSET, T. Y. T. WU and S. W. DOROFF, Office of Naval Research, Dept. of the Navy, Arlington, Va., pp. 147–187.

URSELL, F. (1953), The Long-Wave Paradox in the Theory of Gravity Waves. Proc. Camb. Phil. Soc., Vol. 49, pp. 685–694.

WHITHAM, G. G. (1974), Linear and Nonlinear Waves. Wiley - Interscience, New York, 636 pp.

ZIENKIEWICS, O. C. and P. BETTESS (1976), Infinite Elements in the Study of Fluid-Structure Interaction Problems. In Computing Methods in Applied Sciences, Edited by R. GLOWINSKI and J. L. LIONS. Lecture Notes in Physiscs, Vol. 58, Springer-Verlag, Berlin, pp. 133–172.

Notation

- c Phase celerity (ms^{-1})
- c_g Group celerity (ms^{-1})
- d Rubble diameter (m)
- D Mean water depth (m)
- h Water depth, $\equiv D+\zeta$ (m)
- h_* Measure of the water depth (m)
- k Wave number (dimensionless, except in 7, 8)
- l Characteristic length (m)
- L_* Characteristic horizontal length of a fluid surface profile (m)
- n Pore volume
- N Number of computational points per wave length
- p x-volume flux density, or alternatively x-horizontal momentum level per unit density, $=uh$ (m^2s^{-1})
- P Amplification factor, a measure of the amplification error defined by eq. 14.
- q y-volume flux density, or alternatively y-horizontal momentum level per unit density, $= vh$ (m^2s^{-1})
- Q Celerity ratio, a measure of the phase error defined by equation 15
- r Reflection coefficient
- R Ratio of wave amplitude to still water depth
- S Ratio of wave length to still water depth
- t Time (sec)
- u Velocity in x-direction (ms^{-1})
- U_* URSELL number, $= \zeta_* L_*^2/h_*^3$
- v Velocity in y-direction (ms^{-1})
- x horizontal coordinate (m)
- y horizontal coordinate, direction orthogonal to that of x (m)
- z Bed elevation (m)
- α Laminar flow resistance coefficient for a permeable breakwater
- β Turbulent flow resistance coefficient for a permeable breakwater
- Δ Increment
- ζ Water surface elevation (m)
- ζ_* Measure of the wave amplitude (m)
- ζ^* Complex wave amplitude (m)
- ν Kinematic viscosity (m^2s^{-1})
- ω Angular frequency (s^{-1})
- ϱ Density (kgm^{-3})

The following article has been reprinted from *Journal of Hydraulic Research*, Vol. 29, No. 5, pp. 581–600, with permission from the copyright holders, IAHR International Association for Hydro-Environment Engineering and Research.

The fourth generation of numerical modelling in hydraulics

La quatrième génération de modelès numériques en hydraulique

MICHAEL B. ABBOTT
International Institute for Hydraulic and Environmental Engineering, Delft, The Netherlands

KARSTEN HAVNØ
Danish Hydraulic Institute, 2970 Hørsholm, Denmark

STEN LINDBERG
Danish Hydraulic Institute, 2970 Hørsholm, Denmark

SUMMARY

First-generation modelling was concerned only with computer adaptations of established, human-friendly methods. In second-generation modelling, on the other hand, methods were used that were more specifically "computer friendly", even if "human-unfriendly" (e.g., finite-difference and finite-element methods). In third-generation modelling, these methods were incorporated into modelling systems, or "shells", with which a model of any area could be constructed and run automatically by the system when provided with a suitably-formatted description of the area and its ancillary conditions. During the course of the second- and third-generation developments, however, modelling became an increasingly specialised and hermetic activity, confined for most practical purposes to computational-hydraulics experts. The fourth generation then builds on this earlier development, but in such a way as to provide modelling systems that can be used by professional engineers who are not computational-hydraulics experts.

The paper describes a typical system for fourth-generation modelling, it outlines the manner of transformation of a third-generation system into a fourth-generation system, and it discusses some of the limits of fourth-generation modelling. It is emphasised that fourth-generation modelling demands a serious research commitment that is, however, not scientific research in its usually-understood sense.

RÉSUMÉ

La première génération de modèles concernait seulement l'adaptation aux calculateurs de méthodes "manuelles" bien établies. Dans la 2ème génération par contre, on a utilisé des méthodes plus spécifiquement destinées aux calculateurs, dont l'utilisation par "manuelle" était quasiment exclue (par exemple: les méthodes aux différences finies et aux éléments finis). Dans la 3ème génération de modèles, ces méthodes ont été incorporées dans des systèmes de modélisation ou "coquilles", avec lesquels un modèle de quelque étendue que ce soit peut être construit et exploité automatiquement par le système à condition de lui fournir une description adéquate de la zone et des conditions aux limites. Au cours des développements des 2ème et 3ème générations cependant, la modélisation devint une activité de plus en plus réservée à des spécialistes, experts en hydraulique numérique. La 4ème génération se construit ensuite sur les développements précédents, mais dans une voie qui vise à fournir des systèmes de modélisation exploitables par des ingénieurs hydrauliciens qui ne soient pas des experts en hydraulique numérique.

Revision received December 13, 1990. Open for discussion till April 30, 1992.

L'article décrit un système typique de cette 4ème génération de modèles, montrant la transformation d'un système de 3ème génération en un système de 4ème génération et discutant des limites de cette 4ème génération. Il est en particulier insisté sur le fait que la 4ème génération de modèles demande un effort de recherche important, qui n'est cependant pas de la recherche scientifique au sens strict.

Introduction

The notion of the "generations" of modelling has been described at some length by Abbott (1989) and Cunge (1989), so that only a short recapitulation is necessary here. Models of the first generation were directed to performing the same functions, following the same methods, as human computers, such as finding the roots of algebraic equations and calculating values of analytic functions. The first generation was thus restricted to the programming of well-established, "human-friendly" procedures. In the second generation of modelling, on the other hand, methods were introduced that, although better suited to the ways of working of the digital machine, were usually decidedly "human-unfriendly". The principal methods of this kind were those of finite differences, finite elements and boundary elements. Correspondingly, second generation modelling was the first to pose the problem of translating between statements that we humans could best understand (e.g. differential equations) into statements that digital machines could accommodate (e.g. sequencers of operations necessary to solve difference equations), and then of translating back from the very extended sets of numbers that the machine produced as output into results that we humans could, again, understand. This process of translating back and forth between what are, in effect, statements in essentially different *languages,* was the principal concern of *computational hydraulics* (e.g. Abbott, 1979; Abbott and Basco, 1989; Cunge, Holly and Verwey, 1980; Stelling, 1983). Computational hydraulics began its life as the science of second-generation modelling. At this stage it was mainly concerned with operations, such as those that realise difference schemes in code.

The distinction between second-generation and third-generation modelling is one that is usually made between the construction of each model by the human modeller in a one-off, area-customised way and the construction of each model by the computer itself when provided by the human modeller with a suitably-formatted description of the area that has to be covered by the model. Whereas the accent in second-generation modelling was concentrated on the operational elements of the code, in third-generation modelling the control elements became more prominent, even as the operational elements were further refined. It is common to date the era of predominantly second-generation modelling to the period 1960–1970, and that where most modelling became third-generation modelling to the period 1970–1980. Throughout this period, from 1960 to 1980, numerical modelling advanced from a predominantly academic exercise into a valuable, and in many cases indispensable tool for solving engineering problems. Similarly, in both instances the benefits of numerical modelling in this field were made available through the provision of *results* from modelling studies, so that the clients for such results usually had no direct contact with the modelling work itself.

The rapid increase in the use of models, however, led to an increasing demand on the part of the users of results to participate in the modelling work, and even to conduct such work themselves. At the same time, computer hardware systems and their associated systems' software had developed to such an extent that means to conduct modelling studies had become much more widely accessible. Thus the notion arose that third-generation modelling need not be restricted to specialist centres but could be moved closer to the end-user of results, and that it might possibly even be transferred completely to this end user. The principal obstacle to realising this objective was that numerical modelling and its science of computational hydraulics had become so extended and so specialised that persons closer to the end user, and almost certainly the end user himself, were not longer able to comprehend the problems and methodologies of work in this area. Correspondingly, such persons, even though they might be entirely competent in hydraulics *per se,* were not able to set up and run a model using a third-generation system in a professional manner. The natural response in this situation was to develop a new generation of modelling tools which, even as they incorporated most of the scientific capabilities of the third-generation modelling systems, were altogether more "user friendly".

These then become *fourth-generation modelling tools.* The fourth generation of modelling then has the following special features:

1. The modelling system becomes a design and management tool suited to use by persons who are not computational-hydraulics specialists.
2. In fourth-generation modelling a differentiation correspondingly occurs between tool makers, who are computational-hydraulics specialists, and tool users, who usually are not.
3. The fourth-generation tool is a product, so that computational-hydraulics specialists have to transform, at least partially, from performers of projects to producers of products.
4. The principal line of development of fourth-generation modelling is in the direction of making computational hydraulics much more useful to a much wider range of end-users.
5. Research in this area moves correspondingly away from the direction of scientific research into that of research on the application of science. It is thus a technological research that is not scientific research in the usually-understood sense.
6. When seen from the point of view of computer-science, research in fourth-generation modelling is concentrated on the control aspects of the code, with the operational aspects developing rather little and the data-base aspects only beginning to become significant.

A basic architecture for a fourth-generation modelling system

In view of the development sketched above, it is still usual to regard the computational parts of the system as the "core" of the system. In the overall architecture, however, this computational part is only one of several major components that effectuate the simulation. In the example illustrated in Fig. 1, there are only two such other components, namely the data base containing the time-invariant description of the physical system and the data base containing the time-varying inputs. The example illustrated in Fig. 1 is for an early and correspondingly simple package, intended for simulating flows in urban drainage systems.

It is seen from Fig. 1 that the operational and data base components are actuated, or "driven" through a *user interface,* or *man-machine interface* (MMI), which in this case is composed only of a menu system and a system that drives, in its turn, the necessary printing and plotting facilities. The main features of the system illustrated in Fig. 1 can best be described separately, as follows.

Menu system

All communication between the system and its user takes place interactively through a number of screen menus, and input and output is presented either in tabulated form on a printer or graphically on a screen or a plotter.

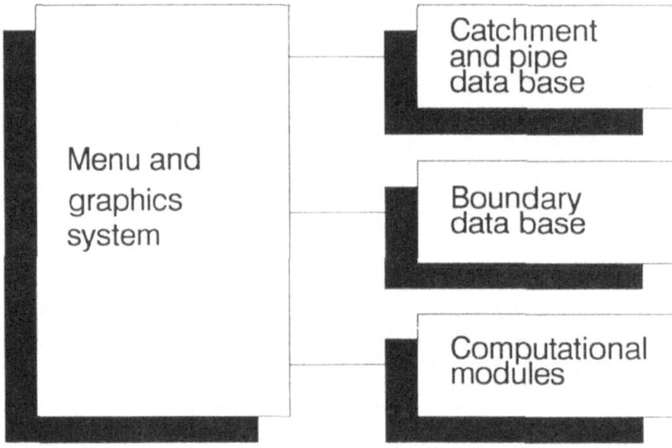

Fig. 1. The main elements in a simulation package for urban sewer systems.
Les éléments principaux d'un code de simulation de systèmes d'assainissement urbain.

Application of the system requires no knowledge of programming or operating systems. The menus guide the user through the application; input data are entered into standard forms on the screen and checked by the system for consistency and correct orders of magnitude. A user's manual is integrated in the system and "help" menus can be displayed on the screen, when required (see Fig. 2).

A.5 S U M M A R Y

	Form name	type	No. of nodes
1	CATCHMENTS	D	154
2	CIRCULAR MANHOLES	KG1	144
3	STRUCTURES........................	KG2	10
4	STRUCTURE GEOMETRY	KG3	10
5	WEIR FUNCTION	KF1	4
6	PUMP FUNCTION	KF2	5
7	CONTROL FUNCTION	KF3	
8	CRITICAL LEVELS	KK	0
9	OUTLETS	KU	0
			No. of branches
A	CONDUITS–PIPES....................	L1	184
B	CONDUITS–TRAPEZOIDAL CANALS........	L2	0
C	CONDUITS(ARBITRARY CROSS-SECTION) ..	L3	0
			lines
T	Text		4

Enter form nr : <Esc> return to A <F1> help

NODAL POINTS – CIRCULAR MANHOLES KG1 – form

144	1 Node–number	2 Coordinates X (m)	Y (m)	3 Levels Bottom (m)	Top (m)	<4>	5 Dia–meter (m)
1	2	736.0	108.0	2.56	7.31	2	2.00
2	3	862.0	117.0	3.62	6.48	2	1.00
3	4	822.0	139.0	2.59	6.18	2	2.00
4	5	897.0	174.0	2.62	6.20	2	2.00
5	6	1016.0	232.0	2.70	7.00	2	2.00
6	7	1085.0	312.0	2.80	7.90	2	2.00
7	8	891.0	126.0	3.53	7.01	2	1.00
8	9	974.0	166.0	4.25	7.00	2	1.00
9	10	1060.0	237.0	4.50	7.40	2	1.00
10	11	954.0	66.0	3.70	7.30	2	1.00

<4> – Shape of outlet: 1 = round edged outlet 3 = orificing pipe outlet
2 = sharp edged outlet 4 = no cross section changes

Entr: (E/I/F/D/T/B/L/ESC) Edit Insert Find Delete Top Bottom Line <esc>=retur

Fig. 2. Sample Menu and Data Entry Form.
Exemple de menu et de page d'entrée des données.

Catchment and pipe date base

In the urban-sewer flow simulation system used here as an example, the lay-out and geometry of the catchment and the pipe system is described in 12 standard forms, e.g. one form for circular manholes, one for pipes etc. The data base in this case also contains files describing the composition of the sewage.

Sub data-sets for calculation and plotting are retrieved automatically by the computational modules and the print/plot modules, (see Fig. 3). It should also be explained that almost all output from such systems is in colour, which, although it greatly aids legibility, cannot be conveniently reproduced in publications like this. Similarly, overlaying, scrolling and other such graphical facilities cannot be adequately illustrated here.

The current systems commonly provide links to external formats for facilitating the input of existing digitised data, and in time these facilities may be expected to interface to standard geographical information systems (GIS's), digital terrain models and other such standard, proprietary data-base-like systems.

Rain data base

The system used as an example here has as a description of its flow input a rain data base. This comprises a system for the processing of historical rain events as well as synthetic rain data. Historical rain events are described as tabulated time series, and synthetic events by means of frequency/intensity/duration formulae or tables. Single rain storm series or sub-series are retrieved automatically by the computation modules and the print/plot modules.

The surface run-off component

This component describes the run-off process, and the user may choose between a simplified and a more comprehensive approach. The simplified approach combines a calculation of major initial losses with a "time/area" function. The more comprehensive approach combines a calculation of all hydrological losses (e.g. evaporation, wetting, storage depression and infiltration) with a non-linear reservoir computation of surface run-off. Input data are automatically retrieved from the catchment and rain data bases.

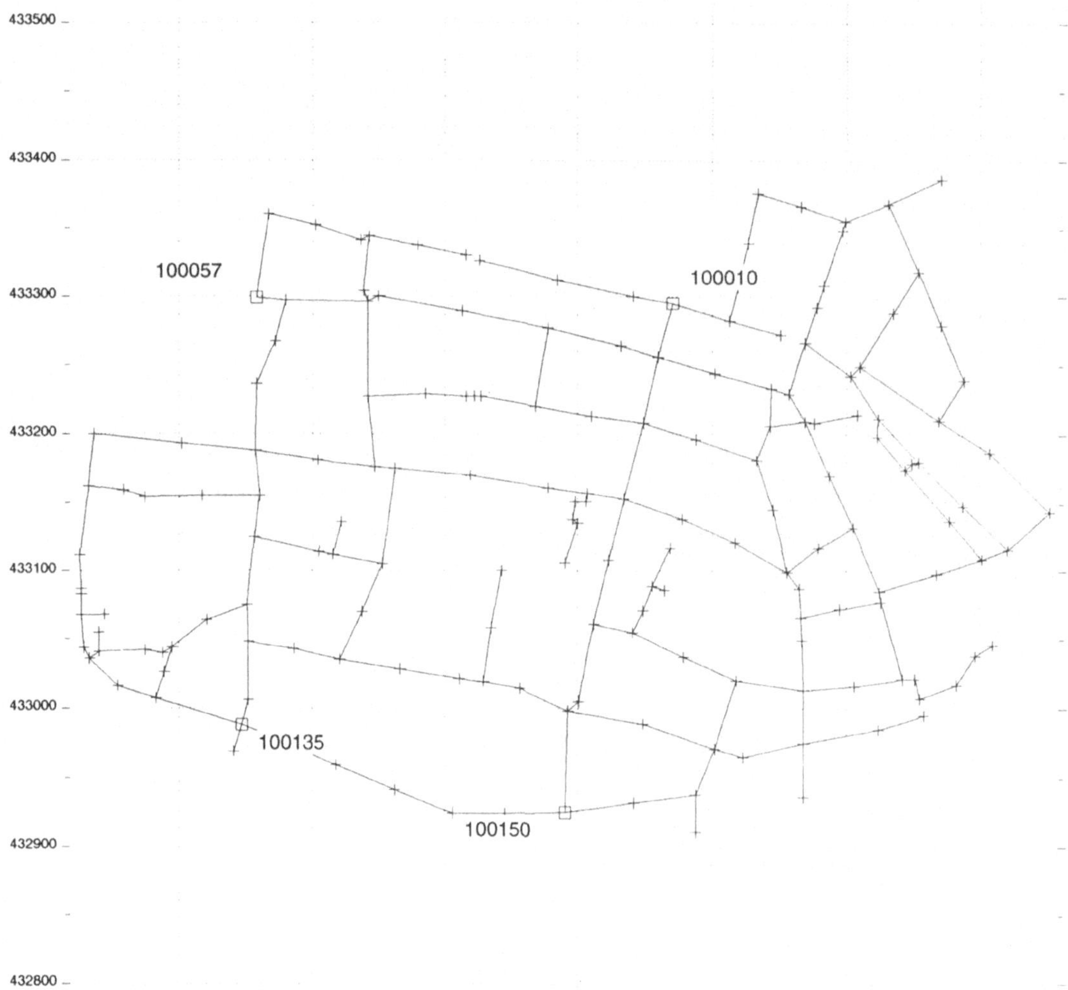

Fig. 3. The typical network configuration shown here is close to the limits of what can be accommodated when using PC's with their usual MS/DOS operating system. Much larger systems (easily twenty times as complicated as the one shown here) can be accommodated using work stations, but even then it is often advantageous to decompose the system into subsystems, both to facilitate the computation and in order to maintain a physical insight into the flow processes.

La configuration type représentée ici est voisine des limites d'utilisation d'un PC travaillant sous MS/DOS. Pour des systèmes plus étendus (facilement 20 fois plus complexes), il faut recourir à des stations de travail, mais il est cependant souvent avantageux de décomposer le système en sous-système, à la fois afin de faciliter le calcul et de maintenir une vue claire de la physique des phénomènes d'écoulement modélisés.

Pipe flow component

The pipe flow component provides the facilities for three levels of hydraulic description, as follows:

- *Kinematic wave approach.* The pipe flow is calculated from the principle of balance between the friction and gravity forces. This simplification implies that the kinematic wave approach cannot simulate backwater effects.
- *Diffusive wave approach.* In addition to the friction and gravity forces, the hydrostatic gradient is included in this description. This allows the user to take downstream boundaries into account, and thus simulate backwater effects as well as pressurised flows.
- *Dynamic wave approach.* Using the complete momentum equation of nearly horizontal flow, including acceleration terms, the user is able to simulate even very fast transients in the system.

Depending on the type of problem, the user can choose the most appropriate description. All three approaches can be used to simulate looped as well as branched systems.

In the system used as an example here, only one, very-general, and indeed "over-general" scheme, is in fact coded, but this contains weighting terms which allow it to correspond to any of the above approaches. As the weighting factors are themselves functions of the flow conditions (i.e. they shape the scheme so as to suit it to the physics of the problem), such a very general scheme can be weighted explicitly from the flow conditions, and it is accordingly called an adaptive numerical scheme, or, more precisely, a *"flow-adaptive"* numerical scheme. Schemes of this type were already originated within second-generation modelling - by Preissmann (see Cunge and Liggett, 1975; Abbott and Basco, 1989) -, albeit initially primarily for "numerical" reasons, but they have long since been extended to other implicit schemes such as the Abbott-Ionescu (1967) scheme used in the present example and its third-generation predecessor, with more hydraulic motivations (Havnø and Brorsen, 1986; see, originally, Hayami, 1951). Some typical results are illustrated in Fig. 4.*

The pipe flow description used in this case accommodates free-surface flows in pipes, as well as surcharges (pressurised flows), flooding and storage on the surface, and flow in multiply-connected systems (loops). Head losses at man-hole inlets and outlets as well as in various flow-controlling structures (valves, weirs, pumps etc.) are taken into account. Combinations of super-critical and subcritical flows, with associated hydraulic jumps, can be calculated by using the dynamic and diffusive wave formulations interactively, where the interaction is organised automatically by the system.

In this system, the organisation and structuring of the numerical grid is automatized. Thus, for example, boundary conditions are assigned at their appropriate locations while the solution algorithm itself is automatically "tailored" for optimal operation. In other systems, (such as open-channel systems), data may be interpolated automatically in time and space in order to make the best use of the available data base.

Corresponding to the introduction of these automatic facilities, graphical facilities are introduced to provide immediate visual checks of input data, and these also can vary from very detailed descriptions to rather gross overviews. Similarly, provisions are made to present standardised overviews of key parameters (e.g. maximum and minimum water levels, outflow

* This approach has been strongly critizised by Samuels (1990), who speaks of a "falsifying of the physics of the flow simulation" and a "crude reduction of the physical basis" and proceeds further to prophesy "for Abbott's scheme... a catastrophic instability in the computation near critical flow conditions". A few such simple computations suffice however to show that these criticisms are not properly founded.

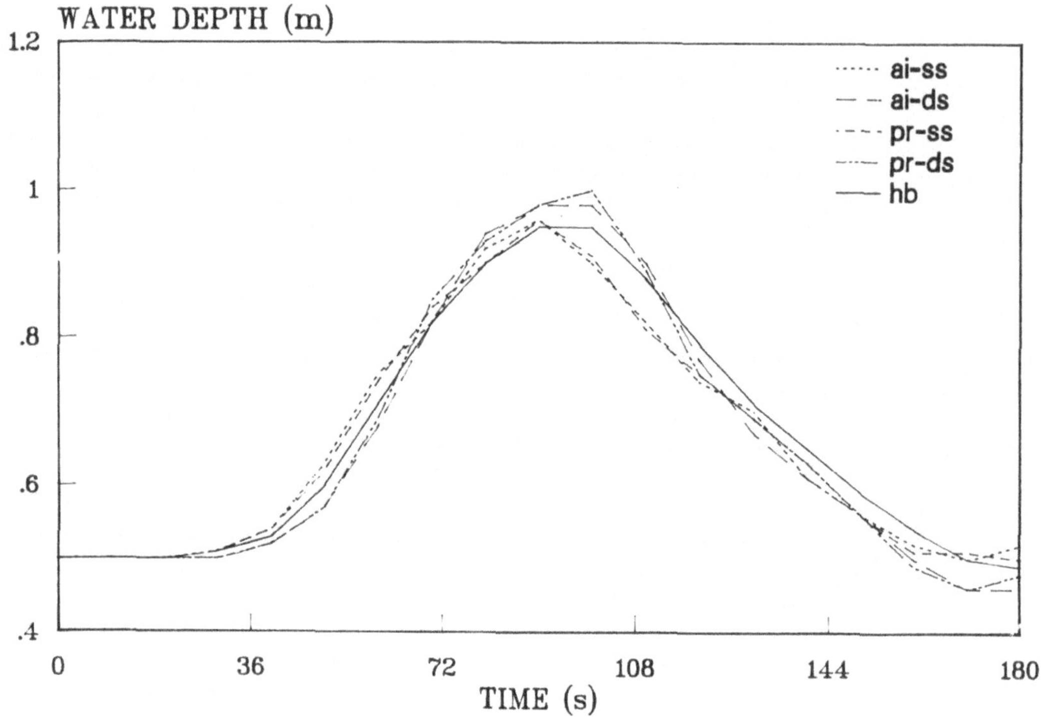

Fig. 4. Comparison between results obtained for a case of supercritical flow using a strictly-appropriate single-sweep (ss) algorithm, the double-sweep (ds) approximation obtained by extracting the convective momentum terms (for both Preissmann (pr) and Abbott-Ionescu (ai) schemes) and the Havnø-Brorsen (hb) approximation. Here a triangular wave of height 0.5 m distributed over 120 s has propagated some 200 m in a uniform channel with slope $i = 0.1$, Chézy $C = 50\ \mathrm{m}^{1/2}\ \mathrm{s}^{-1}$ and, for the single-sweep algorithms, with $u = c\sqrt{hi}$ at the upstream boundary. Δx per function value = 10 m, $\Delta t = 10$ s.

Comparaison entre les résultats obtenus pour un cas d'écoulement super critique en utilisant un algorithme à simple balayage (ss) et à double balayage (ds) avec l'approximation résultant de l'extraction des termes de quantité de mouvement convectifs (à la fois pour les schémas de Preissmann (pr) et d'Abbott, Ionescu (ai)) ainsi que pour l'approximation de Havnø-Brorsen (hb). Le cas concerne une onde triangulaire de hauteur 0,5 m avec une durée de 120 s, propagée sur 200 m dans un canal uniforme de pente $i = 0{,}1$ avec un Chézy $C = 50\ \mathrm{m}^{1/2}\ \mathrm{s}^{-1}$ et pour les algorithmes à simple balayage la condition $u = C\sqrt{hi}$ à l'amont. $\Delta x =$ 10m, $\Delta t = 10$ s.

statistics, including exceedence frequencies, and plan overviews highlighting pressurised regions). These provisions include zooming facilities, whereby local areas can be selected to be displayed at a larger scale.

Similarly, vertical cross-sections can be selected along any required path through the pipe network for later graphical presentation, as illustrated in Fig. 5.

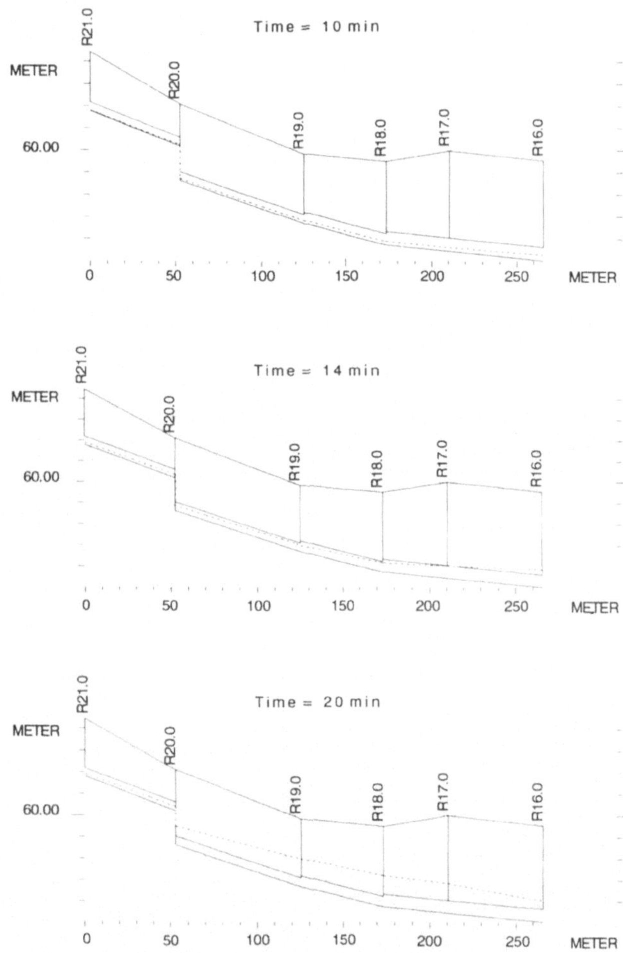

Fig. 5. Instantaneous longitudinal sections along a path chosen by the user through the pipe and manhole network, showing the changing hydrostatic-pressure load line.

Profils en long à un instant donné selon un chemin choisi par l'opérateur à travers les réseau de conduites et de regards (trous d'hommes), montrant l'évolution de la ligne de charge.

Pollution component

The pollution component calculates annual and extreme discharge volumes from overflow weirs on the basis of long-term rain series. The description used here is especially suited for analysis of the following problems:

- calculation of annual pollutant loads to receiving water bodies;
- calculation of extreme BOD loads on lakes and rivers.

A time/area formulation is combined with simplified routeing routines for surface run-off and pipe flows. Relatively simple but efficient routines are used so as to allow for long-term simulations. Thus, for example, the calculation of pollutant loads covering several years is feasible. The results provided by a simulation are presented in the form of a number of loads (volumes, kg BOD, phosphorous etc.) from each overflow weir.

Printing and graphics

Numerical input data as well as output data may be presented either in the form of tables projected on screen, or in the form of hard-copy print-outs. Similarly, graphical presentation may be on screen or plotted. The lay-out of the pipe system and the longitudinal profiles can be retrieved from the Catchment and Pipe Data Base.
Time series as wel as longitudinal profiles may be retrieved from the output data set, and indeed all the facilities detailed for the pipe flow component can be made available in the output form. By way of an example, a sample longitudinal profile consisting of pipe and water levels in a surcharging situation is shown in Fig. 5 and other results in Fig. 6.

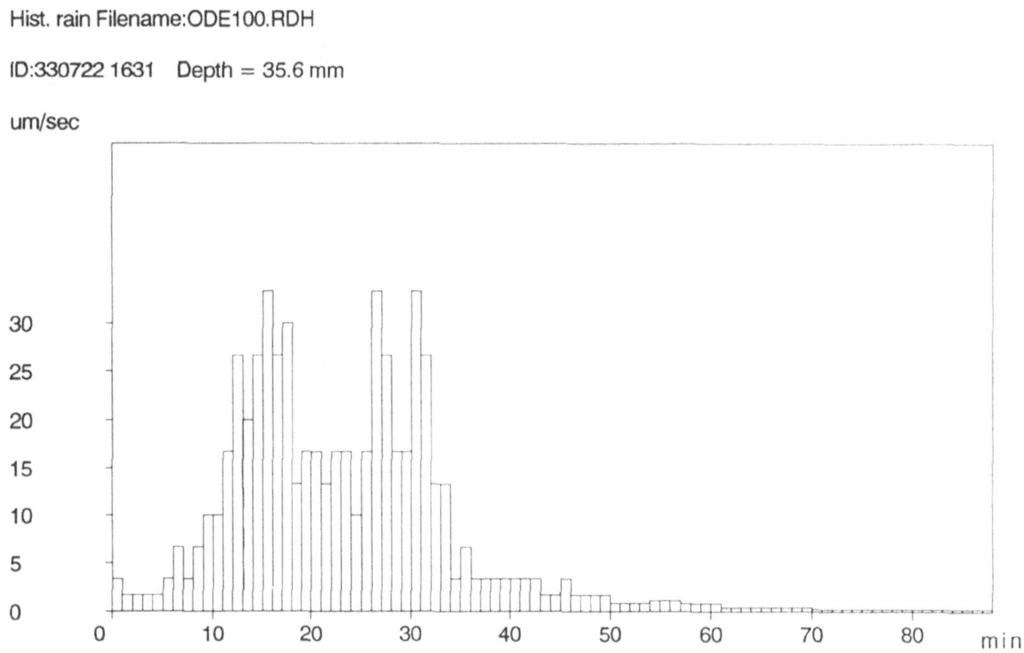

(a)

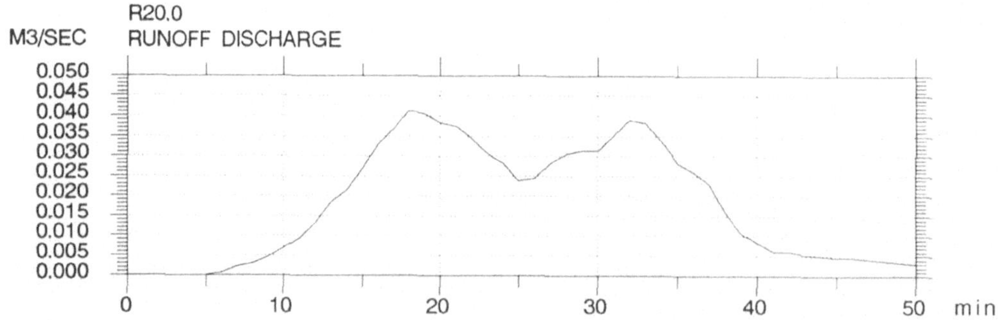

(b)

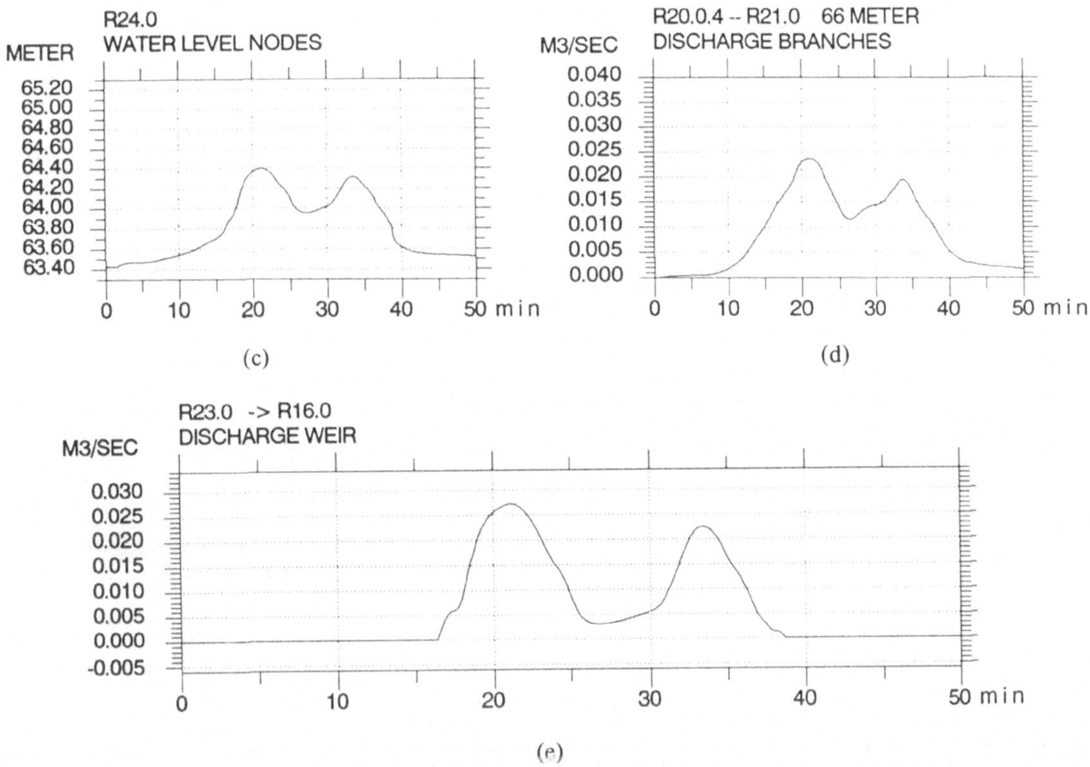

Fig. 6. Typical outputs from the sewer-system simulation package used as an example. The mean areal rainfall (a), occuring over time, provides a runoff at each intake point in the storm-sewer system (b). The system provides water-level variations at all nodes (c) and discharge variations in all branches (d) as well as flow quantities at collecting structures (e). These results can of course be augmented by morphological and water-quality-superstructure computations.

Sorties typiques du code de calcul de réseau d'égouts utilisé dans cet exemple. La pluie (a) en fonction du temps, engendre un débit à chaque point d'entrée du réseau d'évacuation des eaux d'orage (b). Ce système fournit les variations du niveau d'eau à chaque noeud (c) et les débits dans chaque branche (d), de même que les débits dans les structures collectrices (e). Ces résultats peuvent évidemment être complétés par des calculs sur la morphologie et la qualité de l'eau.

Hardware requirements

Earlier systems of the type described here were designed for running on 16- and 32-bit MS-DOS-based micros with fixed disk storage, mathematical co-processor and a minimum RAM of 512 kbytes. A printer had to be available, and a plotter could also be accommodated. Later systems have migrated to more advanced personal systems and to work stations, running on UNIX and UNIX-like operating systems and with practically no limit on RAM requirements. In principle, such a system is independent of the specific host-machine operating system and does not require any external service programs. The system illustrated here was originally made available on floppy disk for a wide range of 16-bit micros. Modifications for installation on more advanced personal systems, work stations, mini-computers and main frames have since followed.

A simulation system of the present type typically contains some 50,000 (FORTRAN, Pascal, C or C++) instructions, of which not much more than 15–20% are normally allocated to the numerical algorithms (Cunge, 1991). Two-dimensional flow systems exceed 100,000 instructions even without water quality, morphological and other such superstructures, but in these systems some 35% is given over to computation. The man-machine interface has also advanced beyond the simple example shown here, as exemplified in Fig. 7.

The results of a simulation can also be extracted for processing by other systems again. Thus, for example, by linking the simulated results to a GIS, the core of an integrated management system is provided. In the same vein, output can be processed for the purpose of on-line control optimization (e.g. Lindberg et al. 1990; Larsson et al. 1990; and references given in these papers). In control-system applications, the simulation results do not figure as the primary concerns of the user, and that which is called "the model" in simulation becomes only a means of "capturing", or "encapsulating" the knowledge of the physical domain the behaviour of which is to be optimized. Thus what was "the model" comes to function more as a *domain knowledge encapsulator*. In fourth-generation modelling, therefore, we already start to make a distinction between "simulation products" and "encapsulation products" – a distinction that becomes much more pronounced again in fifth-generation modelling (Abbott, 1989, 1991).

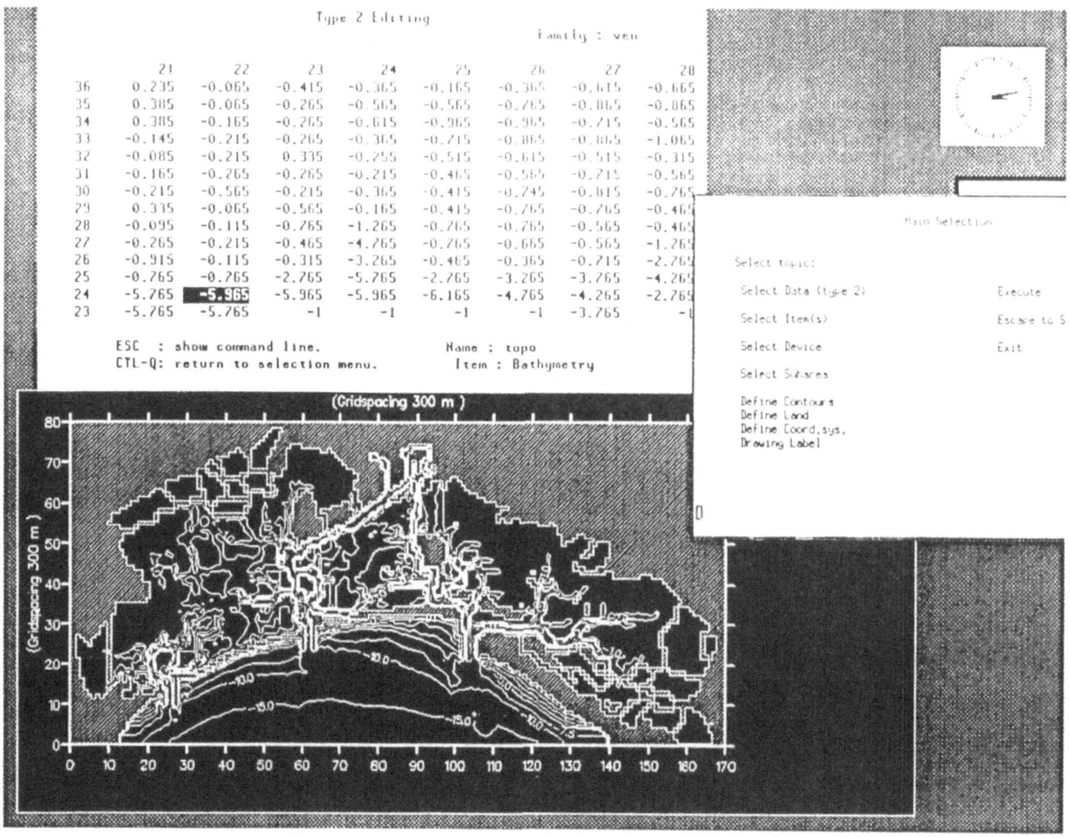

Fig. 7. A typical working environment is here shown for a two-dimensional modelling system. The user opens windows to provide space for menus, help facilities, access to parameter sets and other working features. For the purposes of system development, windows can also be opened for source code display, so that the system can also serve as a tool for its own development.

Vue typique d'un calcul avec un modèle bidimensionnel. L'utilisateur ouvre des fenêtres pour obtenir de l'espace pour les menus, les utilitaires, l'accès aux paramètres et autres éléments nécessaires au calcul. Pour le développement des systèmes, on peut ouvrir des fenêtres donnant le source de telle sorte que le système peut être utilisé pour son propre développement.

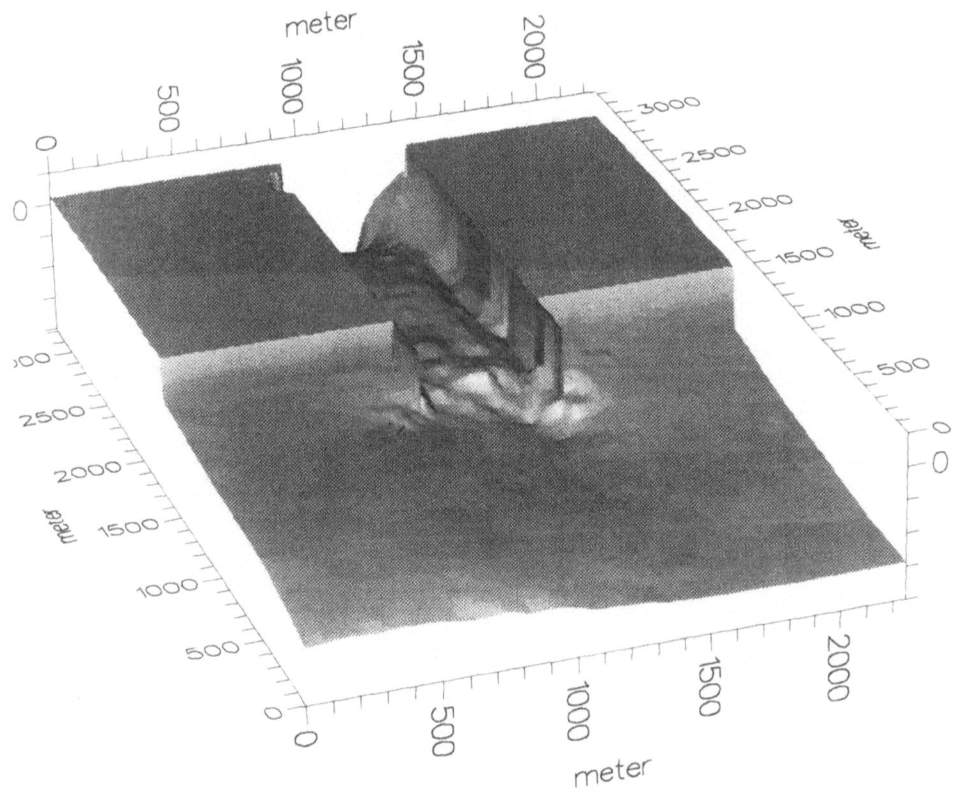
meter
0
500
1000
1500
2000
3000
2500
2000
1500
1000
500
0
meter
Z (m)
0
2500
2000
1500
1000
500
0
meter
0
500
1000
1500
2000
meter

(a)

(b)

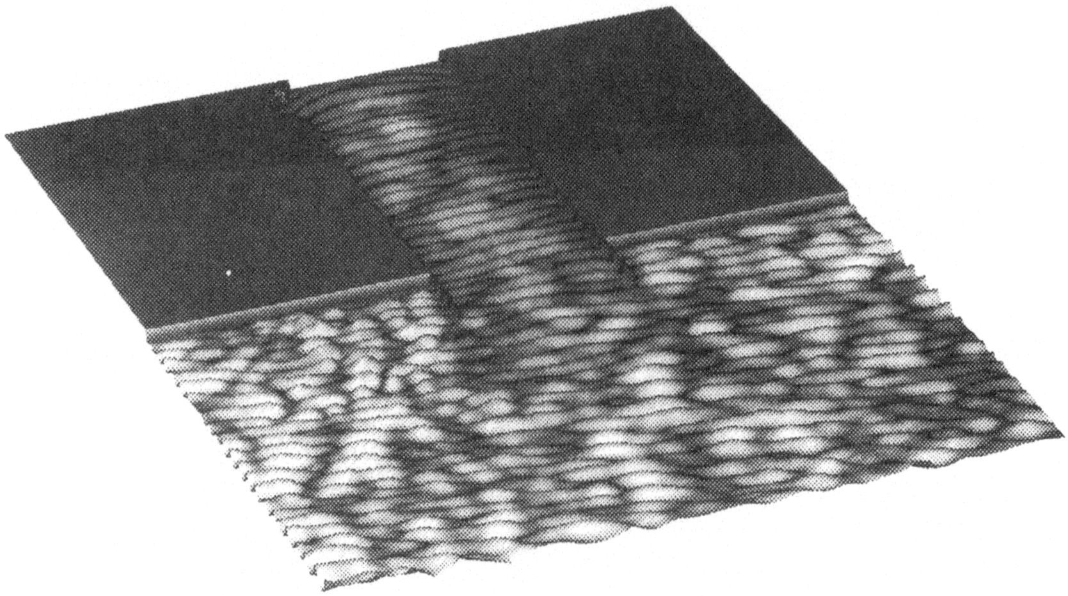

(c)

Fig. 8. By way of an illustration of the use of the graphics facilities made available by using proprietary subsystems, a perspective plot of waves entering a lagoon inlet, (a), are shown here: (b) shows waves with frequency spreading only and (c) waves with both frequency and directional spreading. Models of this kind are quite large (about 150,000 grid points, typically running over some 2,500 time steps) but they run comfortably on standard work stations. The vertical scale is of course greatly exaggerated in these illustrations.

Au moyen d'une installation de l'utilisation des aides graphiques disponibles dans des sous-structures, on montre une vue en perspective d'ondes entrant dans une lagune (a): avec de la houle aléatoire (b), puis avec de la houle aléatoire et multidirectionnelle (c). Les modèles de ce genre sont assez gros (environ 150000 noeuds sur 2500 pas en temps), mais ils passent sans difficulté sur station de travail. Sur les illustrations, l'échelle verticale est évidemment très exagérée.

The transformation of a third-generation system into a fourth-generation system

The existence of a sound and well-proven third-generation system is the *conditio sine qua non* of every successful fourth-generation modelling enterprise. This is because such a third-generation modelling system will have had a long history of practical application behind it, so that its design team will have learnt from practice, and the essentials of what they have learnt will have been incorporated into the system software itself. This last process is commonly called "learning at the level of the code". Moreover, the practice of third-generation modelling, if at all successful, will have led to a system that has the features that its end-users require, so that it will have "tracked the market", so as to have "learnt from the market". At the time of transformation of the third-generation system into a fourth-generation system, it is necessary that the code and its design team have all of this development phase behind them: because of its relation to the market, this is a development that cannot be simulated or otherwise reproduced economically outside of a third-generation environment.

It is often convenient - although incomplete - to regard the *total development* of the third-generation system as the *scientific development* of the fourth-generation system. As already introduced above, the further development of the fourth-generation system does not proceed primarily along the axis of science, but much more along the axis of the application of science to the changing "outer world" of the users of the system, who are not usually computational-hydraulics specialists. It is most important to emphasize, however, that *this further development still constitutes hydraulic research,* even though it is not primarily scientific research. It is, in a very essential sense, technological research (Heidegger, 1949/1977; Abbott and Basco, 1989; Abbott, 1991). This is research into how hydraulics works are currently designed, and how they might be better designed in the future through the use of the tool; of how hydraulics controls are currently operated and how they might be better operated in the future; and of how water resources systems generally are managed and controlled, and how these also might be better managed and controlled in the future. This is to say that the axis of advance of hydraulics through fourth-generation modelling is primarily in the direction of the social application of hydraulics. This kind of research is conducted both by the producers of the system and by various classes of user also. It is currently a particularly interesting area of university research.

The nature of the technological research that underlies fourth-generation modelling can perhaps best be described by starting out from the "front end" of the system, which in the case so far followed is its menu system. In most current systems the menus themselves are seen to occupy certain "levels" in the system, so that they constitute a hierarchy. The ordering of this hierarchical structure must then reflect the orderings of all possible uses of the system, so that it must reflect the ordering of those operations through which works are designed and operated, water systems are managed, and so on. We have here to do with *decompositions* of a large number of work processes, the representation of these in appropriate symbolic forms, and the *recomposition* of all such representations into an integrated design, operation or management tool. At the moment, most menu systems allow only "vertical movements" through the menu hierarchy: the user has to go back to a higher-level menu before descending again. The more the user becomes experienced, however, the greater the need for "lateral thinking", and so for "lateral movements" in the system. As the need commonly varies in this case with the type of application, such "lateral-movement facilities" are difficult to implement. Thus the design of this part of the fourth-generation system alone necessitated extensive research into "the way things are done", over a great many different types of application and over a wide variety of general-social and specific-institutional environments.*

It then necessitates a parallel research into the representation of all these processes in symbolic form, synthesised into specific menus and relations between menus, and then in such a way that certain "completeness conditions" over the field of possible applications are fulfilled. Moreover, as already outlined, this process of successive analysis and synthesis must take account not only of how things are now done (the "initial conditions" of the application of the system), but also of how they may be much better done when the system is in use (the "subsequent" conditions of the application of the system). In effect, the fourth-generation system is not only a tool for solving a given preconceived set of problems, but it is also *a tool for leading its user into new ways of doing things,* and even into new ways of thinking about what this user-he or she, individually, - is in fact doing. From this point of view, the purpose of the fourth-generation modelling system can be characterised as one of "raising the level of the discourse of average intelligibility of its user" (Abbott, 1989; Abbott, 1991).

From the point of view of product planning, and of the strategic thinking that underlies this

* It is indicative in this respect that one producing organisation is currently employing a psychologist in one of its design teams!

entire enterprise, this characterisation (which derives from the "philosopher of technology" of our era, from Heidegger, 1927/1962, e.g. pp. 212–213) is entirely central.

The menu system itself is, of course, only the "front end" of the total system, reflecting its modes of operation as these are initially constituted and as they may be expected to change under the impact of the use of the system. However, these menus have themselves to "drive" the core of the system, its computational component, together with its data-base and input-output environments. It is then often convenient to think of the menu system as driving "downwards" orthogonally from the surface of the screen, or paper. The menu drives then proceed, in the example used here, to the controls of the three computational and data-base components. Clearly then, the original third-generation system must be completely reconstructed on its control side in order that it can work interactively with the user of the total system. This is to say that a great deal of research has to go into the way in which control is passed through the system, working interactively with its user, and as its mode of operation changes with increased user familiarity.

Many of these new ways of passing control have, in their turn, certain influences upon the computational processes and data-base-access processes of the total system. In many cases, the computational and data-base components are altered so that explicit control functions are avoided: there occurs, so to speak, an automatic passing of control through the "central", or "core", computational and data-base components. For example it would be possible to follow the development of subcritical flow to supercritical flow and back, in space and time, through the setting up of a control function (Abbott, 1979, p. 200) and by the introduction of "if-then" control-statements and more general sweep algorithms. However, it is much more convenient, just to keep to the example already illustrated, to reduce the convective-momentum effects so as to provide a parabolic-wave-like mode of propagation, and so simultaneously to introduce a forward and upstream weighting within a flow-adaptive implicit scheme, whereby subcritical flow and supercritical flow are comprehended entirely within the one computational component (Havnø and Brorsen, 1986). By these means a great simplication is attained on the control side and the topological properties of the network remain inviolate, at the cost of only insignificant errors. Although in the case of this particular example the control simplification was already effected at the level of the third-generation system, in other cases such simplifications are best introduced when transforming a third-generation into a fourth-generation system.

In this kind of representation, there occurs a trade-off between the local accuracy of the simulation and the control complexity of the total system, and the nature and effects of such trade-offs have also to be extensively researched.

Alongside such changes on the computational side, other changes occur on the data-base side. The first and simplest of these is the introduction of such facilities as "help menus", scanning devices and warning notices that locate and identify possible errors, facilities for providing default values and settings, and various mnemonics and other aids to system use. A further and more complicated class of facilities have to do with minimising the consequences of improper applications of the system. It is well known that, despite the provision of introductory courses and the supplying of extensive documentation, many users persist in making certain common operational errors. (Indeed, the more extensive the documentation, the greater is the temptation to omit the reading of some sections of it, following the maxim that: "only when all else fails, read the manual"!). For example, unless appropriate measures are introduced, the novice-user may compile an extensive data base describing a particular water body and then come to erase this entire data base through a single improper command. Facilities have then to be introduced that will reduce the risk of failures of this kind.

The overall tendency in fourth-generation systems is for data bases to become more "professional", which is to say, for the most part, that they become more *structured*. To the extent that they take on the standards of data base types used more generally (e.g. structured-query-language (SQL) data bases: see for example, Date, 1989) so it becomes increasingly advantageous to use "bought-out", proprietary data bases in fourth-generation systems. This tendency in data-base technologies, of using more standardised products that have been developed for a much wider market as components in a fourth-generation modelling system, reinforces a general tendency in the direction of using standardised, proprietary components. Thus menu systems tend to be replaced by standard windowing systems, "home-made" post-processors by standard graphics facilities, and so on. On the one hand this greatly reduces the cost of a system for a given level of specification, but on the other hand it presents difficult problems of interfacing hydraulics-specialised components with a great variety of standard, bought-out, proprietary components. These problems are compounded by differences in standards in the machine-operating environments upon which the various configurations of components have to be run (e.g. ZENIX, AIX and OS/2 operating systems). Once again, research has to be directed into finding solutions to these problems, and for the most part this is not scientific research of the kind normally associated with hydraulics.

It follows from all this that the task of transforming a third-generation modelling system into a fourth-generation system is a long, complicated and expensive one. The elaboration of a fourth-generation system requires other talents, backgrounds, educations, organizations, marketing and management practices than those appropriate to a system of the third generation. (See also Abbott et al. 1987).

The limits of fourth-generation modelling

Fourth-generation modelling does not necessarily replace third-generation modelling; it only extends the range of application of modelling far beyond that which was formerly available. Indeed, for some computational-hydraulics experts, the third-generation device remains the preferred tool, as it appears to offer superior opportunities for personal initiatives and, generally, for the free play of fantasy of the modeller. In fact it is often thought that the user-friendliness of the fourth-generation system can only be bought at the cost of a marked reduction in the flexibility and range of its capabilities, but this is not necessarily the case. It is indeed a measure of the *value of a fourth-generation system* that it combines a high level of user-friendliness (including reliability) with a high level of flexibility and generality, (including overall accuracy). This value may be very high when it allows such a level of flexibility and realiability that it makes possible a much superior design or management of works, while it will be very low *(and it can very easily be negative!)* if it is inflexible and unreliable.* In this connection it is observed that a third kind of organisation already begins to intercede between the system builders and the end-users of the

* The computational-hydraulics expert will usually be able to get acceptable results in an acceptable time from even a very mediocre fourth-generation system, but the average user – who, though a professional hydraulician, rarely possesses a computational-hydraulic background – may easily be led into serious errors when using this same system. Indeed, such errors can then only be avoided when this user expends such an effort and time to attain to an acceptable level of computational-hydraulic expertise that any initial "saving" obtained by acquiring the mediocre system is quickly lost again. There is thus a real economic incentive to use thoroughly-engineered and reliable systems, despite their necessarily higher first cost – a cost that is still usually only a very small fraction of the costs of errors.

system. Organisations of this kind are directed towards customising systems for more special end-users applications, and even towards "personalising" them to particular application environments.

The value of a system is raised further, then, to the extent that the user of the fourth-generation system is provided with facilities that make it possible, with increasing familiarity and scientific insight, to use it with the flexibility of a third-generation system. These facilities may be then include "porting" facilities that enable its user to construct "add-on" features, thus customizing the system to the needs of the users or specific groups of users (e.g. such as a urban drainage system may be customised for use in airport environments, so as to treat specific de-icing and fuel-spill problems).

On the other side, a fourth-generation system is also not a fifth-generation language, environment, or tool. Characterisations of fifth-generation devices have been given elsewhere (Nielsen et al., 1987; Abbott, 1989; Cunge, 1989; Abbott, 1991; Lindberg et al., 1991; see also Fuchs et al., 1987), so that only the barest outline is in place here.

Very briefly, unlike the general fifth-generation device, the fourth-generation modelling system does not incorporate intelligent expert-advice-serving facilities and neither does it provide equivalent intelligent interrogation facilities (or *introspective* capabilities: see Genesereth and Nilsson, 1988, and Thayse, 1989). The difference between these generations devolves upon differences in the extent to which they make use of those branches of computer science that are subsumed under the one rubric of Artificial Intelligence (AI). Fourth-generation systems make very little use of this; fifth generation systems come to depend upon it.

From the side of practical applications, fourth-generation modelling has to do primarily with equipping the hydraulic engineer and some other users with specific "closed" tools and toolsets: in effect the engineer-user only "parameterises" these tools, so as to provide modelling and domain-encapsulating facilities. Fifth-generation modelling, on the other hand, has more to do with providing the engineer, chemist, biologist, sedimentologist or whatever other user with one or another, more general, "open" *working environment,* within which this user is able to assemble his or her own tools and tool sets, both for an own use and for the use of others. Corresponding to this, fourth-generation systems are written in only one, or at most two languages, whereas fifth-generation environment link together many diverse language-like elements (e.g. various shell-built expert systems, SQL data bases, objected-orientated data bases and lists and list processors, windowing systems and graphics facilities, i.e. elements which may be regarded as constituting statements written in a considerable variety of different languages). It follows that fifth-generation systems must employ quite other architectures than those currently used in fourth-generation modelling systems.

More generally still, both fourth-generation and fifth-generation modelling together constitute just one part of that area of technology for which the name of *Hydroinformatics* has been proposed and already quite widely accepted (Abbott, 1991; see also Larsson et al., 1990, for one of the first integrated applications). Hydroinformatics has to do with the whole gamut of information technology applied in hydraulics and water resources, and in this total picture simulation and encapsulation themselves constitute only relatively limited parts.

Conclusions

The direction of advance of hydraulics that proceeds through fourth-generation modelling proceeds along the axis of the social applications of hydraulics. When seen from this point of view, however, it constitutes the greatest advance in hydraulics that has been accomplished in recent times, in that the benefits of hydraulics' knowledge, and so of hydraulics research, are made far more widely available within society. By these means, a town engineer, a factory superintendent or a water resource planner comes to have at his or her disposal a body of knowledge and expertise that was formerly available to only a very few specialised research and design teams. Thus, for example, in Denmark and Sweden alone, the number of organisations with advanced, general-purpose urban drainage system and channel-flow modelling capabilities has increased from about 6 to more than 150 in just 4 years.

From the point of view of traditional hydraulics' research, on the other hand, this advance may at first appear much less interesting, and especially so since it appears to introduce so little that is of a scientific nature. Indeed, to the extent that scientific principles are invoked at all, they may appear to be used in a retrogressive way, such as exemplified earlier for the case where a fully-dynamic flow situation is reduced to a simpler parabolic-wave approximation when supercritical flow occurs. The essential point here, however, is just this: that research in this area remains *bona-fide* hydraulics research even when it is not scientific research in the usual, to-date-established sense. Ultimately, hydraulics has to do with technology - and research in fourth-generation modelling is *technological research in its most essential sense.*

What the fourth-generation system does call for, however, is a much improved education of hydraulic engineers in the principles of unsteady-flow hydraulics, turbulence theory and other such "advanced" topics, so that these engineers can better appreciate and use the tools that are now being made available to them (Abbott, 1991; Cunge, 1991).

References / Bibliographie

ABBOTT, M. B. (1979), Computational Hydraulics, Elements of the Theory of Free-Surface Flows, Pitman, and subsequently Longman, London.

ABBOTT, M. B. (1989), Review of recent developments in coastal engineering, in Falconer, R. A., Goodwin, P. and Matthew, R. G. S., Hydraulic and Environmental Modelling of Coastal, estuarine and River Waters, Avebury Technical, Aldershot, pp. 3-39.

ABBOTT, M. B. (1991), Hydroinformatics: information technology and the aquatic environment, Avebury Technical, Aldershot.

ABBOTT, M. B. (1991), A central issue of teaching within the hydroinformatics paradigm, in Colloque, Société Hydrotechnique de France, La Formation de l'Ingénieur Hydraulician Européen de l'an 2000, La Houille Blanche, 3/4, pp. 257-262.

ABBOTT, M. B. and BASCO, D. R. (1989), Computational Fluid Dynamics, an Introduction for Engineers, Longman, London, and Wiley, New York.

ABBOTT, M. B., HAVNØ, K. and LINDBERG, S. (1987), Reliability aspects of micro-based modelling systems, Proc. XXII Congress, IAHR, Lausanne, pp. 414-418.

ABBOTT, M. B. and IONESCU, F. (1967), On the numerical computation of nearly-horizontal flows, J. Hyd. Res. 5, 2, pp. 97-117.

ABBOTT, M. B., MCCOWAN, A. D. and WARREN, I. R. (1981), Numerical modelling of free-surface flows that are two-dimensional in plan, in Fischer, H. B., (editor), Transport Models for Inland and Coastal Waters, Symposium on Predictive Ability, Academic, New York.

BACH, H. K., BRINK, H., GLESEN, K. W. and HAVNØ, K. (1989), Application of PC-based models in river-water-quality modelling, in Falconer, R. A., Goodwin, P. and Matthew, R. G. S., Hydraulic and Environmental Modelling of Coastal, Estuarine and River Waters, Avebury Technical, Aldershot, pp. 568-577.

CUNGE, J. A. (1989), Review of recent developments in river modelling, in Falconer, R. A., Goodwin, P. and Matthew, R. G. S., Hydraulic and Environmental Modelling of Coastal, Estuarine and River Water, Avebury Technical, Aldershot, pp. 393-410.

CUNGE, J. A. (1991), Informatique, hydraulique numerique, hydroinformatique, in Colloque, Société Hydrotechnique de France, La Formation de l'Ingénieur Hydraulician Européen de l'an 2000, La Houille Blanche, 3/4, pp. 253-256.

CUNGE, J. A., HOLLY, F. M. and VERWEY, A. (1980), Practical aspects of computational river hydraulics, Pitman, London and subsequently Institute of Hydraulic Engineering, Univ. Iowa, Iowa City.

CUNGE, J. A. and LIGGETT, J. A. (1975), Numerical methods of solution of unsteady flow equations, in Mahmood, K. and Yevjevich, V. (eds.), Unsteady flow in open channels, Water Resources Publications, Fort Collins.

DATE, C. J. (1989), A guide to the SQL standard (second edition), Addison-Wesley, New York.

FUCHS, L., MULLER, D. and NEUMANN, A. (1987), A learning production system for the control of urban sewer systems, in Beck, M. B. (ed.), Systems Analysis in Water Quality Management, Pergamon, London, pp. 411-423.

GENESERETH, M. R. and NILSSON, N. J. (1989), Logical foundations of artificial intelligence, Morgan Kaufmann, Palo Alto.

HAVNØ, K. and BRORSEN, M. (1986), A general mathematical modelling system for flood forecasting and flood control, Proc. Int. Conf. Hydraulics of Floods and Flood Control, Cambridge, UK, BHRA, Stevenage.

HAYAMI, S. (1951), On the propagation of flood waves, Kyoto University Disaster Prevention Research Centre, Kyoto.

HEIDEGGER, M. (1927/1962), Being and Time, trans. Macquarrie J. and Robinson, E., Blackwell, London.

HEIDEGGER, M. (1949/1977), The Question Concerning Technology, and Other Essays, trans., Lovitt, W., Harper and Row, New York.

LARSSON, J., LINDBERG, S. and STAHRE, P. (1990), Mars - a real time operation and on-line optimization in the city of Malmö, Sweden, in Proc. in Proc. Fifth Int. Conf. on Urban Storm Drainage, Osaka.

LINDBERG, S., AMDISEN, L. K. and BABOVIC, V. (1991), Fault diagnosis in real-time operation of sewer systems, Proc. XXIV Congress, IAHR, Madrid, pp. D403-D409.

LINDBERG, S., NIELSEN, J. B., PETERSEN, S. O., AMDISEN, L. K. and HARREMÖES, P. (1990), Real time optimization of a sewer system as part of a larger municipal prototype project, including treatment plant operation, in Proc. Fifth Int. Conf. on Urban Storm Drainage, Osaka.

NIELSEN, J. B., LINDBERG, S., AMDISEN, L. K. and JENSEN, N. E. (1987), Towards the fifth generation of numerical modelling systems, Fourth Int. Conf. Urban Storm Drainage, Lausanne.

SAMUELS, P. G. (1990), Cross-section location in 1-D models, Int. Conf. River Flood Hydraulics, Editor White, W. R., Wiley, London.

STELLING, G. S. (1983), On the construction of computational methods for shallow water flow problems, Thesis, Delft.

THAYSE, A., Editor (1962), From standard logic to logic programming, Wiley, New York.

The following article has been reprinted from *Journal of Hydroinformatics*, Vol. 1, No. 1, pp. 03–19, with permission from the copyright holders, IWA Publishing.

3 © IWA Publishing 1999 Journal of Hydroinformatics | 01.1 | 1999

Introducing Hydroinformatics

Michael B. Abbott

ABSTRACT

Hydroinformatics is the name of a new way of applying knowledge as this knowledge is utilised in the worlds of the waters. This new way of applying knowledge, which is developing generally within our present-day societies, is concerned with ways to access and employ electronically encapsulated information, which itself becomes knowledge just to the extent that it is genuinely accessed and authentically employed. The knowledge of how to apply knowledge in the new way is thus itself a certain kind of 'metaknowledge'.

Key words | hydroinformatics, knowledge application, information, metaknowledge

Michael B. Abbott
International Institute for Infrastructural, Hydraulic and Environmental Engineering—IHE, Westvest 7, 2601 DA Delft, The Netherlands
E-mail: *mba@ihe.nl*

INTRODUCTION

This first issue of the *Journal of Hydroinformatics* marks a new and important step in the progress of this subject. It marks the point where hydroinformatics becomes accepted as a viable discipline in its own right, where it comes to presence in a more connected and homogeneous way, and where it attains to a certain permanence of form. It announces the fact that hydroinformatics has arrived, that it is already standing right here, and that it has come to stay.

Hydroinformatics carries into the world of the waters, the *hydro* of its name, all the developments in information, including communication, technologies that our present-day societies provide, thereby giving it the *information* of its name. We here take the word 'information' in the common colloquial sense of 'that which has the capacity to impart knowledge', rather than in the narrower but more strictly scientific sense in which it is used in classical information theory. From the point of view of the societies in which these developments occur, information and communication technologies are only the means to realise what is essentially a social transformation, namely one in which *knowledge becomes a kind of product*, to be produced, configured, brokered, marketed, transmitted, further transformed and ultimately consumed in new ways. It is then usual nowadays to associate this change with a transition from a *modern* to a *postmodern* condition within the societies so affected. In studies of postmodernism, these new ways in which knowledge comes to function are often associated through one common factor, namely a change from an emphasis upon *knowing* to an emphasis upon the *consumption of knowledge*. Thus, for Appignanesi & Garrett (1995, p. 107; see, more originally, Baudrillard 1963; Lyotard 1979//1992):

> The irreversible change from *knower* to *consumer of knowledge* is the cornerstone of postmodernity. This is the real historic change which legitimises postmodernism.

This general tendency within societies that enter into the postmodern condition then leads to the formation of a new way of employing knowledge, which in turn necessitates a new kind of knowledge, which is a knowledge of how to access, absorb and apply an electronically encapsulated knowledge, ubiquitously called information, which itself truly becomes knowledge precisely to the extent that is genuinely accessed, authentically absorbed and properly applied. This new kind of knowledge is therefore a certain kind of 'metaknowledge'. Thus, within the postmodern context, hydroinformatics is the name of this new kind of knowledge as it is applied to the worlds of the waters.

In the case of hydroinformatics, this general postmodern condition first made its appearance during the development of the fourth generation of modelling tools in the mid-1980s. These were numerical simulation packages that could be set up and run by persons who, although competent in fields to which the solutions were applied, had little knowledge of the way in which these functioned at the level of their numerics, their graphics, their other codings, and indeed all else of this technical-enabling kind. Every aspect of the operation of these tools – their input of site-specific data, their appropriate instantiation with this data for a wide variety of map projections, their calibration, their operation and the forms of their output – were automated. A division then already arose between the producers of the resulting 'packages', who thereby encapsulated their knowledge of the numerics and the control codes that realised the automation process, and the users of these packages, who made use of this encapsulated knowledge without needing to possess such knowledge personally (Abbott 1993). We accordingly nowadays commonly make a distinction between 'tool makers' and 'tool users' within this field. Naturally, a good tool-making organisation is simultaneously one that is a user of its own productions, but the number of the end users of the tools and the variety of their applications usually causes the tools to be applied over a much wider range of applications than are normally available to any one individual tool maker-user. For this reason, institutional arrangements have to be set up to link users of any specific tool to the makers of that tool, usually generalised over classes of tools. This introduces a new kind of social entity which links together local and regional user groups, local-language-service and other (regional) knowledge centres to an on-line service centre at the premises of the tool maker that is linked further to all relevant in-house specialists and their facilities. The on-line service centre may itself be constituted as a multilingual centre using standard commercial Computer Integrated Telephony (CIT) and Customer Relationship Management (CRM) systems. We observe correspondingly that even this first stage introduces new institutional arrangements, so that it already exhibits a *socio*technical dimension.

The first and most immediate consequence of the introduction of such fourth-generation simulation modelling tools and their social-institutional infrastructure has been that the number of organisations and individuals making use of the highest level and most comprehensive range of knowledge in hydraulics and water resources has increased by about an order of magnitude every five years. This development is shown as a graph in Figure 1, which also projects this development further forward in time on the basis of expectations that will be introduced presently.

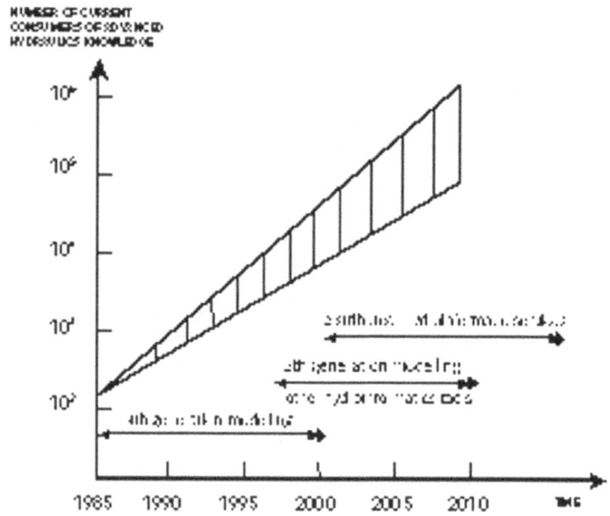

Figure 1 | Historical and predicted growths in the numbers of consumers of high-level knowledge in hydraulics, hydrology and water resources.

Alongside the elaboration of fourth-generation modelling tools, we may identify the following developments that have evolved directly within this first stage of the hydroinformatics paradigm:

- The assimilation of measured data into numerical models to provide fast and reliable data-assimilation capabilities.
- The analysis, design, installation and operation of combined measuring and numerical-modelling facilities.
- The further integration of such facilities with geographical information systems, and thus with a wider range of social/infrastructural services.
- The introduction of flood- and other early-warning systems.
- The introduction of real-time flood-management systems.

- The construction and operation of real-time control systems, especially for urban drainage networks and wastewater treatment facilities.
- Several enhancements to existing products and services, such as the on-line control of marine operations.

In order to realise these developments, several enabling technologies have been adopted and adapted, such as those of artificial neural networks (ANNs), genetic algorithms (GAs) and a host of enabling tools, environments and internet development technologies such as ActiveX, Java and CGI. Throughout this development we observe an increasing use of knowledge not only by persons who do not possess the knowledge itself, but increasingly by artefacts for which knowledge, as such, can have no meaning.

We further observe that, in almost all of these developments, persons and tools appear in close interaction, thus providing the *material overlaps* of sociotechnical studies (see, for example, Law 1986, 1991). Further to this again, most applications involve combinations of interacting persons and tools, and so interacting combinations of material overlaps.

THE TWO MAIN LINES OF CURRENT DEVELOPMENTS

Against the backdrop of the above developments, we currently discern two main lines of advance within hydroinformatics. The first of these is that of *data mining for knowledge discovery*, which promises to revolutionise the way in which knowledge is produced, represented and applied. Data-mining-for-knowledge-discovery environments are composed of combinations of persons and tools with strong material overlaps that transforms raw data into representations of knowledge that are accessible to the end-users or consumers of this knowledge. The second main line of development is often nowadays fitted under the one rubric of *knowledge management*. As we use the term here, this covers all the ways in which given bodies of knowledge are prepared for consumption so as to be easily distributed and turned to various uses within society. It is also concerned with the effects of changes in knowledge relations within societies upon power relations within these societies, and vice versa. Knowledge management in this more general sense is nowadays particularly directed to the distribution of knowledge as information over electronic networks, such as the Internet, and the corresponding social networks that then come into being.

The first of these two lines of development, that of data mining for knowledge discovery, has one singular advantage for the purposes of this introduction. This is that, as it is currently pursued, it is almost entirely technical, with few if any immediate social implications. The social aspects do not disappear, of course, but their consideration can usually be suspended, to be taken up later in applications, such as at the level of knowledge management. It is essentially this feature that makes this line of developments so much more congenial and correspondingly more readily acceptable to engineers and other professionals.

Reinforcing this tendency to professional acceptance again is the perception that data mining may provide a certain correction to many of the misuses and abuses of fourth-generation modelling tools, and especially to the practice of 'mindless calibration' of models from measurement that has unfortunately characterised so many of the applications of these tools. All too many examples exist where the calibration procedures have been used to cover over and mask gross misrepresentations of the physical realities of a modelled situation, so that the resulting model, although apparently 'well calibrated' is of little or no use for predictive purposes. In view of the now widespread use of simulation modelling tools and the frequency of their misuse, this corrective or at least palliative role of data mining finds a sympathetic reception in many places, thus making it acceptable already within present-day engineering practice. The real significance of data mining passes way beyond such immediately pragmatic aspects, however. Indeed, the question concerning the essential significance of data mining is critical in hydroinformatics.

'Questioning', observed Heidegger, 'is the piety of thought', so that in fact every such fundamental development in technology as this necessarily starts with a searching question. In the present case this question concerns the way in which knowledge is produced and the means used to represent it relative to a particular kind of consumer. Within the modernist paradigm, now already fast receding behind us, it was the 'knower' who postulated and formulated 'the laws of nature', as a kind of reflection of the world of nature experienced within his or her own mind. The mental reflection at a moment of time was provided with a certain permanence through the practices of certain ways of expression, and in the first place through *writing* (Eco 1976). The modern era was introduced and marked by one very special way of writing, which was a quite new kind of symbolic writing. This is to say that it was, and still is, a writing in which certain marks ordered on a place surface – ink marks ordered on a sheet of paper, chalk marks ordered on a blackboard, etc. – *replaced* certain isolated experiences of our world of nature for the purposes of facilitating our own, otherwise unaided, mental operations (Abbott 1999). Each such mark thereby constituted a *token*, as that which is given up in place of a something else, and in this period the predominant tokens used in science and technology became those that not only replaced, but effectively *effaced* that which they betokened, so that they functioned as *symbols*. In effect, *symbols are tokens that point to themselves*. We accordingly call this period in the history of human activity, with all its corresponding means of representing the matters of science and technology, *the symbolic era*. It reached its apogee in the setting up and playing of language games, its *mathesis*, in an extension and reformulation of an earlier mode of representation using the devices of geometry (Galileo 1638; Abbott 1992; see also again Abbott 1999). Its counterpart was a *taxonomia* which, although it employed symbols extensively, had little or no use for an algebra of the kind that was so essential to this kind of *mathesis*. We situate the period of vigorous growth of the symbolic era in the nineteenth century, although of course its practices have continued and proliferated much further into our own times.

When seen within a wider historical perspective, this development can be seen as providing a certain completion of the programme of 'Enlightenment', increasingly associated with a humanism in which a new kind of subject called 'man' became 'the centre of all things' (Horkheimer & Adorno 1944/1969//1972). As such, it was analysed and questioned critically even before the nineteenth century by such as Locke, Hume and Kant. The great achievements of the nineteenth century itself in science and technology largely drowned out even these voices of questioning, however, with only a few dissenting voices making themselves heard, such as those of Kierkegaard, Schopenhauer and Nietzsche.

It is really only in our own time, but then through an apparently quite other process, through the introduction of computer-based tools in which symbols are not employed explicitly at all, that we arrive at a new emphasis upon signs, understood as tokens that *point away from themselves* towards other objects, situated in the world of nature. Since so many of these tools – finite state automata, artificial neural networks, evolutionary programmes, phase-space reconstruction techniques, etc. – make little or no explicit use of symbols at all, we commonly describe them as *subsymbolic*. Obviously, the symbols still subsist at the level of the descriptions of the ways of working of these devices and precisely for the purposes of facilitating, and even enabling, our own understanding, but they now already come to live a more shadowy, secondary existence, at least one step removed from the immediate experience of the tool-using technologist or scientist. We speak of this new period in technology, and increasingly in science also, as *the postsymbolic era*.

The essential point about the symbolic era is that it locked its proponents into specific kinds of language games in which the systematic manipulation of symbols became the principal instrument for guiding human understanding. However, one cannot employ any language as an instrument without becoming, at least some extent, an instrument of that language. The scientist, and the technologist increasingly also in the symbolic era, framed the laws of nature (in the words of Kant (1787)) after his or her own design, where this design was expressed within the rules of symbol manipulation and in the very first place within the rules of algebra.

The postsymbolic era in technology is then characterised in the first place by a retreat from the position that it is the 'knowing human' who frames the 'laws of nature' after his or her own design. Instead, the person concerned, who is now more of 'a seeker after truth', sets up the means for nature herself to speak of her own designs, but then in a language which we humans can understand. Instead of setting up as a judge presiding over the affairs of nature, handing down judgements like a Kantian grand inquisitor to be executed by means of the physical interventions of humans and their machines in nature, the technologist becomes one who mediates between the ways of nature and those of humans. In order to do this, he or she must necessarily use devices that allow nature to express so far as possible her own ways, her own interests and her own desires. And then of course these devices that the technologist must necessarily use in this new mediating role are in the first place those for mining data collected with the specific purpose of providing insights into the own ways of nature and for making the products of this mining comprehensible to humans. Thus data mining for knowledge discovery is a first and essential requirements if technologists, and especially hydroinformaticians, are to perform this new mediating role between the world of nature and the world of human societies. Its precept is: *Let Nature speak!*

It should be clear again that those directly involved in data mining for knowledge discovery will not normally adduce these reasons for their activities, but will point to much more immediate and pragmatic objectives. And indeed it is essential that they should do this within the Heideggerian 'mundane and average world' with all its imperatives of financial viability. All ages have remarked on this discrepancy between the reasons that practitioners adduce for their actions and the deeper socio-historical reasons that have been subsequently identified to justify these actions, and indeed the adduced reasons and the historical grounds rarely, if ever, coincide. (Consider, for example, the observations of Tolstoy in literature, of Keynes in economics, of Heidegger in philosophy and of Jung in psychology.)

Every such change of paradigm naturally brings with it a new round of misapplications of the technology and a new crop of abuses within the social applications. Although examples of these are instructive and are unfortunately already plentiful enough, this does not seem to be the appropriate place to regale the reader with 'horror stories' about misapplications of these tools and technologies. The *Journal of Hydroinformatics* may perhaps still publish some 'Cautionary Tales' in due course and with appropriate circumspection.

KNOWLEDGE MANAGEMENT

The transformation in the role of the technologist generally and the hydroinformatician in particular to that of a mediator is necessarily associated with another transformation, but now one that is proceeding quite generally within present-day societies regardless of whether hydroinformatics participates or not. We have already introduced the initial stage of this transformation as a transformation in the role of technologists, and increasingly of scientists also, from 'knowers' to 'consumers of knowledge'. Now, however, we have to extend the range of persons involved in this process to include those who, although consumers of technology and scientific knowledge, are themselves neither technologists nor scientists, or at least they are not persons who are normally working professionally in the areas of concern here. We could then try to introduce this process as one of a *democratisation of knowledge*, but, as we shall shortly see, this is not entirely satisfactory, and from several points of view.

On the other hand, like so many half-truths, it does make for a nice slogan!

In the period that proceeds that which we are advancing as a 'hydroinformatics era', decisions influencing relations between the worlds of the waters, as the common property of nature and of human societies inseparably, as the 'Goods of the Earth' of Thomas d'Aquinas, are made by only a relatively few persons, who are described correspondingly as 'decision makers'. In one or another instance of last resort in democratic societies, these persons are elected, although in practice their decision-making powers are devolved upon other, appointed, unelected persons who are deemed to be 'experts'. These 'experts' are or are supposed to be 'knowers' in the original, modernist, sense. The aims of tools and tool users are still widely seen as those of helping the 'experts' to draw up proposals for human interventions in the world of nature that the 'decision makers' will subsequently implement or cause to be implemented. Even at the present stage of development, however, the 'decision makers' are left with very little room to make their own 'decisions', being constrained on the one side by the representations of their 'experts' and on the other side by the expectations of their constituents. Indeed they are often little more than ciphers in the real decision-making process. Moreover, even the part of the 'experts' in this paradigm is being increasingly subverted in practice by the first wave of hydroinformatics tools as introduced earlier, for several studies have already shown how in practice the 'experts' are in their turn often reduced to mere ciphers by the productions of measuring and modelling studies, whereby nothing much remains for these 'experts' to do than to approve the consequences that flow inexorably from the studies (e.g. Teknologiraadet 1995).

Further to this again, just as data mining might be regarded initially as a reaction to the abuses of calibration and other modelling procedures within an existing social context, so the development of network-distributed environmental impact assessment and decision-support systems can be regarded initially as a reaction to abuses of existing sociotechnical decision-making arrangements. It is widely understood, even it not so widely accepted institutionally, that a considerable part of all investments in the water sectors of human economies has been wasted, while the cost to the natural economy has often been immeasurable, and is in many cases irreversible. Indeed in some sectors, and especially in the case of irrigation schemes, it has been argued that the losses often outweigh the benefits; in the case of irrigation, the loss of fertile land owing to salination, the disruption of long-established social arrangements and the increase in water-borne diseases may well have outweighed the benefits of increased food and cash crop production. If we add to this the damage done to the natural economy, then the consequences of such interventions often appear to be very negative indeed.

In many such cases it has been recognised by at least some of the organisations responsible for these projects that the primary reason for this kind of situation is the absence of participation on the part of the local population (World Bank 1992–98). This population is the immediate repository of knowledge about the region itself and of its own social arrangements, whether explicit or implicit. If this population is to be involved, however, a whole new raft of tools is required that will help a great variety of persons to explicate their own positions and elaborate their own judgements about a number of proposed interventions. On this basis, they may then go on to make counterproposals, to organise themselves in support of specific positions and otherwise engage themselves actively in the relevant assessment and decision-making processes. Tools that promote the explication of positions and on this basis facilitate the making of judgements are called *judgement engines*. These tools must necessarily draw upon facts that are presented in ways that this new class of judgement-makers can best understand, such as may be provided already to some degree by fourth-generation modelling systems, measuring networks and data assimilation systems, GIS integration and other such facilities. Such fact-providing tools then becomes *fact engines*. One attempt to provide one component of a network distributed judgement engine is presented in this first number of our *Journal*, but several other such developments are already proceeding to the reporting stage. Such judgement engines will be complimented by other tools designed to facilitate the processes in which concerned persons organise themselves into interest groups and interact with the processes of an increasingly electronic governance of society by applying political pressure and proceeding to litigation and otherwise engaging in social–institutional activity. We may observe here already that, whereas simulation modelling systems and other fact engines are essentially instruments of enlightenment, judgement engines and their associated tools are essentially instruments of empowerment. The purpose of judgement engines and thereby their supporting fact engines in this area is to empower interested persons as genuine stakeholders in water resources.

Thus the notion that 'everything can be left to the experts' is being subverted further again by the second wave of the informational revolution, by the introduction of electronic networks for transporting and distributing knowledge, as initiated through the physical medium of the Internet. As the theory of semiotics tells us, any a such far-reaching technical development, working as it does at the level of signs, must have the most profound social consequences. The immediate result is the need for a new class of tools that can be characterised in the first place as networked-distributed environmental-impact-assessment and decision-support systems.

The management of the knowledge resources of persons and networks of persons and of the knowledge encapsulated in tools and network of tools, as these persons, tools and their respective networks interact, is, then, the proper business of knowledge management. Knowledge management has to do with the design of social environments and events, new business processes and organisational development initiatives. It is far wider than information management, even as it draws heavily upon the skills and products of information management. Further to this again, within the Anglo-Saxon business community the term 'knowledge management' is usually restricted to the knowledge resources of a particular company or other individual organisation, while we are using it here in a much broader sense. The same may be said also of the term 'information management'. Similarly, the Anglo-Saxon practice has led many organisations to institute a position that they call a 'Chief Knowledge Officer (CKO)' specifically to develop the knowledge core of their business.

Some observations need to be made about the position of papers about sociotechnical developments of this kind within the *Journal of Hydroinformatics*. The first of these is that such developments may appear as rather uninteresting from a traditional engineering, including software engineering, point of view: the technology does not appear to present anything like the same challenges as does that of data mining for example. In point of fact, however, even the technical challenges are still present, but they are of a quite other kind. It is in this case very much as in philosophy as described by Wittgenstein, in that the software industry has managed to produce any number of tools and even complete environments with little or no mutual compatibility, so that anyone who tries to produce integrated products using these tools soon gets tied up in the most agonising knots as they try to link the elements together. The technical difficulty is, in effect, to untie the knots that the software industry has (unwittingly) introduced. It is precisely because the knots have been tied in such complicated ways in the first place that one is obliged to go through such complicated motions, and even downright contortions, to disentangle everything.

The second general point that has to be made at this juncture is that, whereas most people will not claim to know a great deal about such enabling technologies as ActiveX, Java or CGI, almost all will consider themselves fully conversant with such main-line processes as those of gathering experience and making judgements. Accordingly, although most persons will understand the need to study the enabling technologies on the software side, they will not necessarily see much purpose in studying, let alone researching, the human-cognitive and social sides of the subject. Moreover, whereas there is a fair measure of agreement in a still quite restricted literature about the nature of ActiveX, Java or CGI, there is an immense spread of meanings and correspondingly an overwhelming confusion in an (indefinitely) extended literature even about such inference chains as are introduced elsewhere in this issue as:

$$\text{(beliefs, facts (data))} \rightarrow \text{attitudes} \rightarrow \text{positions} \rightarrow \text{judgements} \rightarrow \text{decisions} \rightarrow \text{actions.} \qquad (1)$$

The difficulty confronting the hydroinformatician in this area is that of making some sense out of this confusion, and so in this case of disentangling the complicated knots that have been tied by countless opinions over the millennia. Of course, hydroinformatics does not stand alone and unsupported in this situation, for it is as well sustained by the immense efforts, that have been made, and also over the millennia, to create some order out of this confusion.

By way of analogy, if we may compare the task of the data miner with that of cutting tunnels through deep and difficult geological strata in order to reach a vein of valuable minerals, we may compare the task of the hydroinformatician working in the sociotechnical area with that of hacking a path through a dense jungle of opinions, preconceptions and biases. And moreover, just to complicate this situation still more, this is a jungle that, for most of its denizens, does not even exist! Thus, that part of hydroinformatics that often appears from the outside as the 'softest' part, and therefore by implication the easiest part, in fact turns out to be the hardest of all. Most people simply use the words and relations expresses in (1) without reflection, with the self-assurance that they understand their working without further consideration. Heidegger expressed this situation to perfection already in his first major work, as follows (Heidegger 1927, pp. 168–169//1962, pp. 212, 213, see also Abbott 1991, p. 95):

> 'We do not so much understand the entities that are talked about; we already are listening to what is said-in-the-talk as such – what is said-in-the-talk gets understood; but what the talk is [really] about is understood only approximately and superficially.
>
> '. . . What is said-in-the-talk as such spreads in wider circles and takes in an authoritative character. Things are so because one says so. The average understanding . . . will *never be able* to decide what has been drawn from primordial sources with a struggle and how much is gossip. . . . The average understanding, moreover, will not want any such distinction and does not need it, because, of course, it understands everything.'

The third cause of the difficulties that we have to face here, and one that already prefigures the more contentious issues that we shall introduce next, is that this background to our sociotechnical studies can probably no longer be produced. For the most part, it can nowadays only be studied at a distance in time. So far as we can ascertain, our present-day societies can no longer produce a *Kritik der reinen Vernunft*, a *Sein und Zeit*, an *Erfahrung und Urteil*, or even a *Les mots et les choses*, any more than it can produce a Rembrandt self-portrait or a Beethoven symphony. The societies which made these peaks of creative activity possible have passed irrevocably from the face of the earth, just as surely as have the tectonic forces that created the Alps or the Himalayas. Of course, we can now do many other things that these societies could not. For example, these lines are written while looking across the waters of the Caribbean at a visiting American nuclear aircraft carrier which concentrates the greatest mobile potential for man-made destruction ever known. No previous society could have produced that! The point here is that every era presents its own challenges and provides its own means to meet these challenges. The task of the hydroinformatician is to meet the challenges facing nature and humanity in the world of the waters with the means that are placed at hand in his or her own times. And these *necessarily* include the means that have been bequeathed to us by earlier societies, even (and in fact precisely because!) these societies have themselves passed beyond recall.

Even as the brilliant instant of clarity of impression and thought became diffused and suffused through the rendering of that impression and thought in a more permanent form, it was only through this rendering that it could be carried into its futures, and thus into our own times. This is particularly true of the thought expressed in writing:

> *The word is the crucifixion of the thought.*

THE BUSINESS DIMENSION OF HYDROINFORMATICS

This brings us to the first of the more difficult and more contentious issues that face our *Journal* and its subject generally, which are those of the true purpose of hydroinformatics and thus the uses to which it is properly to be put. This issue is often presented as one of 'ethics', but that can be a misleading designation within a postmodern setting (see, for example, Bauman 1993). A better point of reference is that of the influences of changes in knowledge relations upon power relations, transforming in time into the influence of changes in knowledge structures on power structures. We then speak generally, following Foucault (e.g. 1966//1970) about the problematics of 'knowledge/power'. As introduced earlier, it is just to the extent that hydroinformatics realises its programme through the provision of knowledge to very many persons over electronic networks, and provides also the capabilities for people to express their fears, concerns and aspirations over these same networks, and thence to interact and organise themselves, that it must change power relations within society as a whole. The sign vehicles which the new communication technologies allow us to employ for the transmission, processing and exchange of knowledge lead already to changes in the nature and function of signs within these societies. Thus, just to the extent that these signs succeed in functioning as social forces, so they cause social changes, and then not only in power relations: they proceed further to change power structures.

The hydroinformatician cannot possibly remain indifferent to these changes, but must analyse and research them alongside his or her other analyses and researches of the means with which they are effected. The new means now being mobilised for hydraulic and related environmental knowledge to be produced, encapsulated, marketed, brokered, leased and purchased, distributed, and everything else of this kind, cannot be properly analysed and researched without a simultaneous and closely co-ordinated analysis and research of the social implications and repercussions of such a mobilisation.

So far, this analysis and research has led to three main conclusions that already influence the way in which hydroinformatics is developing in this direction. The first of these is that hydroinformatics can best be developed and nurtured within a commercial or business environment. The second is that the analysis and design of the new tools of knowledge management and empowerment must proceed on the basis of an investigation of the intentions of their users within appropriate physical and social contexts, so that they must be grounded in the field of science that is called *Phenomenology*. The third, which may appear superficially as conflicting with the first, is that the driving forces of hydroinformatics, the ultimate sources of its creativity, of its *poieses*, must also be investigated, but this brings us into a whole subject area of 'motivations' which must often appear as quite highly irrational. Such forces as drive the quasi-religious zeal of truly creative activity do not proceed on the basis of rational expectations and well-informed calculations, even though these may be employed by way of self-justification, but they proceed from a quite other place. We really have no alternative but to relate this place to the Kierkegaardian 'level of the religious' even though we are not then using the word 'religious' at all in its conventional sense. Since these conclusions are so contentious and the discussion of them is so far removed from the normal scope of a 'technical' journal, some explanation is necessary.

The relation between these theses, passing through the essential reciprocity of the first and third, was explicated already by Heidegger (1963//1977; see also Abbott 1991, pp. 78–79 and Abbott 1999):

> 'The flight from the world of the suprasensory is replaced by historical progress. The otherworldly goal of everlasting bliss is transformed into the earthly happiness of the greatest number. The careful maintenance of the cult of religion is relaxed through enthusiasm for the creating of a culture or the spreading of civilisation. Creativity, previously the unique property of the biblical God, becomes the distinctive mark of human activity. Human creativity finally passes over into business enterprise'.

The essential point concerning its business dimension is that hydroinformatics is a creative activity, but it is creative in a new kind of 'metaknowledge-producing' way, at least one part of which, roughly that of knowledge encapsulation and distribution, is unfamiliar to most persons already working in hydraulics, hydrology and water resources generally. Indeed, this new way of thinking is in several respects quite foreign to that which is followed within most of the established sciences and technologies, which are still dominated by the ethos of an all-exclusive 'knowing'. Hydroinformatics correspondingly encounters a strong resistance in most existing organisation, which are strongly stratified in the sociotechnical sense that it is not so much that persons-as-such are stratified by their education and experiences or that tools-as-such are stratified, but that the material overlaps between persons and tools employed within one stratum of the organisation differ considerably from those employed within other strata (Abbott 1996). Hydroinformatics is often seen accordingly as introducing a new stratum which is intrusive upon established ways of thinking, established procedures and established institutional hierarchies. It may even be perceived as something that is potentially destabilising within an already-established organisation. Like almost every other kind of new technology, it often appears as an unwelcome guest, supported only on sufferance, only as a way of 'staying in business' or of 'keeping up with the current fashion'. Thus, from the standpoint of many organisations, hydroinformatics is something that has to be carefully 'kept in its place', and preferably at a place as far as possible removed within the organisation from the point of application of the technology. The most difficult problems of applying hydroinformatics in practice are rarely if ever of a technical nature, but they are of a socio-institutional nature (see, for example, Abbott & Refsgaard 1998).

At the same time, of course, hydroinformatics has the potential to confer immense benefits upon society, and indeed it is already demonstrating this in many areas and in many ways, such as will be increasingly apparent as this *Journal* progresses. The basic problem of hydroinformatics at the level of its application is then that of changing or bypassing or otherwise getting around the socio-institutional roadblocks that are erected in its path.

The most immediate instrument that falls to hand for this purpose is then that of business enterprise. Hydroinformatics has to avoid the imposition of inappropriate organisational arrangements, it has to overcome uninformed (and often uninformable!) interventions in its activities and it has generally to escape from as many as possible of those constraints that are already only too familiar to practitioners in this field. Hydroinformatics has problems enough in meeting its responsibilities to society as a whole without having to fight strings of time- and energy-consuming battles within its own organisations. This it can best do by meeting the demands of the users of its products as directly as possible, through the setting up of new business arrangements and by practising business enterprise generally.

Unfortunately for this process, the notion still persists among many scientists and engineers that business is only about 'making money'. Now for some persons – financiers, business lawyers and other such opportunists – it may in many cases be no more than that. But for the hydroinformatician-become-entrepreneur it is really about something quite other again, which is *autonomy*. And then not just autonomy in relation to *how* things are done, but, as a consequence of an own kind of research and development, an autonomy in relation to *what* things are done. For the creative spirit, business enterprise is the means to sustain independence of mind and action, such as is the most conducive to the freedom to create in a responsible, truthful and thus pleasurable environment. Business enterprise, and even the money that may go with it, are only the means – even though still the necessary means – to provide the required creative environment.

It follows that hydroinformatics must be very much occupied with the researching, establishing and developing of its business arrangements. This aspect of hydroinformatics must accordingly find a place also in our new *Journal*. (See, by way of an earlier example, Thompson (1998).)

THE PHENOMENOLOGICAL DIMENSION OF HYDROINFORMATICS

In order to approach the most contentious issue of all in this area – and so by way of a halfway house on our way to the formulation of the purpose of hydroinformatics – we can best proceed through the consideration of the nature of phenomena in this subject. Our starting point is that specific direction within philosophy that studies how objects give place to phenomena within our minds, which study is called *Phenomenology*. Thus, in philosophical terms, phenomenology is the systematic study of our ways of thinking about our possible worlds. As a 'thinking about ways of thinking' it necessarily leads to circularities, but these have long since been identified and the precautions required to avoid their potentially vicious consequences have been largely established.

The need for some kind of study of phenomenology arises already in the design of any user interface. For example, even the elementary processes involved in the schematisation of, say, an urban drainage system in a simulation package necessitates a thinking about the ways in which the user of the package may be thinking or come to think about the urban drainage system itself. In the design of the user interface more generally, the designer has to make as systematic a study as possible of the various likely trains of thought of the various users of the package as they apply it in various situations and with a variety of intentions. A large part, if not all, of this process may receive a graphical representation, such as in the form of a directed graph as already exemplified in very general terms in (1) above.

Of course we know of no designer of a simulation package in hydroinformatics who has actually studied Phenomenology as a strict science (and so with an upper-case 'P') within philosophy for such purposes, although we would undoubtedly have much better user interfaces if these studies had been made! It is in fact still possible to manage without such systematic studies at this level. Clearly it becomes more difficult to manage on such an *ad hoc* basis when we arrive at the design of environmental impact assessment and decision support tools, negotiation environments and other such facilities with more varied and complicated human interactions, and for these purposes some study of Phenomenology is surely desirable, and may well be necessary.

For this purpose we may draw upon the great and irreplaceable studies in this area that were initiated in the nineteenth century by such as Bolzano and Brentano and which were brought to fruition by the twentieth century school of Phenomenology that was established in the first place by Husserl (in 1900/1901//1970). In general, these studies become increasingly important the more that we pass from primarily technical systems to essentially socio-technical systems. This is because it is these studies, more than any others, that enable us to escape from the morass of vapid opinion that otherwise threatens to drag down our thinking and drown us in empty speculations. Phenomenology in this sense enables us to make much more definitive statements about situations, events and phenomena generally and it is for this reason that it is commonly accepted as an 'exact science'. For example, in the design of knowledge management facilities it does not require much reading of Phenomenology to understand that any attempt to provide facilities using technical means alone will be at best suboptimal, and more likely unsuccessful. Indeed, a study of the logical requirements of a purely technical approach, following for example Husserl, should be sufficient to show the impracticability of such an approach. The notion of a 'knowledge centre' then comes to the fore, as a place, physical and virtual, where a variety of humans and their constructs work, together and interactively, while communicating with quite other activities of humans and tools outside the centre, situated at one or more 'peripheries' (Abbott & Jonoski 1998; Jonoski & Abbott 1998; Thein & Abbott 1998).

In the same vein, not only is some background in Phenomenology highly desirable in the design of such processes as (1) as these proceed during the operation of judgement engines, but they are essential when the operations of these engines by other persons must be made 'transparent' (Findley 1961). This occurs when the process that is dual to (1):

$$\text{actions}\rightarrow\text{decisions}\rightarrow\text{judgements}\rightarrow\text{positions}\rightarrow\text{attitudes (beliefs, facts, (data))} \quad (2)$$

has to be explicated with, for example, only 'actions' and 'facts [data]' as observables.

The consequence of the neglect of Phenomenology as a strict science are currently experienced most clearly in the failure to meet their objectives of many research and development programmes, whether at the local, national or international level. Even more to the point, the great damages that these programmes have caused and continue to cause in more general business-industrial terms, as exemplified by the virtual elimination of the computer hardware and basic-software industries in Europe, can all be traced to failures at the phenomenological level of understanding in research and development programmes. Our *Journal* can scarcely avoid giving some attention to these matters, albeit within an historical context and exercising the utmost restraint and the greatest decorum in order to avoid unnecessary pain and embarrassment. Done is done, and, after all, in the case of most of the research programmes 'it was only the taxpayers' money', while the distortion of competitiveness relations occasioned by these programmes could have been and in some cases was compensated by other strategies on the part of the disadvantaged enterprises.

In the case of development programmes also, the recognition of the social aspects on the one hand and the neglect of the overall phenomenology of the proposed development on the other hand, leads to situations in which the technical and social aspects do not complement one another at all adequately, and indeed may never come together properly at all. Indeed, on the basis of sociotechnical-historical studies generally we may propose the following principle:

The more experienced and brilliant the persons on the technical side of the project and the more experienced and brilliant the persons on the social side of the project, the more complete will be the failure of the project if these two sides are not properly co-ordinated and connected together.

As a particularly pertinent, because topical, example we may point to the stock market valuations placed upon several Internet companies, and specifically to search-engine providers. It is commonly observed that the enabling technology is really quite mundane and the social-application side often appears confused, ill-informed and poorly organised, but in fact it is precisely because of this mediocrity on both sides that it is possible to hold the two sides together and thereby make such financial-valuation successes of these enterprises. As probably the most widely available reference describing the highest level of ability on both the technical and social sides held together by a person of exceptional sociotechnical experience and brilliance, we cannot do better than refer to the semi-fictional figure of Jack Aubrey in the 18-volume series of Patrick O'Brian on the role of naval power and intelligence in the Revolutionary and Napoleonic wars of 1800–15. Indeed, the very fascination of these works, leading to their sales of millions in several languages, derives precisely from the pleasure they provide from their depiction of sociotechnical excellence. They illustrate the thesis, advanced fully in the spirit of Heidegger's teachings, that *a fascination with technology is a fascination with truth*.

Whether and to what extent our *Journal* can allow itself to comment at all critically on current programmes, and especially on certain development-aid programmes involving hydroinformatics components with sociotechnical dimensions, is more questionable, and remains for the moment undecided. It is a common experience that critical comments in such areas have little impact and rarely change anything, and it is usually better to make the best of the programmes as they stand and to proceed to other initiatives accordingly.

THE TASK OF HYDROINFORMATICS

The two main lines of hydroinformatics, of data mining for knowledge discovery on the one side and of knowledge management on the other side, clearly present great differences in the kinds of difficulties that they present, in the nature of their applications and in the manner in which they can be employed. And yet we have again to insist that hydroinformatics must proceed along both of these lines if it is to realise anything like its full potential. Without a data mining capability, the sociotechnical side will be severely restricted in its field of applications, while without a sociotechnical development data mining will be just as surely constrained in its scope. The first and primary task of the *Journal of Hydroinformatics* at the moment is to keep these two lines of study of our subject together.

It is when we come to ask *why* we should do this at all, however, that we arrive at our most significant, contentious divide. What is the true purpose of hydroinformatics? And, given that we can identify this purpose, what has that to say about our procedure in any particular case?

We have already introduced the level at which the creative drive of hydroinformatics, as of any other creative activity, must properly derive, as the (Kierkegaardian) 'level of the religious' (e.g. Abbott 1991). We have again at once to add that by this we do not refer to any social religions or their combination or their absence, but to a level of experience and activity that is so far removed from normal rational behaviour that, although we do not like to condemn it as 'inhuman', we can no longer strictly predicate it as 'human'. We must then identify it in the strict and proper (and theological) sense of the word as 'superhuman', and indeed it is quite common to speak of a 'superhuman effort' when referring to the corresponding exertions. This is a place where the division between the possible and the impossible is no longer clearly discerned, and where probability has no currency at all. This level of experience appears to be present in all humans whether they know it or not and whether they like it or not. Its manifestation is *faith* even when this is experienced within contexts that are apparently far removed from those associated with social religion. And then, of course, whatever the context, 'faith is a miracle, otherwise it is not faith'. It is indeed the lesson of all ages that all creation springs from this level of internal experience. Hydroinformatics as a technology and so as an act of creation, as a place where, in the words of Heidegger, '*aletheia*, truth, happens', must be founded at this level (Abbott 1991).

The creative act may here rise to the level of a 'passion' in the exact, and again theological, sense. We accordingly have to do here with all manner of behaviour which appears to be highly irrational within the social and specifically institutional context of the Heideggerian 'mundane and average world'. It is driven by forces that are usually hidden even to the individual so possessed and which frequently lead that individual into conflict with established mores and ethics: we then have to do here with the (again Kierkegaardian) 'teleological suspension of the ethical'.

Of course none of this can possibly be admitted in an institutional context, and least of all within a conventional business context. Thus, to take an example that is particularly topical at this time of writing (Wolff 1998, p. 101):

> 'You can't say to investors: I have a problem. A big problem. You can't say, I need your money to feed the mouths I have to feed. I need the money to pour down the maw. You can't say, Hey, what do you think is going on? There's a fire burning like crazy that we have to keep throwing dollar bills on.
>
> And that was, unmistakably, what [the other CEO] was saying. And while that was true of this business and of every other business in the new Internet industry and while everyone knew it was true - that is that cash was being consumed at a rate and with an illogic that no one could explain, much less justify - you must never, never admit it'.

Now two things must be said about this that are of vital significance to our *Journal*. The first of these is that although our *Journal* cannot possibly become involved in any discussion of these matters as such, it cannot avoid considering their influence within specific sociotechnical projects or situations. Moreover, these apparently 'irrational' influences may intrude not only through the individual experience of this 'level of the religious' in the creative process, but also at the social-application level as well. For example, a negotiation platform designed to assist in the settlement of disputes between partners who subscribe to Judaism in the one side and to Islam on the other side cannot fail to take account of the different kinds of values that are placed upon land within the respective social religions (Bany-Mustafa 1998). Similarly, the value of water as a social-unifying force in Buddhism must be taken into account in a decision support system in a Buddhist community (Thein & Abbott 1998). In the same vein, a network-distributed impact assessment system can best use different social nodes for communicating between its inner and its outer peripheral segments, such as schools and microbanking institutions in some Islamic communities and monasteries in certain Buddhist communities (Thein & Abbott 1998). The technical side then has to adapt correspondingly.

The second matter that unavoidably presents itself here is that any such discussions as may arise at all in this area can themselves only be conducted within a specific tradition. Thus not only does the mode of application of hydroinformatics change as we pass from a community with the one tradition to one with another tradition, and with this the technical means that are employed also, but our very way of writing about this must change also. We are often moving from a place where the waters of the world are experienced in one way to a place where they are experienced in a very different way, and our writings as well as our actions must reflect this difference.

Both of these aspects obviously lead to great difficulties in presentation, and these are exacerbated again by differences in languages and the uses of these languages between traditions. Our *Journal* must endeavour to overcome these difficulties within its pages.

THE TECHNOLOGIES OF PERSUASION

It will now be clear that hydroinformatics is not just concerned with the way in which man changes his outer world, the world which he shares directly with nature, but it is also concerned with changing man's inner world by providing the means for men to persuade one another in more equitable ways. The technologies that are used to change the inner worlds of individuals are known collectively as *the technologies of persuasion* (e.g. Norris 1993). These are the traditional technologies of advertising, politics and other activities that may appear at first sight to be far removed from the interests of hydraulicians, hydrologists and environmentalists generally - and which probably in the view of most of these professionals should be kept at as great a distance as possible from their list of concerns! And yet hydroinformatics is unavoidably drawn into this area also and once again regardless of its so-far-established inclinations.

In fact, these professional inclinations often have a sound foundation, in that the devices that pass as the technologies of persuasion are often morally indefensible and even downright obnoxious because of the evident wrong-mindedness of their applications. But we shall have to insist, and again with Heidegger, that a technology is defined as 'a place where *aletheia*, truth, happens' and the technologies of persuasion, if they are truly technologies at all, must 'tell the truth'. 'Being true to the technology' automatically equates to 'telling the truth'. That which is used for the purposes of misrepresentation cannot then be an authentic technology, but only *an imitation of a technology*. As Karl Barth explained so convincingly (1938–1950//1961) imitation is the hallmark of the inauthentic, of nihilism, of 'nothingness' (*das Nichtige, le néant*) as the vehicle of the lie. The one who is true to an authentic technology is thus the one who is automatically 'in the service of the truth', and those who have no use for truth have no use for a genuine technology.

The hydroinformatician whose drives originate 'at the level of the religious' is then obliged to expose deceptions, but even for this purpose he or she must resort to the technologies of persuasion to expose the productions of the psuedo technologies. We thus have to do with applications of the technologies of persuasion which have the purpose of combating and ultimately prevailing over other such (mis)applications. Phenomenology, however, being an exact science, cannot meet this kind of challenge, and indeed was never meant to do so. Although Phenomenology pays the greatest attention to intentions and intentionality within its own field of study, it cannot itself be the subject of specific intentions, and least of all to conflicting intentions. For these purposes something quite other than an exact science is required.

The whole purpose of hydroinformatics is to persuade, even if only to persuade local governments to proceed along one path of action rather than others, or to persuade investors to invest in one way rather than in others, or to persuade contractors to build in one way rather than others, and so on already indefinitely. As hydroinformatics proceeds further into its own version of the communication revolution, extending the range of its influence to millions of interested citizens and empowering them as genuine stakeholders in water resources, so the nature of its persuasive activity changes also. Its task now becomes one of, so to say, *persuading people to persuade people*, rather than leaving them as helpless spectators or, possibly even worse, as the victims of one or the other form of duplicity or coercion. The hydroinformatician has then to provide the means for facilitating processing of persuasion, which processes are always multidirectional. Since persuasion can only proceed through the agency of signs, this necessitates the research and analysis of a range of persuasive activities and their semiotics, and this in turn necessitates a serious study of the technologies of persuasion on the part of hydroinformaticians. Clearly a study of Phenomenology is essential to this purpose, but in this case it cannot be sufficient. It cannot serve as a moral foundation, and it is not well suited as a conceptual foundation either in this area.

Now it has in fact been argued, and by authors as otherwise differing as Karl Jaspers and Emanuel Levinas, that although Phenomenology-as-such cannot properly serve the purposes of insinuating the lie, it does have the capacity, *when suitably extended*, to serve the power of truth and thereby to expose the lie. Experience with other such extensions of Phenomenology, and especially the critical analyses of the extensions proposed by Heidegger and Sartre, do little, however, to support this argument (Barth 1938–1950//1961).

In effect, as the 1998 Encyclical Letter *Fides et Ratio* of John Paul II proclaims (p. 36):

> 'We face a great challenge at the end of this millennium to move from phenomenon to foundation, a step as necessary as it is urgent. We cannot stop short at experience alone; even if experience does reveal the human being's interiority and spirituality, speculative thinking must penetrate to the spiritual core and the ground from which it arises.'

Appeals to psychology, and specifically clinical psychology as an empirical science, or to psychiatry as a set of technologies in its own right, appear equally unproductive, with the possible exceptions of certain parts of the works of Carl Jung and his school, which then, however, do not really subscribe to psychology as an empirical science.

More potentially fertile are certain areas of literary criticism and especially those that cover the field of *pastoral* (often rendered as 'pastorale' in English), understood as the rendering of specific mental experiences with a minimum of signifying resources (Empson 1966). By understanding pastoral in this way, we place it within a general theory of semiotic economy. Thus, in the case of a graphical user interface, we may pose the problem of how we can express a certain fact, belief, position or whatever else of that kind, with the minimum number of pixels and the minimum of effort on the part of any specific user or class of users.

More fundamentally, such studies teach us, among other things, that 'understanding' occurs not so much in the signs themselves but in the 'spaces' between the signs, such as between two lines in a sonnet or between two adjacent illustrations in a strip cartoon. Associated with this feature, in the technologies of pastoral we increasingly meet instances of trains of inference that are not logical in any 'standard' sense at all – neither predicate, temporal, deontic or whatever. Instead, we enter into trains or strings of inference which subvert the 'standard' forms, even as these subversive forms do seem to have some kind of 'inner logic' that appears to defy any explanation in 'standard' logical terms. We accordingly say of these inference strings that they are *paralogical*. There are any number of current examples of paralogical devices, of which probably the most widely known are those employed in advertisements and in sequences of advertisements where again, in many cases, 'the message is in the spaces between the images'.

Current studies of paralogical thinking take many of their cues and much of their vocabulary from the work of Derrida. In her English-language edition of certain of Derrida's key essays, Kamuf (Derrida 1991) correspondingly exhorted her readers to 'read between the blinds', observing how the ordered sequences of signs through which knowledge is necessarily transmitted simultaneously obstruct the transmission process itself, so that (Kamuf, in Derrida 1991) 'these . . . could be thought of as slats of a venetian blind, of a jalousie which partially obstructs the view'.

The same notions currently arise also in applications of social value theory to knowledge management processes, where the social value of a working group's knowledge is observed not to combine as a simple sum of the knowledge of a collection of individuals and nor again only as an additive augmentation of these with the 'knowledge content' of their tools, but as well and essentially in the 'social space' between the individuals and their tools that come together to form the group. In the case of commercial organisations, this may be extended further again to encompass the social spaces between the organisation and its clients. These notions are essential ingredients in any attempt to estimate the social value and thence the money value of any organisation in 'the knowledge industry'.

Another set of technologies, used to great effect by the marketing organisations that nowadays run many political campaigns is that of *aporia*, understood as the induction of mutually inconsistent beliefs into individual minds and into collectivities of minds. All of these arcane technologies have of course been rather completely transformed and greatly strengthened by overall developments in 'the media'. The position of our *Journal* here is difficult, but it would seem best to accept some work on pastoral techniques while drawing the line at paralogical and aporiaic devices. In support of this position we may draw upon all traditions, which uniformly regard these devices as potentially dangerous and therefore to be avoided. As Jung so succinctly expressed the danger involved: 'One cannot possess this kind of knowledge without being possessed by it', and that is something that we would not wish on anybody (including ourselves!). Clearly this limit on the sociotechical side of hydroinformatics is far removed from the traditional concerns and interests of professionals in hydraulics, hydrology and water resources; but it is one of the principal duties of our *Journal* to wean these professionals over to new sources of nourishment and inspiration. Equally clearly, our task will not be easy. As a reviewer of a recent work on the playwright Samuel Beckett observed, 'In the 20th century it is axiomatic that the avant-garde is to be misunderstood'. On the other hand, one essential part of our purpose here is to ensure that the next century and millennium will understand the need for such a breadth of view and will come to embrace this range of interests.

CONCLUSIONS

Hydroinformatics is creative, poietic; it is a place where '*aletheia*, truth, happens'. It is therefore a technology in the full Heideggerian sense. We have now seen, however, that it is a technology that itself draws upon, combines and co-ordinates a considerable number and variety of quite other technologies, and even of some sciences. The hydroinformatician is thus every bit as much of a consumer of knowledge as is everyone else in the postmodern condition. Hydroinformatics is thus a kind of 'technology of other technologies, and sciences', and so a kind of 'metatechnology'. We have now seen how its knowledge content is directed on its input side towards an imbibing and consuming of quite other forms of knowledge than its own, and on the other, output, side to transforming, or 'refining' this knowledge into very different kinds of knowledge again: it consumes knowledge only in order to provide this quite other kind of knowledge. For the moment hydroinformatics does this mostly within its own frontiers, so to speak, in that it is the hydroinformatician who personally performs this work. However, with the more widespread and deeper-going application of electronic information-transmission networks, hydroinformatics aims much more to provide the equipment for other persons than hydroinformaticians to carry out this task. Hydroinformatics itself thus passes from a rhetoric of expertise to a rhetoric of persuasion, or from constative to performative modes of functioning. This is to say that hydroinformatics becomes directed also to providing the means for persons who are by no means hydroinformaticians, and who may have little or no knowledge of most of the knowledge that the hydroinformatician consumes, nonetheless to apply this knowledge in a responsible and valuable way. It is here in point of fact that the added value of hydroinformatics increasingly accrues, both in business terms and in terms of its meeting its human responsibilities.

It is in its role of a consumer of knowledge that this subject becomes 'postmodern', coalescing into a more general postmodern condition of society. It is essentially from this point of view that hydroinformatics may be regarded as a 'postmodern technology'.

Now, of course, many things that pass for 'postmodern' in our present-day societies are uncongenial to many of us: we would much rather that such things were not going this way, and indeed that these changes did not occur. The fact none the less remains that most of our current societies are moving in this direction whether we like it or not, and certainly whether we want it or not. Correspondingly, most of the best writing on postmodernism is by way of reporting on what is actually happening, and not on promoting it. This writing is primarily constative, and not prescriptive.

It is a dominant theme in every tradition that not only does each age present its own challenges to mankind, but each and every age provides its own means for mankind to meet these challenges. Hydroinformatics is placed firmly within such a tradition. The challenges facing mankind in its relation to the waters of the world appear to us as unprecedented, but at the same time the means that are made available to us, both from the past and in our own time, are just as unprecedented also. Hydroinformatics is all about meeting these new challenges by employing these new means. The works on postmodernism are concerned in the first place to present the challenges as these arise in our present-day societies, but it is the task of practitioners in all fields, and not least in hydroinformatics, to meet these challenges. To the extent that we succeed, so we change also the nature of these societies by changing the very way that they think and behave towards the worlds of the waters. From this point of view, not only is hydroinformatics a postmodern technology, but it belongs to a process of redefining the postmodern condition as a whole.

Some concluding words should be said also about the manner in which this *Journal of Hydroinformatics* should contribute to the development not only of hydroinformatics as a discipline, but to the body of hydroinformaticians as a community. A primary aim of this *Journal* is to promote this community; to provide a 'home' to those who have a community of purpose in this area, whatever the specific nature of their immediate interests, means and objectives. This aim of 'building community' remains as a beacon towards which we must constantly steer.

REFERENCES

Abbott, M. B. 1991 *Hydroinformatics: Information Technology and the Aquatic Environment*. Ashgate, Aldershot, UK, and Brookfield, USA.

Abbott, M. B. 1992 The theory of the hydrologic model, or: The struggle for the soul of hydrology. In *Advances in Theoretical Hydrology, a Tribute to James Douge* (ed. O'Kane, J. P.), pp. 237–254. Elsevier, Amsterdam.

Abbott, M. B. 1993 The electronic encapsulation of knowledge in hydraulics, hydrology and water resources. *Adv. Wat. Resour.* **16**, 21–39.

Abbott, M. B. 1996 The sociotechnical dimension of knowledge. In *Hydroinformatics '96* (ed. Mueller, A.), pp. 3–18. Balkema, Rotterdam.

Abbott, M. B. 1999 Forchheimer and Schoklitsch: a postmodern retrospection à la recherche du temps perdu. In *Memorial Symposium P. Forchheimer and A. Schoklitsch, Proc. 28th IAHR Congress, Graz* (ed. Graf, W. H.).

Abbott, M. B. & Jonoski, A. 1998 Promoting collaborative decision-making through electronic networking. In *Hydroinformatics '98* (ed. Babovic, V. M. & Larsen, L. C.), pp. 911–918. Balkema, Rotterdam.

Abbott, M. B. & Refsgaard, J. C. 1998 (eds) *Distributed Hydrologic Modelling*. Kluwer, Dordrecht.

Appignanesi, R. & Garatt, C. 1995 *Postmodernism for Beginners*. Icon, Cambridge.

Bany-Mustafa 1998 *The Role of Hydroinformatics in the Oslo Peace Process*. IHE internal seminar, summer semester, unpublished material.

Barth, K. 1938–1950 *Die Kirkliche Dogmatik*. Evangalische, Zollikon-Zürich//1961 *Church Dogmatics* (trans. Bromiley, G. W. & Ehrlich, R. J.). Clark, Edinburgh.

Baudrillard, J. 1963 *Le Système des Objets: la Consommation des Signes*. Gallimard, Paris.

Bauman, Z. 1993 *Postmodern Ethics*. Blackwell, Oxford.

Bolzano, B. 1837/1989 *Wissenschaftslehre, Bernhard Bolzano-Gesamtausgabe* (ed. Berg, J.). Fromann, Stuttgart.

Brentano, F. 1862/1960 *Von der Mannigfachen Bedeutung des Siendes nach Aristoteles*. Herder, Freiburg in Breisgau/Olms, Hildesheim.

Derrida, J. 1991 *Between the Blinds* (ed. Kamuf, P.). Harvester, New York/Wheatsheaf, London.

Eco, U. 1976 *A Theory of Semiotics*. Indiana Univ., Bloomington, USA.

Empson, W. 1966 *Some Versions of Pastoral*. Penguin, Harmondsworth.

Findley, J. N. 1961 *Values and Intentions: A Study in Value Theory and Philosophy of Mind*. Macmillan, New York.

Foucault, M. 1966 *Les Mots at les Choses*. Gallimard, Paris//1970 *The Order of Things*. Routledge, London.

Galileo Galilei 1638 *Discorsi e Dimostrazioni Matematiche Intorno a Due Nuove Scienze Attenenti alla Meccanica*. Elzevier, Leiden.//1914 *Dialogues Concerning Two New Sciences* (trans. Crew, H. & de Salvio, A.). Macmillan, London/1954 Dover, New York.

Heidegger, M. 1927 *Sein und Zeit*. Neimeyer, Tuebingen//1962 *Being and Time* (trans. Macquarrie, J. & Robinson, E.). Blackwell, Oxford.

Heidegger, M. 1963 *Die Technik, und Die Kehre*. Neske, Pfullingen//1977 *The Question Concerning Technology and Other Essays* (trans. Lovett, W.). Harper, New York.

Horkheimer, M. & Adorno, T. W. 1944 *Dialektik der Aufklaerung*. Soc Studies, New York/1969 Fischer, Franfurt am Main//1972 *Dialectic of enlightenment* (trans. Cummins, S.). Harder, New York.

Husserl, E. 1900,1901/1913 *Logische Untersuchungen*, Niemeyer, Tuebingen and Halle//1970 *Logical Investigations* (trans. Findlay, J. N.). Routledge, London.

John Paul II 1998 *Fides et Ratio*. Libraria Editrice Vaticana//Veritas, Dublin.

Jonoski, A. & Abbott, M. B. 1998 Network distributed decision-support systems as multi-agent constructs. In *Hydroinformatics '98* (ed. Babovic, V. M. & Larsen, L. C.), pp. 1219–1226. Balkema, Rotterdam.

Jung, C. G. 1944/1952 *Psychologie und Alchemie*. Rasch, Zuerich//1953 *Psychology and Alchemy* (trans. Hall, R. F. C.). Routledge, London.

Kant, I. 1787 *Kritik der Reinen Vernunft*, second edition/1979, facsimile edition, Reclam, Leipzig//1929 *Immanual Kant's Critique of Pure Reason* (trans. Kemp Smith, N.). Blackwell, Oxford.

Law, J. (ed.) 1986 *Power, Action and Belief: A New Sociology of Knowledge*. Routledge, London.

Law, J. (ed.) 1991 *Sociology of Monsters; Essays on Power, Technology and Domination*. Routledge, London.

Lyotard, J.-F. 1979 *La Condition Postmodern; Rapport sur le Savoir*. Minuit, Paris//1992 *The Postmodern Condition; A Report on Knowledge* (trans. Bennington, G. & Massumi, B.). Manchester University Press, UK.

Norris, C. 1993 *The Truth about Postmodernism*. Blackwell, Oxford.

O'Brian, P. 1970–1997 *Master and Commander*, through 16 other titles to *The Yellow Admiral*. Harper Collins, London.

Teknologiraadet, 1995 *Magt og Modeller: Om den Stigende Anvendelse af edb Modeller I det Politiske Beslutninger* (Danish Technology Council, *Power and Models; On the Increasing Application of Computer Models in Political Decision Making*, in Danish). Teknologiraadet, Copenhagen.

Thein, K. N. N. & Abbott, M. B. 1998 Internet-based management of water resources for a new Burma. In *Hydroinformatics '98* (ed. Babovic, V. M. & Larsen, L. C.), pp. 957–963. Balkema, Rotterdam.

Thompson, G. 1998 Hydroinformatics: a cornerstone to information technology in business. In *Hydroinformatics '98* (ed. Babovic, V. M. & Larsen, L. C.), pp. 9–14. Balkema, Rotterdam.

Wolff, M. 1998 *Burn Rate, or How I Survived the Formative Years of the Internet*. Weidenfeld, London.

World Bank 1992–98 *http://www.worldbank.org/wdr/*

The following article has been reprinted from Proceedings of the Memorial Symposium, with permission from the publisher and distributor, TU Graz Institute of Hydraulic Engineering and Water Resources Management.

GRAZ Forchheimer & Schoklitsch Memorial Symposium 1999
AIRH – CONGRES - IAHR

FORCHHEIMER AND SCHOKLITSCH : A POSTMODERN RETROSPECTION

Michael B. Abbott
International Institute for Infrastructural, Hydraulic and Environmental Engineering
Delft, The Netherlands

SUMMARY: This truncated archaeology of the meaning and understanding of the hydraulics of Forchheimer and Schoklitsch is used to bring out the changes occurring in such a technology as it passes from the modern to the postmodern condition of society.

Mais les instants d'extase, où les hasards d'une sensation présente permettent la renaissance du passé et nous donnent le sentiment joyeux de notre permanence, sont peu nombreux dans une vie.

André Maurois, Preface to the Pléiade Edition of Proust's ***A la Recherche du Temps Perdu.***

Introduction

The working lives of Philipp Forchheimer (b.1852) and Armin K. Schoklitsch (b.1888) together covered a period of hitherto unprecedented technological development, spanning the era between the first introduction of electrical power generation and distribution systems, through the thereby enabled developments of telephone, radio and television systems, into the era of digital computing machines. Among their many achievements, both made important contributions to the design and analysis of hydroelectric power plant and both left behind methods that contributed greatly to those first employed to apply digital machines to the solution of hydraulic problems. And then, of course, both taught at Graz.

In the modern, or rather modernist, tradition, Forchheimer and Schoklitsch might well stand as symbols of a particular order in technology, which we might then try to characterise as 'modern' technology. Their works would in that case serve to *symbolise* a 'modern era'. From the point of view which is to be adopted here, on the other hand, which is essentially postmodern — or perhaps we should rather say postmodernist — the works of these innovators and writers must stand in another relation to their age, serving more as *signs*, and thereby *pointing the way* towards an era, which they thereby *signify*. In the first, modernist, case we should have to refer to the works for what they still have to teach us, in order that we might come to know them better or attain to a deeper understanding of the matters that they treat. We should then approach these works in our capacities as *knowers*. In the second, postmodernist, case that is adopted here, however, we are concerned essentially with how they encapsulated knowledge and how they induced understanding. We thus now approach these works in our capacities as *consumers of knowledge*. For, as Appignanesi *et al* (1995, p.107) emphasise before all else: "The irreversible change from *knower* to *consumer of knowledge* is the cornerstone of postmodernity. This is the real historical change which legitimises postmodernism." Or, as the present author has observed in another place (Abbott, 1992): "We could previously navigate our way through the world, whether as individuals or as a profession, simply by building our models of the world; from now on, however, we must increasingly navigate our way through a world of models. But still, of course, *navigare necesse est.*"

It follows from this position that we shall have rather little to do here with what Forchheimer and Schoklitsch actually said; our attention will be directed almost exclusively to the way in which they said it. For this purpose a certain amount of preparation is then required, and so it is to this that we shall first turn.

The eras of 'modern' and 'postmodern' technology

The world of Forchheimer and Schoklitsch is gone forever; it has passed irrevocably: it is beyond any recall in our present-day experience. And yet many of us are in some part still of that world: we lived it even as it happened; we can still recall its music just as we can recall a performance of a Brahms concerto or a Mahler symphony. And yet we no longer exist, whether physically or mentally, in that world. We exist in and experience a world that has a quite other constitution, that is of a completely other composition. Those of us who span the space between paradigms must be forever torn between the anguish of the present experience and the pain of past memories, with each of these engendered and reinforced by the other. At best, we perch precariously between the endeavour and the regret.

(Even if only by way of an aside, we should observe that there is an essential simultaneity about these experiences. While I was reading an introduction to *Java* and its various programming languages, my wife asked me the function of a strange object that she had found after our last move. It was my old slide rule, now yellow with age. It appeared to me as the very embodiment of the abandoned order and I rather ostentatiously placed it on top of the glossy cover of the *Java* compendium so as to compose a still life in the manner of the old Dutch masters, a *vanitas, sic transit gloria mundi*. We are in the one world and in an entirely other world at one and the same time, and at each and every instant of that time.)

If we should put the first and most readily available of labels on these two coexisting and conflicting conditions, we might say that Forchheimer and Schoklitsch lived and taught the modern experience, as we did too, whereas we now also live and teach the postmodern experience. To live through the first was to live the *Heldenleben*, the excitement and elation of shaping certain isolated forms of the material world within a passive natural setting, while remaining ourselves in a state of spiritual stability. To live through the second is to live an entirely other kind of experience, which we might characterise as an uncertain search for an equilibrium between a hyperactive, complex and highly interactive artefactual world and an often strongly reactive natural environment, a search which is further complicated by our own closely-related states of spiritual turmoil and always threatening instability. Those of us who span the space between these paradigms live through the twilight but still comforting glow of the one experience and the lightening but uncertain dawn of the other, occupying a space that, although often dark enough, is still, so to say, lit from its two extremities.

Forchheimer and Schoklitsch constructed, and we constructed after them in emulation; now we just as necessarily also deconstruct, following paths that they would mostly not recognise – and just as little accept even if they could recognise them. The present study is just such a deconstruction, conducted as a search for a relation, and even a certain equilibrium, between our modern and our postmodern situations. Since from a post-modernist point of view this is always a search for an equilibrium between our social and individual situations as knowers, and in this case of technologies, and our social and individual situations as consumers of knowledge, as provided by these same technologies, it is necessarily an equilibrium that is situated between the modern and the postmodern *conditions*. But, what are we to understand in the first place by these expressions of 'modern' and 'postmodern' conditions?

The modern and the postmodern conditions

It was of the essence of modern man, and especially of later 'enlightened' man, that he trained his mental, and especially his logical, faculties to control his 'more primitive' emotions, even at the cost of his immediately intuitive faculties. The thereby-controlled person was that much better equipped to take the process of control further, and principally to social entities, and then at larger and larger social scales (*e.g.,* Horkheimer and Adorno, 1947 // 1973). These social entities could in their turn again take over an increasing control of the world of nature.

(We speak here exclusively of 'man' because, as a number of postmodernist studies have indicated, a woman could usually only enter into this modern world by renouncing an essential part of her womanhood, at least within the social-economic sphere.)

Postmodernist writers, on the other hand, often portray the modern condition as one of an increasingly socially-induced amnesia and self-delusion, whereby nature can no longer satisfy the truly essential needs of societies, societies can less and less satisfy the real needs of individuals, and the individuals, both men and women, themselves become ever more profoundly, and ultimately depressively, dissatisfied with themselves and with one another. The further convulsions of society as it grapples with this situation are then portrayed from the postmodern standpoint as expressions of *hypermodernity*.

To speak rather more abstractly from this point of view, modernism *objectified* the world: nature was seen increasingly as just so many objects that were available for control and manipulation, ultimately proceeding at all scales, even down to the level of the manipulation of genetic codes. Similarly, the social organisms that realised this programme were in their turn treated as collections of objects – as material and human 'assets' – that could be controlled and manipulated as required for some, usually one-sided, 'performative' purpose. Finally, the very individual treated him- or herself also as a collection of objects to be controlled and manipulated as required, and usually for some similar 'social' purpose. The whole world, from the totality of the creation down to the single individual, became, in the language of Heidegger, a *standing reserve* (*Bestand*): a collection of objects that are 'just waiting there' to be manipulated for use, available 'at the flick of a switch' for serving some further, and further-objectified, purpose. Indeed, when reduced to this level, as Heidegger has also explicated, these things are not even objects in the strict sense any more, since they are no longer present to consciousness, but are quite lost to view and forgotten, being entirely replaced by their simple functionalities.

All of this, of course, has its own enthusiasms and its own ecstasies, and it even has its own aesthetics. It has, correspondingly, its own vocabulary and associated thought world, as represented by such words as 'conquest' and 'domination'. It has just as much its own exhileration, which becomes social and spreads downwards even to the exhileration of the mob. At this lowest and most basic level of objectification, its aesthetic expression is perhaps epitomised by the hunting scene in the *Götterdämmerung* that surrounds the killing of the demi-god Siegfried. It is this, however, which unleashes all manner of untoward consequences, and specifically the overthrow of the old gods, as the repositories of traditional values, and the reclaiming of the gold of nature by nature through the flooding waters of the Rhine.

From the particular kind of postmodernist position that is adopted here, which, as Bruno Latour in particular has observed, has also many pre-modern precedents, one cannot even begin to function as a human being unless one treats other persons as human beings. The 'self-control' now becomes that of controlling our own traits of selfishness; the domination is now directed to overcoming our own narrow egoistic drives. This in turn leads to our acceptance that any control that we may have over others is inseparable from our responsibility for these others, and thus from their control over us, and that this can best occur through our own volition. More formally, no one can become a *subject*, as one who can develop an own individual character and thereby participate freely in the construction and deconstruction of objects, unless he or she treats other persons, and indeed ultimately all other forms of the creation, as subjects. Thus, in the words of Gabriel Marcel (1950; see Troisfontaines, 1968, Part 3, p.8): "To be a subject is not a fact or a point of departure, but a conquest and a purpose of life."

This particular postmodernist position (which is, however, by no means the only one of its kind!) extends naturally through social organisations, whereby the element of human consideration and solicitude must become ever more pronounced, if only by way of compensation, as individuals spend more and more of their time working through artificial media. The so-called 'caring society' thereby becomes also increasingly "a point of departure", and indeed *hydroinformatics*, as the present-day mode of application of hydraulics in society, is primarily concerned with providing the socio-technical means to realise this caring process (Abbott, 1991). This caring then proceeds equally inevitably to encompass the world of nature itself. The equilibrium that is sought by hydroinformatics is necessarily a sustainable equilibrium that is maintained between human economies and natural economies, both of which are water-based and therefore bound together by their sharing of the waters of planet Earth. Thus, for Matthews and Grabs (1994, p.15):

> Fresh water is the vital natural resource which supports all environmental activities, that is, the natural economy, and all human social activities, that is, the artificial economy. Therefore life on this planet is essentially an aquaculture living in a hydro-economy.

But, if we are ever to attain to such an equilibrium, or even advance towards it, we must question what has prevented us so far from doing so. What is there, for example, about the hydraulics that was established at the climax of the modern period by such as Forchheimer and Schoklitsch that allowed us to drift so much off-course from this goal? And what is there in or about this hydraulics that may still place us on-course again? Let us now accordingly open our first investigation of this question.

The geometrical period in technology

It has become usual since the works of Duhem (1913-1959//e.g. 1985) to start the clock of the era of modern science on March 7, 1277, with the condemnation by Etienne Tempier at the University of Paris of Aristotle's theses on infinity, space and time. This condemnation allowed scientific thinking to follow new paths, and these paths were initially traced through the use of geometry in a new way — just as the geometrical methods were themselves used increasingly to trace the actual paths of the heavens. From the point of view that asks *how* knowledge is and was recorded and transmitted, this new realm of application of geometry is of central importance. Knowledge is always transmitted between humans through *tokens* and the principal kinds of tokens are *signs* and *symbols*. A change in paradigm, whether in science or in technology or in whatever other field of human endeavour, is always marked by a change in the way in which signs and symbols are employed. Thus, that which Michel Foucault (*e.g.*, 1966//1970) has characterised as *an archaeology of knowledge* always proceeds through a study of the functioning of signs and symbols in knowledge transmission processes. From the point of view of the theory of the ways of functioning of signs, which is the *theory of semiotics* (*e.g.*, Eco, 1967) the most significant of the changes that mark paradigm shifts are recorded by the choice of the *sign vehicles* that are employed. Sign vehicles are, roughly speaking, the physical means that are employed in order to effect the signification process. Thus the application of existing geometrical methods in a new way marked a paradigm shift and one that can be traced in science as well as in technology. Thus, for Addis, speaking from the technological side (1990, p. 126):

> Although we may never know how the people of the time perceived what was happening, nor why it happened, there is absolutely no doubt that something extra ordinary occurred in the art of building design and construction in the 12th century. The nature of the 'Gothic design revolution' is apparent from the buildings themselves. After several hundred years of gradual development since the end of the Roman empire, the design of large buildings took a sudden and sharp change of direction.

Addis's studies leave no doubt about the impetus for this design revolution: the translation of Euclid's *Elements* in 1120, just seven years before the construction of the trend-setting basilica of St. Denis was begun. For want of a better name, the Abbot of St. Denis called this 'new look' an *opus modernum*, and this was subsequently called *Gothic* as something synonymous with 'barbaric', and thereby intended originally as a term of abuse. Addis explained:

> Geometry had, of course, survived as a practical art throughout the Middle Ages, but the appearance of Euclid did improve the level of geometrical knowledge that could be learnt. Improved geometry facilitated the more accurate 'description' of proposed building designs and was of great practical use in the construction process, in setting out the building and enabling the finished parts and their relative dispositions to be checked within better tolerances. Such an improvement alone would have enabled builders to contemplated larger and taller buildings.
>
> However, it was in its capacity to provide 'justifications' of designs that geometry probably had the more profound effect. Euclid introduced a crucial new ingredient — the notion of the geometrical proof. ….. Just as occurred 600 years later, philosophers put the new theoretical tool to use in every conceivable way and created, quite literally, a new type of geometry — *'geometria theorica'*.

This influence of the theoretical proof upon the practical analysis, design and construction processes was in fact emphasised already by Victor (1979, p.53: see Addis, 1990):

> The use of theoretical methods in practical geometry seems to have increased between the twelfth and the fourteenth centuries. At first their role was ancillary to the purposes of practical geometry. Once proofs had found a place in practical geometry, their role increased and changed. Theoretical proof became the goal even of practical geometry.

Thereby, as Addis observed (*loc. cit.*, p. 132), "simultaneously not only was a new type of architecture born, but also a new world view."

For our own part we may add that, just as the great Gothic cathedrals symbolised in their own time the new power of the Roman church, whereby salvation itself could proceed only through the intercession of the Church, *extra ecclesiam nulla salus,* so for us they may now serve to symbolise a change in the means of expression — the sign vehicles — that became available to justify and to realise their construction. We observe that this birth of a new paradigm in technology predated the pronouncements of Etienne Tempier by some 150 years, so that we may well suppose — as Duhem also suggests — that these were 'pronouncements after the fact' with the usual academic lag time. In any event it can be established that a new paradigm was initiated in technology a considerable time before the announcement of the beginning of the era of modern science.

We should now observe however that the geometrical construction persisted through the entire intervening period into the works of Forchheimer (1931, *e.g.* pp. 417, 452, 455,) and Schoklitsch (*e.g.*, 1923), albeit augmented and combined with another kind of sign vehicle, namely that of the curve of a graph. Figures 1 to 4 show typical geometrical constructions as these appear in the works of Galileo, Descartes, Newton and Forchheimer, with the last two of these introducing curves of graphs representing the behaviour of objects situated in physical space and time. Thus, we may establish that the works of Forchheimer and Schoklitsch carry the marks of an era that preceded that of the era of modern science by a considerable margin.

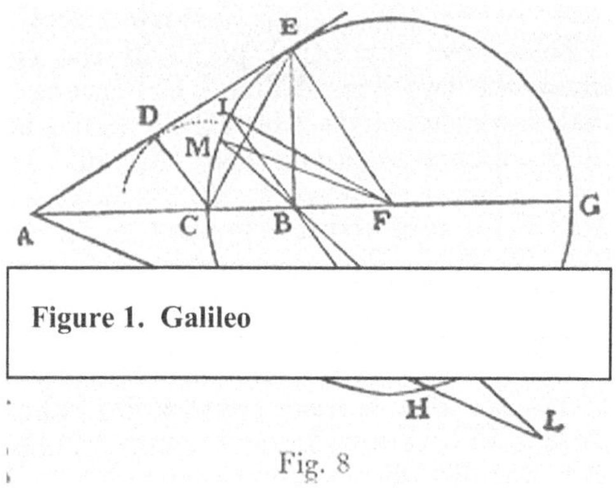

Figure 1. Galileo

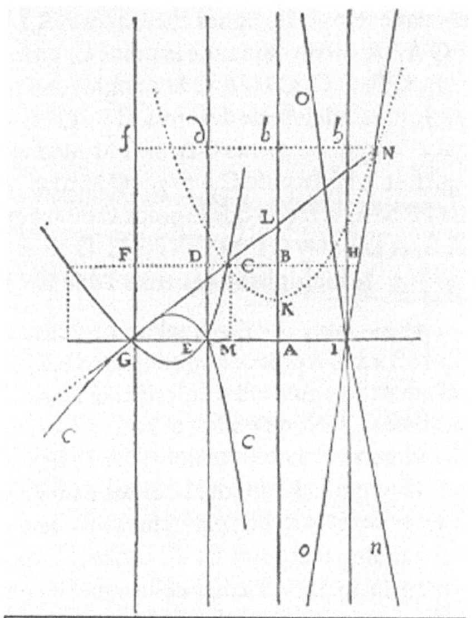

Figure 2. Descartes

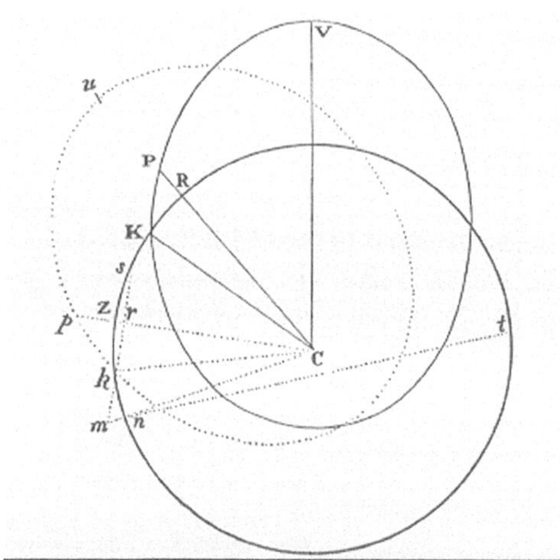

Figure 3. Newton

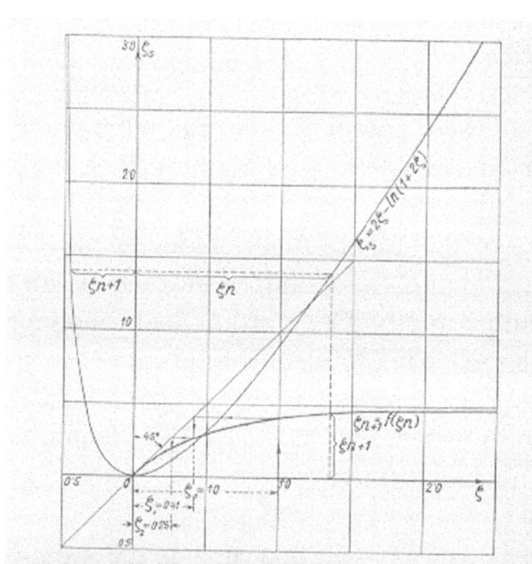

Figure 4. Forchheimer

The symbolic era in technology

The symbolic era that followed, apparently inevitably, from the "new world view" of applied geometry is commonly supposed to have been initiated by Galileo Galilei. It can be marked semiotically by a comparison between two earlier and lesser-known of Galileo's works, *de Moto Accelerato* (1590 +/-, *e.g.*, pp. 332/104) and *Le Meccaniche* (1600 +/-, *e.g.* -/p. 159 *et seq*), and Galileo's later and celebrated *Two New Sciences*, of 1638. Although Galileo thereby laid the foundations for the representations of the laws of dynamics in the algebraic-symbolic form that we still use today, the general process of representing the knowledge-content of the sciences in an algebraic-symbolic form was greatly accelerated by the almost-simultaneous appearance of the algebra of Descartes (1637//1925/1944). Descartes demonstrated, at least in principle, how the quasi-totality of problems that were expressed in geometrical terms could be translated into problems expressed in algebra, and of course *vice versa*. In semiotic terms, Descartes provided the means to transform the knowledge transmission process from one that was realised using the sign vehicles of geometrical constructions to one that was realised using the sign vehicles of ordering strings of symbols subject only to a very small number of fixed rules of algebra.

The geometric representations of motions, as exemplified by the representations of the Keplerian orbits of the planets as ellipses, had already transformed problems of motion to problems of geometry, and the algebraic representations of the resulting geometrical figures led in their turn and apparently just as inexorably to the mechanics of Newton. These were necessarily expressed most generally in symbolic form, even if only for discrete increments of space and time. We then usually date the full-blown symbolic representation of continuous processes from 1750, with Euler's formulation of what is still commonly called 'Newton's' second law of motion. Since it usually takes at least half a century for notions that are current in mathematics to penetrate into technology, it was only in the beginning of the last century that the description of processes in algebraic-symbolic forms became an accepted norm in technology, but then with such impetus and to such an extent as to attain to the status of an all-inclusive 'mathematical analysis' *per se*. The negative aspects of the reorganisation of the *Ecole Polytechnique* of 1816 in the spirit of Laplace, Poisson and Cauchy were clear enough to Olivier (1857: see also Weiss, 1982):

> These men, who knew no language but algebra, who thought that one was ready for anything when he knew algebra, who esteemed a man only to the extent that he knew algebra, who were incapable of rendering services to the country other than in algebra, destroyed from top to bottom the original organisation of studies at the *Ecole Polytechnique*....

This transformation naturally did not proceed unopposed even at the time, and indeed one opponent at the *Ecole Polytechnique* acerbically remarked that it would reduce that already venerable institution to an *Ecole Monotechnique*! Despite such opposition, however, as the military academies and other such institutions transformed into the *Technische Hochschulen* in emulation, so the symbolic representation became the principal *desideratum* in technical-scientific studies of processes generally, and indeed it came to represent the very hallmark of 'scientific-technical' advancement. Entirely consentaneously, the introduction of symbolic methods into a subject like hydraulics – which in this respect was accomplished by hydraulicians such as Boussinesq, beginning already in the middle of the 19th century – established the scientific legitimacy of hydraulics, this use of symbolic methods being the only acceptable 'mark of respectability' of a scientific- technical subject. Indeed, it was only by these means that hydraulics could be admitted within the university-level educational system. Again correspondingly, even quite closely related subjects such as hydrology had a much more difficult time in acquiring such credentials, treating as they did matters that were less susceptible to algebraic representations.

From this point of view, it was the remarkable property of algebra that it came to legitimise certain extreme forms of intellectual deprivation, whereby it became acceptable for certain academics to be ignorant of almost everything outside of a narrow area of specialisation, just so long as they could do algebra. Nowadays we often refer to such persons as 'mere symbol shufflers', but it may be less denigrating to speak here of a very particular kind of *intellectual asceticism*. Corresponding to this process, teaching and research organisations were created or adapted to accommodate this kind of activity, so that we may speak of an institutionalised ignorance, or a specific form of institutionalised asceticism.

Let us now however go back a step, to recall that a symbol constitutes only one particular type of token, where we now understand by a token any thing that is offered up to circumspection in place of another thing. We have already seen that in the case of the symbol the token replaces in our minds that which it betokens. In the case of the sign, on the other hand, it is that which is betokened, the significant of the sign, as our mind's view of the ultimate referent, which takes over our attention as soon as the sign has fulfilled its signifying function. To use a common figure of speech, through the symbolisation process the symbol takes over the stage, allowing that which it symbolises to bow itself off the stage, while in the case of the sign it is the other way around, in that it is the signified object, the signified and with this its referent, that takes over the action, while it is the signifier that bows itself off.

Following the theory of semiotics it is of the essence of the symbolisation process that there is a *symbolisation function* that maps the symbolised object into the symbolising object, while it is of the essence of the signification process that there is a *sign function* that maps the signifier into the significant, which is our mind's or whatever other's reflection of the outer-world or otherwise externalised referent. In the mathematical notation of category theory, which at once provides a convenient means of representing these process:

$$\text{Symbolised object} \xrightarrow{\text{Symbolisation function}} \text{symbolising object ('the symbol')}$$

or: $a \xrightarrow{f} b$ or: $f{:}a \rightarrow b;$

and: signifying object (or 'sign') $\xrightarrow{\text{signification (or sign) function}}$ signified object

or: $c \xrightarrow{g} d$ or: $g : c \longrightarrow d.$

One and the same object can commonly appear as: *a, b, c or d*, in the above relations. Thus, when Forchheimer (*Hydraulik*, 1930, p.47) makes verbal-descriptive distinctions between:

Das Druckgefälle $\gamma \dfrac{\partial h}{\partial x},$

Trägheitskräfte von Typus $\dfrac{\gamma}{g} u \dfrac{\partial u}{\partial x},$

Reibungskräfte von Typus $\frac{\gamma}{g} \nu \frac{\partial^2 u}{\partial x^2}.$

he does so in his subsequent discussion by referring to the properties of the streaming fluid, which ultimately provides the referent, so that he is using the tokens $\gamma * \partial h / \partial x$; $\gamma / g * u * \partial u / \partial x$, and $\gamma / g * \nu * \partial^2 u / \partial x^2$ as signs. When, however, he shortly later (p. 65) differentiates

$$t + \frac{1}{m}(e^{-mt} - 1) = -\frac{m}{g}\left\{H \log nat\left(1 - \frac{h}{H}\right) + h\right\} \qquad \text{Forchheimer's (37c)}$$

in which:

$$H = -\frac{n}{2g} = \frac{6\sigma}{\gamma d_\omega}\frac{1-\mu}{\mu} = 0.446\frac{1-\mu}{\mu}\frac{1}{d_\omega} \qquad (37d)$$

to provide:

$$1 - e^{-mt} = -\frac{m}{g}\left\{-\frac{H}{H-h}\frac{dh}{dt} + \frac{dh}{dt}\right\} = \frac{m}{g}\frac{h}{H-h}\frac{dh}{dt}$$

or:

$$\frac{1}{m}\left(e^{-mt} - 1\right) = -\frac{1}{g}\frac{h}{H-h}\frac{dh}{dt}$$

or:

$$\frac{dh}{dt} = \frac{1-e^{-mt}}{\frac{m}{g}\frac{h}{H-h}} = \frac{1-e^{-mt}}{\frac{m}{g}\left(\frac{H}{H-h} - 1\right)} = \frac{1-e^{-mt}}{\frac{m}{g}\left(\frac{n}{n+2gh} - 1\right)}$$

to obtain, by introducing the required initial conditions, the expression

$$\left(\frac{dh}{dt}\right)_{t=h=0} = \sqrt{-\frac{n}{2}} = \sqrt{gH} = \sqrt{0.446\frac{1-\mu}{\mu}\frac{g}{d_\omega}}$$

he is using the tokens H. $-n/2$, etc., as symbols, in that we no longer give any attention to the physical interpretations of these tokens, but we treat them only as mathematical objects, to be manipulated solely by the rules of mathematics, through the practice of algebra. In this last case these symbolic objects and symbolic processes have entirely replaced in our minds their 'corresponding' objects and processes in the material world. Of course, there must come a moment in any technology when we are again led to read some meaning into the expression obtained by these devices, but throughout the intervening argument we forget completely about such meanings. As a result, to the extent that we regard the rules of mathematics as 'laws of thinking' and we apply these laws to our mathematical representations of the 'laws of nature', so we may come to regard the rules of mathematics as 'laws of the laws of nature'.

We may already observe here, that the use of symbols allows us to erase from our minds all thoughts about the nature and reality of the 'objects' that are so symbolised. The shear facility with which we can order and reorder and act upon the symbols quite covers over the reality of the corresponding processes of ordering, reordering and operating upon the 'objects' themselves. However, these only become 'objects' through our own thought processes, and may in reality be, for example, living creatures that in no way deserve to be so treated in the world of nature. The use of symbols constitutes one of the principal means of objectifying the world in the manner of the Heideggerian *Bestand* (Abbott, 1991).

It is of the essence of a technology, as the process of circumspective creation in the world that its tokens are primarily those of signs, while those of its tokens that are employed as symbols are in fact only used to link together its signs. In the more 'pure' sciences, on the other hand, and especially in mathematics, the tokens are employed primarily as symbols, and indeed, in the limit, in an 'absolutely pure' mathematics, there would be no signs at all: we should have "a language that refers entirely to itself" (Foucault, 1966//1970).

Hydraulics, for Forchheimer and Schoklitsch, and through them and with them for other hydraulicians, is thus a technology, even though it may utilise certain parts of science, including certain fragments of mathematics. Hydraulics in this tradition is that which brings together and partially orders a range of descriptions of the behaviour of fluids, and then particularly of fluids in motion, that takes up our primary reflections of this physical world, which re-presents these through its use of tokens, first as signs but subsequently in some situations as symbols, which manipulates (or 'shuffles') these symbols as it does so take them up, and which finally interprets the results of its manipulations or shufflings again in signs that point, in their turn, to outer-world, and indeed primarily material-world, events. We may also observe already here that, as a result of their following this kind of approach, the technologies that were taught in the *Technische Hochschuler* could be distinguished semiotically from the sciences taught at the traditional *Universitäten* by the differences in the range of the sign vehicles that these two classes of institutions introduced and employed.

(As we shall later indicate, this difference in the use of sign vehicles reflects in its turn a deeper difference at the level of the social legitimisation of these two classes of institutions. Similarly, from our present postmodern point of view, only an age as completely confused and destabilised as our own could have produced the oxymoron of a *Technische Universität*. At the same time, and as we shall explain in conclusion, even this strange, and indeed paradoxical, construction does in fact open up new, and essentially postmodern perspectives, but then these are surely entirely different from anything envisaged by those who 'instituted' this term).

Among the many and essentially semiotic innovations introduced by Forchheimer were those based upon the total energy line that, although implicit in hydrodynamic studies from the time of David Bernoulli in the eighteenth century, only came into widespread use in the twentieth century (see, for example, Jaeger, 1956, p.76). It was Forchheimer similarly who extended the energy-line concept to unsteady flows in a thoroughly systematic way (1931, pp. 246 *et seq.*), but then making extensive use of graphical, and essentially geometrical methods, while it was Schoklitsch (1917) building on the pioneering work of Ritter (1892) who extended these notions to dambreak fronts and waves analogously. Such graphical methods reached their apogee in studies of water hammer of the kind that supported the design and operation of those hydropower plants which were coming into increasingly widespread use at that time. The graphical method of Schoklitsch for the analysis of pipeline and surge tank oscillations may serve as a paradigm in this respect.

It is then clear that 'modern' technology as constituted in hydraulics and as exemplified in the works of Forchheimer and Schoklitsch never embraced the symbolic paradigm to anything like the same degree as did the physical sciences. Forcheimer and Schoklitsch were definitely *not* intellectual ascetics. At the same time, any reading of the learned societies or, from the 1950s onwards, of the *Journal of Fluid Mechanics*, shows that these were given over almost exclusively to 'symbol shuffling'. Hydraulics instead wove together geometrical, other-graphical and symbolic methods into an emergent web of relations between physical observable. It combined and juxtaposed its sign vehicles to its own immediate technical advantages. Hydraulics, like several other technologies besides, was *semiotically heterogeneous*.

The significance of sign vehicles that were more specific to hydraulics

Of course, there was no 'grand design' or indeed anything 'ideological' at all about this: the hydraulician just solved his problems with whatever tools he had available and he had little time for, or could not afford much in the way of, ideological conformity. Hydraulics, again like most other technologies, sought its social legitimacy, even if not its academic status, primarily in the successes of its applications in the material world, and thus in its value to society as a whole. This 'performative' property was the essential ground for the legitimacy of the *Technische Hochschulen* alongside the science-for-science's-sake legitimacy that was established in the spirit of Humbolt and Schliermacher in the Germanic *Universitäten*. Ideological conformity was good for scientific, and even general intellectual, legitimacy, which certainly had its place in maintaining an equivalent status with the *Universitäten*, but a show of conformity was often about as far as it really went in practice.

Beyond these more pragmatic aspects there are others again that, although related to them, are in fact much more specific to the hydraulics of this period. The first of these was that certain sign vehicles, and specifically the geometric-graphical methods that were expounded in the works of our two authors, were not only intended to convey knowledge as such, but to provide devices that could be used to produce knowledge. They thus already catered to some degree to users of knowledge at the level of their signs. Thus, for example, Schoklitsch's graphical procedure was intended to be constructed by practising engineers for each application again, simply by following the processes that he described. There was no question of a once-and-for-all-times result, but rather the prescription of a working procedure, or algorithm.

We should further observe, however, that hydraulics made extensive use of a quite other class of sign vehicles again, which was that of the *physical-hydraulic models*. Moreover, this second specific class of sign vehicle existed almost entirely apart from the written texts of hydraulics. Using the terminology of semiotics in the broadest sense, a model is "a collection of signs that serves as a sign" (Abbott, 1992, 1993). A collection of visual observations and measurements, even when each one is individually only indicative of some small part of a flow, so as to constitute only an *indicative sign*, when taken together may point towards a quite other kind of sign again, whose significant is ultimately a deeper state of understanding of the phenomenon so modelled, which phenomenon thereby becomes its referent. It then comes to constitute an *expressive sign* (Husserl, 1900/1970; Derrida, 1967,1968//1973). However, the physical-hydraulic model is special in that it is a collection of signs that points to a sign by virtue of its sensual (and in this case purely visual) *resemblance* to its referent. Such classes of sign vehicles that function through a sensual resemblance between signifier and referent belong, however, to an even earlier and even more 'primitive' semiotic substratum. Thus, for Foucault (1966, p.32//1970, p.17):

> Up until the end of the sixteenth century, resemblance played a constructive role in the knowledge of Western culture. It was resemblance that largely guided exegesis and the

> interpretation of texts; it was resemblance that organised the play of symbols, made possible knowledge of things visible and invisible, and controlled the art of presenting them.
> ...... The semantic web of resemblance in the sixteenth century is extremely rich: *Amicitia, Aequalitas (contractus, consensus, matrimonium, societas, pax et similia), Consonantia, Concertus, Continuum, Paritas, Proportio, Similtudio, Conjunctio, Copula.* And there are a great many other notions that intersect, overlap, reinforce, or limit one another on the surface of thought.

(Let us observe in passing that this mode of signification through resemblance is essentially *metonymic.* Very similarly nowadays, our user interfaces also work, at least on the surface of things, entirely with metonyms, so that the design of a user interface is confined, at least at first sight, to a study in metonymic devices and metonymic structures. On the other hand, however, as soon as a user interface is actually 'used', it may in fact introduce all manner of *metaphores* in the minds of its 'users'. In effect the metonyms serve as surrogates for elements of our outer world, operating within our field of (visual) cognition, while the metaphores are essentially products of our inner world. Thus, although metonym is commonly regarded as the poor relative of metaphore, it is the metonyms that serve to generate the metaphores. It follows that the design of a user interface is in point of fact an exercise in the production of metaphores – and the theory of the design process must take account of this. In the words of de Man (see Norris, 1993, p. 190) writing specifically of how these processes function in Proust's masterpiece, "The figural praxis and the metafigural theory do not coincide...The assertion of the power of metaphore over metonymy owes its power [in fact] to the use of metonymic structures". At this level of retrospection, the same could then of course also be observed of the physical-hydraulic models)

The physical-hydraulic models that also enter, albeit only in passing, into the works of Forchheimer and Scholklitsch thus refer semiotically back to the praxis of this earlier age. But then, we must ask: what was the essential *socio*-technical purpose of these most 'primitive' of all semiotic devices, such as were generally eschewed by a contemporaneous science whose 'experiments' were usually of a quite other kind? Taking the example of physical-hydraulic model constructed at *Delft Hydraulics,* a storm-surge barrier situated at the entrance to the *Rotterdamse Waterweg* that was actually completed only in 1997, Latour (*loc cit*, p. 230; see also Abbott, 1996) observed of this most basal class of sign vehicles, but in the full-blown language of 'Enlightenment', that:

> Sure enough, another 'Copernican revolution' has taken place. There are not that many ways to master a situation. Either you dominate it physically; or you draw on your side a great many allies; or else, you try to be there before anybody else. How can this be done? Simply by reversing the flow of time. Professor Bijker and his colleagues *dominate* the problem, *master* it more easily than [do] the port officials who are out there in the rain and are much smaller than the landscape. Whatever may happen in the full-scale space-time, the engineers will have *already seen it.* They will have become slowly acquainted with all the possibilities, rehearsing each scenario at leisure, capitalising on paper possible outcomes, which gives them years of experience more than the others. The order of space and time has been completely reshuffled.

We see then on the one hand how the semiotic heterogeneity of hydraulics in *its* 'classical' era, as recorded in the writings of Forchheimer and Scholklitsch and accentuated further by the hydraulic model, was still being employed as a *combination of means to control and so to dominate situations*. Entirely in the spirit of 'Enlightenment', as so exactly recorded by Kant, some fragment of nature is asked specific questions, which are the questions that *we* want to ask, and *we* only listen to nature to the extent that it answers just these, *our* questions. We behave here as though we were Kantian grand inquisitors, with nature as our prisoner. At no time do we allow any of our fellow creatures to pose their questions, for we provide no means for them to do so. We consider just one set of fragments that happens to be that which we identify as 'significant' to us and we ignore everything else. Moreover, this attitude pervades not only our texts and our physical models, but it extends further to our direct observations and measurements of nature also. But, on the other hand, we see something else again that resides in these self-same texts: that the means that are mobilised to realise this programme are not at all uniformly those of this modern era, but have proceeded from other, and sometimes much earlier, eras again.

Finally, in this respect, we observe that Forchheimer himself explained how he had arranged the material of the third edition of the *Hydraulik* in the chronological order of a history, just as, for Foucault (1966, p.140 // 1970, p.128):

> In the sixteenth century, and right up to the middle of the seventeenth, all that existed was histories: Belon had written a *History of the nature of birds;* Duret an *Admirable history of plants*; Aldrovandi a *History of serpents and dragons*.... From then on, however, there is the sudden separation, in the realm of *Historia*, of two orders of knowledge henceforth to be considered different.

At the same time, in the case of Forchheimer we may also see this historical approach as an expression of a faith, shared with his readers, in *progress*, and indeed as one inseparable part of the *progress of mankind*. His work was thus conditioned by the metalanguage of a *Grand Narrative*, and was thus in this sense again fully congruent with the modern condition.

So, has there ever been a 'modern' technology?

Could there ever be any doubt about the answer to this question? Has not 'modern technology' been portrayed to perfection long ago by such as Fritz Lang in his *Metropolis* and Chaplin in his *Modern Times*? Was this technology not explicated in the greatest detail long ago by such as F.W. Taylor? Certainly upon this surface of outer appearances there has been a modern technology. But, of course, our whole purpose here is to pose this problem at other levels than those of superficial appearances. For this purpose we turn, as we usually do in these matters, to Heidegger, who, even as he already recognised the common superficial view of technology, saw that this view was essentially misleading. Indeed, technology generally has a quite other nature than that which might at first appear from its outer appearances (Heidegger, 1977, pp 12-13: see also Abbott, 1991):

> This prospect strikes us as strange. Indeed it should do so, should do so as persistently as possible and with so much urgency that we will finally take seriously the simple question of what the name 'technology' means. The word stems from the Greek. *Teknikon* means that which belongs to *technē*. We must observe two things with respect to the meaning of this word. One is that *technē* is the name not only for the activities and skills of the craftsman, but also of the arts of the mind and the fine arts. *Technē* belongs to bringing forth, to *poiesis*; it is something poietic [creative, formative, productive, active].
> ".... Technology is a mode of revealing. Technology comes to presence in the realm where revealing and unconcealment take place, where *alēthia*, truth, happens.

Thus, although the technology of such as Forchheimer and Scholklitsch was and continues to be used for many 'modern' purposes, it need not be itself in any way 'modern'. Indeed, the semiotic heterogeneity of hydraulics indicates already that a technology such as hydraulics cannot be entirely absorbed into the ideology of any such movement. In this respect it differs quite essentially from modern science, so that, again, technology cannot be regarded merely as 'applied science', as indeed Heidegger also explicated convincingly. Any technology can of course be turned to the one or the other purpose of such a modern movement, just as any other social activity may be influenced by the prevailing ideology. But, at the end of the day, the engineer in the mould of a Forchheimer or a Schoklitsch could not be so much interested in conforming to the current scientific ideology by filling the pages of the *Proceedings* of learned societies, but in solving real-world problems. Such persons cannot and indeed cannot afford to be constrained by the 'good taste' of the current ideology. The path of creation of technology, as the loci of places where 'truth happens', as the continuation of the natural creation through human intervention, necessarily transcends the limits of any 'ism', including modernism.

This is then a view of technology which is at one and the same time more modest and more exulted than that accorded to it within a modernist ideology. It is more modest than the role accorded to it within the modernist condition because it does not set out to conquer anything or to dominate anything. At the same time, it is also more exalted in that it places the essence of technology above any prevailing ideology, whether modernist, postmodernist or anything else of that kind.

(To one side again: this situation may well appear as self-evident to those who unquestionably accept the prevailing condition, whether modern or postmodern. For the more questioning – roughly speaking, those who have followed this piece so far – it will however appear as very strange indeed. For the prevailing condition is a state of being-in-the-world (*in-der-Welt-sein*) so that this position of technology makes it in some way extra-ontological, and indeed, in its essence, pre-ontological. This position should indeed astonish us in the same way as Kant was astonished by the strange contiguity between 'the starry sky above me and the moral law within me'. As the first book of Genesis proclaims and as Levinas, in particular, has explained in our own time, this places 'the moral law within me' before human existence itself, as something pre-ontological. To propose a similar place for technology must then appear as a scandal for good sense and indeed bordering upon, if not transgressing, the edge of sanity. And yet, as anyone who has experienced technology at the outermost limits of the possible can attest, it has this feeling about it. This situation is clearly totally different to that which obtains in science. Indeed, sometimes I think that when the deserving technologists and scientists are called before God, the technologists will stand on the right hand side and the scientists on the left; and there will be no one left in between them, at the centre, not even an Augustine, an Aquinas or a Barth. Such is the gulf that I envisage between these two classes of persons).

Thus in this essential sense there never has been a 'modern technology' but only a modernist use, or misuse, of an otherwise timewise-overarching technology.

Can there then be a 'postmodern technology'?

Could not all the above arguments be applied just as well to this question also? Could we not dismiss the notion of a postmodern technology just as readily? In fact, as we shall now show in conclusion, everything proceeds in the contrary direction in this respect, and indeed we now have to show that the answer to this question is again by no means so obvious or so simple as it may at first appear. We shall now have to show, moreover, that this very difficulty follows directly from the postmodern situation of technology itself. Following this, we shall still have to identify what we mean by a 'postmodern technology' as a socio-technical reality.

Technology has to do with creation in 'the real world' and the postmodernist experience is just that of an ever increasing 'destabilisation of reality' and the formation of ever-wider divisions of views about what constitutes 'the world'. The roots of the postmodern themselves express the paradox that is inherent in this situation: *post* (after) *modo* (just now). Thus the postmodern is, so to say, 'something that never arrives but is always arriving'. A postmodern technology would be like everything else, including ourselves, existing in this condition of forever *Waiting for Godot*. And this condition applies every bit as much to the rules of the creative act as to the creation itself. A technology can therefore only be 'postmodern' to the extent that it sets up its own rules of construction at the same time as it constructs — it has to be continuously, and even recursively, poietic.

Let us therefore conclude by placing hydraulics within this postmodernistic context, as a 'postmodern hydraulics'. We may perhaps best present this point-by-point as follows:

1. The hydraulics of Forchheimer and Schoklitsch was 'purely technical' in the sense that it did not treat or research any of the social aspects of its applications in practice. At that time it was possible, in effect, to separate the technical from the social in order to produce what might be described as a *homogeneous engineer*. Of course in the world outside of the teaching and research organisation there had always been *heterogeneous engineers*, as epitomised by Bell and Edison. Thus, for Law (1991, p. 9 , see also Abbott, 1996):

> "Edison was a 'heterogeneous engineer'. He worked not only on inanimate physical materials, but [also] on and through people, texts, devices, city councils, architects, economics and all the rest. Each of these materials had to be moulded to his design if the system as a whole was to work. And, as a consequence, he travelled between these different domains, weaving an emergent web which constituted and reconstituted the bits and pieces that it brought together."

These social functionings of the engineer were scarcely treated at all, however, in the teaching and research of the *Technische Hochschuler*. They were naturally implicit, but never, or hardly ever, made explicit. In this respect teaching and research proceeded in a kind of *utopia* where 'everything had its place and there was a place for everything'. Nowadays, we scarcely meet any purely technical problems at all, and least of all in practice: all our problems are now more-or-less *socio*technical. One cannot any more teach and research in hydraulics, for example, without simultaneously teaching and researching the mode of application of that hydraulics within society. Corresponding to this social aspect, teaching and research in hydraulics now presents more the aspect of a *heterotopia* where (Foucault, 1966, p. 9 //1970, p.xvii-xviii):

> Things are 'laid', 'placed', 'arranged' in sites so very different from one another that it is impossible to find a place of residence for them, to define a *common locus* beneath them all. ...... *Heterotopias* are disturbing, probably because they secretly undermine language, because they make it impossible to name this *and* that, because they shatter or tangle common names, because they destroy syntax in advance, and not only the syntax with which we construct sentences but also that less apparent syntax which causes words and things (next to and opposite each other) to hold together.

As Foucault also explained, however, this experience also announces the arrival of a new kind of relation between the 'order of things', including the 'order of signs', the relation that underlies all our means of representation and associated cognition. A new structure of thought is breaking through the matrix of the old structure, and in this process necessarily creating an heterotopian impression. This event then signalises that which Foucault has so aptly described as a "discontinuity in the history of thought". The domain where this occurs is correspondingly "more confused, more obscure and undoubtably less easy to analyse" (Foucault, 1966 p.12 // 1970 p. xx).

By way of an example, we may observe how in simulations in hydrology a whole rhetoric of metonyms, and indeed a complete metonymic structure, 'interception', 'stream flow', 'infiltration', 'unsaturated zone', 'soil-water tension', etc., - may disappear, only to reappear as a set of 'neurons', 'perceptrons', 'weights', 'input patterns', and other such, quite different, metonyms. In effect, the rhetoric of the descriptor collapses, or even implodes, in the one region of the total semantic field, only to reappear as a quite other descriptor rhetoric in a completely different part of the total semantic domain. The descriptor rhetoric is in effect 'displaced' discontinuously over a large distance in the semantic field. This apparently 'technical' process then has all manner of social consequences, such as those of providing efficient real-time control systems, viable Internet-distributed environmental impact assessment games, working decision support systems, and all sorts of other such facilities.

At the same time, however, if one were to take the new title of *Technische Universität* at all seriously, at its Janus-like face value, then this institution should provide a more appropriate space within which to pursue this new 'dual' approach to sociotechnological research and development with all its semantic discontinuities. The historian A.J.P. Taylor invented the word 'literalism' to describe this strategy of taking seriously that which was never intended seriously in the first place, as a means of attaining certain ends. It is a strategy that the hydroinformatician may well adopt to good effect.

2. Further corresponding to the purely technical nature of the subject matter of Forchheimer and Schoklitsch, there is never a moment's doubt about the *reality* of the world that is treated in their texts. Every sign that is introduced has a referent that is to be found 'out there' in the 'real world'. In the same vein, the relations between the signs and their referents always 'makes sense' in that there are clear logical connections between them. Corresponding to this again, there is a definite, identifiable and univocal channel of communication between writer and reader. Today, on the other hand, (Lyotard, 1988, p. 13 // 1984, p. 73):

> It is not necessarily the same thing to formulate a demand for some referent (and objective reality), for some sense (and credible transcendendence), for an addresee (and audience), or an addresser (and subjective expressiveness) or for some communicational consensus (and a general code of exchanges, such as the genre of historical discourse).

Corresponding to this, our experience of reality, and indeed our very *sens de réalité*, has changed, and apparently irreversibly. We recall that, for the hydroinformatician, "Reality is the name that we give to the interface between our inner and our outer worlds" (Abbott, 1994). This reality now however grows increasingly from the cathode ray guns of our television and computer-display tubes. On the one hand it becomes increasingly labile and at least potentially unstable, while on the other hand it increasingly conforms to the rule enunciated by Lyotard (1978, p. 19 // 1984, p. 17) that:

> There is no reality unless testified by a consensus between partners over a certain knowledge and certain commitments.

The 'realities' of 'Limits to Growth' of the 1960s, of 'Global Cooling' in the 1970s and of 'Global Warming' in the later 1980s and 90s may serve as paradigms here. The conformity of these 'realities' to Grand Narratives shows however that they are not essentially postmodern, but, if anything, hypermodern. But in any event the modernist situation that Heidegger described ironically already in 1927 (p. 168, 169 // 1962, p. 212) is now becoming the established social norm:

> What is said-in-the-talk as such spreads in wider circles and takes on an authoritative character. Things are so because one says so.... The average understanding will *never be able* to decide what has been drawn from primordial sources with a struggle and how much is just gossip. The average understanding, moreover, will not want such a distinction and does not need it, because, of course, it understands everything.

But then, to take the examples further, our present-day science introduces 'ten billion years', 'three thousand light years', 'a big bang', 'black holes', 'quasars' and all manner of other textual signs for which no one can adduce the existence of any referents, but only indications to other signs again. As Baudrillard explained already in 1973, these signs no longer have a 'real' referential function: "the signified and the referent are now abolished to the sole benefit of the play of signifiers, of a generalised formalisation in which the code no longer refers back to any subjective or objective 'reality', but to its own logic." (Baudrillard 1973, p.127). *Along very much the same lines* - and this is the crux of the postmodern *critique* - a so-called 'music-television channel' projects images that are inconsistent with and even contradict the accompanying 'texts', an advertising agency uses images that have no logical relation at all to the products that it promotes, a political party wins elections with patently self-contradictory 'policies' concocted solely by its marketing consultants, and so we can go on. And yet all of this is immensely successful, whether in cultural, business, political or whatever other terms, serving to create senses of 'identification', 'community', 'belonging' and other emotional attachments, and indeed it does this *precisely because of* its application of inconsistent and self-contradictory means. Since these means are so effective even as they are predicately illogical, they must be in some sense paralogical.

There is, on the other hand, nothing essentially new about all this; it is only that it is now so immensely reinforced by new and entirely unprecedented technical means. For one brought up in the tradition of a Forchheimer or a Schoklitsch, a natural reaction may be just to shake one's shoulders and go fishing, take up ballroom dancing or retreat to a monastery in Tibet. For the hydroinformatician, however, no such escape is possible (being anyway clearly illusory). Relative to this 'down-side' of the postmodern or, strictly speaking, hypermodern condition, the hydroinformatician must always be a *résistant*, and in order to succeed in this endeavour he or she must turn at least some of these *technologies of persuasion* around to good use and, if necessary, even against those who so misuse them. This necessitates, however, that the hydroinformatician must study at least some of these technologies, and then not only alongside, but in the closest synergic relation with the technologies that were so advanced by such as Forchheimer and Schoklitsch.

(May I be permitted to interpose one more aside? It is that since earliest times certain technologies of psychic manipulation that are in fact being employed quite extensively, and generally indiscriminately, in modern society, have been treated with the greatest suspicion and have been hedged around with all kinds of warnings, restrictions and prohibitions. Very briefly, and as Jung explained on various occasions (*e.g.*, 1944//1953/1968) one cannot possess this kind of knowledge without being possessed by it. Thus, although we must witness how, for example, certain marketing consultants can snatch an electoral victory from almost certain defeat by employing paralogical and aporiaic means [as did Saatchi and Saatchi for the Conservative Party in the 1992 UK general elections; see Norris, 1993] we hydroinformaticians are best advised, not just to eschew, but to abjure the use of such devices. For a most thorough argument for this renunciation, reference must be made to Karl Barth's *Kirchliche Dogmatik*, and then specifically to the third volume of the third part of that monumental work. At the same time, of course, the hydroinformatician may still be obliged, *in extremis,* to study inconsistency-tolerant logics, even though these might be subsumed under the paralogical, as well as undecidable problems, even though these may be aporiaic, and similar subjects, but then never for such purposes.)

3. The lives and works of technologists like Forchheimer and Schocklitsch were primarily those of creation in the material world, and much less, if at all, those of persuasion. Today, our major efforts in hydraulics are directed to the technologies of persuasion, even though still necessarily building on the foundations provided by that earlier era. For this purpose the writings on postmodernism of such as Baudrillard, Lyotard and Kristeva, together with the *grands maîtres,* Foucault and Derrida, are of inestimable value. We accept – how could we deny? – that their findings are often grossly misused, and that we often experience the hypermodern and postmodern conditions that they describe, analyse and dissect as most uncongenial. But we *are* living through these conditions whether we like it or not, and we must adapt correspondingly.

It follows from this situation that although hydraulicians during the time of Forchheimer and Schoklitsch could restrict themselves to statements of facts without being concerned with judgements of value, we can no longer do this in most cases. Alongside everything that we now constate, there has to be an insight into how this will influence or perform in society: our constative and performative modes of utterance must proceed together. Whereas we could previously speak and treat of objects as and by themselves, today we can rarely speak of objects at all without speaking of values – and indeed it is precisely because we cannot avoid value judgements that we must now think and create so explicitly in an object-orientated way (Abbott, 1994). Thus while Forchheimer was teaching and researching at Graz, Alexius Meinong, the founder of the present-day theory of objects and values, was also teaching and researching at Graz, but the works of the one find no mention whatsoever in the works of the other. So far as we can see, Forchheimer and Meinong never met or communicated at all, and indeed within the context of their times this was not so extraordinary. This kind of institutionalised asceticism was then entirely justifiable. Today, however, such a situation would be, or should be, untenable, and indeed it should be so from both sides.

4. Forchheimer and Schoklitsch researched, taught and wrote for a select group of engineering specialists and they consulted for a small number of corporate clients. Today, we are working for a much wider range of 'end-users' of hydroinformatics' products and services. One European organisation alone already markets 80 hydroinformatics (including modelling) products in 50 countries, with an agent network of 35 companies to support the marketing and to provide local support and consulting services for the installed products. Some 4,000 organisations worldwide use the products of the two major European suppliers alone. Now, however, this process is being extended much further again as we move out 'into Cyberspace' to reach a new and much more numerous kind of consumer of hydraulics knowledge. The new intranetted and extranetted generation of distributed environmental assessment systems, decision-support systems, negotiation-facilitating environments and everything else of this kind are intended to bring the benefits of the most advanced hydraulic and environmental engineering knowledge and data to hundreds of thousands, and ultimately millions of persons. At the same time, these new products and services are all sociotechnical constructions, so that the social aspects of their applications must be accommodated and properly integrated with the technical. Moreover, the new kinds of end-users that are now brought into this process are by no means passive observers, and even less are they mere objects in any grand design, but they are themselves active participants, becoming initiators of projects and creators of objects on a scale hitherto unknown. But - and this is the essential thing in this place - far below the level of the interactive user interfaces, deep within the workings of the code, the thinking, researching, teaching and writing of such as Forchheimer and Schoklitsch, and with them their whole life world in Graz, are still alive, and indeed making their contribution to humanity and to the creation as a whole as never before.

In one sense, their world has departed for ever, and in another sense, by way of electronic encapsulation, it has never been more alive than today, and certainly it has never before had such an influence upon our futures. Those who retrospect, *à la recherche du temps perdu*, must nowadays always look in the direction of the code of the digital machine – and as soon as they do this they inevitably also look into their possible futures. For it is primarily through the intervention of the code that we project and plan our possible futures as a means of choosing the one future that we find most desirable. The experience of these new hydroinformatics devices is thereby an inseparable part of the paradoxical, and often paralogical hypermodern and postmodern condition as a whole, for it is an experience that replaces our present moments in order to represent our possible futures, even as it exists at the present time.

REFERENCES

Abbott, M. B., 1991, *Hydroinformatics: information technology and the aquatic environment*, Avbury, Aldershot, UK, and Brookfield, USA.

Abbott, M. B., 1992, The theory of the hydrologic model, or: the struggle for the soul of hydrology, in *Topics in Theoretical Hydrology: A Tribute to Jim Dooge*, ed. J.P. O'Kane, Elsevier, Amsterdam, 237-59.

Abbott, M. B., 1993, The electronic encapsulation of knowledge in hydraulics, hydrology and water resources, *Advances in Water Resources* **16,** 21-39.

Abbott, M. B., 1994, The question concerning ethics, or: The metamorphosis of the object, in *Hydroinformatics '94, Proc. 1st International Conference on Hydroinformatics*, Verwey *et al* (eds.), Balkema Rotterdam, pp. 3-8.

Addis, W., 1990, *Structural engineering: the nature of theory and design*, Horword, New York.

Barth, K., 1960, *Kirchlishe Dogmatik*, Evangelisher Verlag, Zollikon-Zurich//*Church dogmatics*, trans. Bromiley, G.W., and Ehrlich, R.J., Clark, Edinburgh.

Baudrillard, J., 1973, *Le miroir de la production: l'illusion critique du matérialisme historique*, Asterman, Paris //...... *Mirror of production*, trans. Poster, M., Telos, St.Louis.

Derrida, J., 1967, 1968//1973, *La voix et le phénomène*, Presses Universitaires, Paris; La forme et le vouloir-dire: note sur la phénomènologie du langage, *Rev. Int. Phil. LXXXI*; La différance, *Bull. Soc. FranKaise Phil. LXII// Speech and phenomena, and other essays on Husserl's theory of signs*, trans. Allison, D. B., Northwest Univ., Evanston.

Descartes, R., 1637//1925//1954, *La geometrie,* appendix to the *Discours*, occupying pp.297-411// *The geometry of RenM Descartes*, trans. Smith, D.E., and Latham, M.L., Open Court, New York / Dover, New York.

Descartes, R., 1637//1962, *Discours de la méthode*, Maire, Leiden/Third edition, edited and annotated by Gilson, E., Vrin, Paris.

Duhem, P. 1913-1959, *Le système du monde. Histoire des doctrines cosmologiques* (in 10 volumes), *De Platon a Copernic*, Herman, Paris.

Duhem, P. 1985. *Medieval cosmology; theories of infinity, place, void and the plurality, of worlds*, ed. and trans. Ariew, R., Univ. Chicago, Chicago, Ill.

Foucault, M., 1966, L*es mots et les choses: une archéologie des sciences humaines,* Gallimard, Paris // The *order of things: an archaeology of the human sciences*, Routledge, London.

Galileo Galilei, 1638, *Discorsi e dimostrazioni matematiche intorno a due nuove scienze attenenti alla meccanica,* Elzevier, Leiden//1914/1954, *Dialogues concerning two new sciences* Trans. Crew, H. and de Salvio, A., Macmillan, London/Dover, New York.

Heidegger, M., 1927, *Sein und Zeit*, Niemeyer, Tübingen and Halle // 1962, *Being and time*, trans. Macquarrie, J., and Robinson, C., Blackwell, London.

Heidegger, M., 1977, *The question concerning technology and other essays*, trans. Lovitt, W., Harper and Row, New York.

Horkheimer, M. and Adorno, T.W., 1947, *Dialektik der Aufklärung*, Querido, Amsterdam // *Dialectic of enlightenment*, 1973, trans. Cumming, W. Lane (Penguin), London.

Husserl, E., 1900, 1901/1913, *Logische Untersuchungen*, Niemeyer, Tübingen and Halle // 1970, *Logical investigations*, Trans. Findley, J.N., Routledge, London.

Jaeger, C., 1956, *Engineering fluid dynamics*, Blackie, London.

Jung, C.G., 1944, *Psychologie und Alchemie*, Rascher, Zurich // 1953 /1968, *Psychology and alchemy*, trans. Hall, R.F.C., Routledge, London.

Law, J., 1991, *A sociology of monsters: essays on power, technology and domination*, Routledge, London.

Latour, B., 1991//1993, *Nous n'avons jamais été modernes*, La DMcouverte, Paris // *We have never been modern*, Trans. Porter, C., Harvester, London.

Lyotard, J-F., 1988, *Le postmoderne expliqué aux enfants (correspondence 1982 – 1985*), Galilée, Paris.

Lyotard, J-F., 1979, *La condition postmoderne: rapport sur le savoir*, Minuit, Paris // 1984 *The postmodern condition: a report on knowledge*, Manchester Univ., Manchester, U.K.

Marcel, G. 1933//1946//1946//1947, *On the ontological mystery // existence and human freedom // testimony and existentialism // An essay in autobiography* // collected in 1995 as the *The philosophy of existentialism.* Trans. Manya Harrari, Carol, New York.

Mathews, G. and Grabs W., 1994, Hydro-economic development and tools for decision aiding, in *Journal of hydraulic research*, **32**, 15-19.

Norris, C., 1993. *The truth About postmodernism*, Blackwell, Oxford.

Olivier, T., *Mémoires de géométrie descriptive, théorique et appliquée*: see Addis,1990.

Ritter, A., 1892, *Zeitschr. D. V. deutsch Ing.*, **36**, p.954.

Schoklitsch, A., 1917, *Wien, Ber.,* **126**, p.148

Schoklitsch, A., 1923, *Graphische Hydraulik*, Leipzig-Berlin.

Troisfontaines, R., 1968, *De l'existence à l'être: la philosophie de Gabriel Marcel*, Nauwelaerts, Louvain and Paris.

Victor, S. K., 1979, Practical geometry in the high middle ages, *Mem. American Philosophical Soc. 134.*

Weiss, J.H., 1982, *The making of technological man: the social origins of French engineering education*, MIT, Cambridge, Mass.

The following article has been reprinted from *Journal of Hydraulic Research*, Vol. 38, No. 4, pp. 339–349, with permission from the copyright holders, IAHR International Association for Hydro-Environment Engineering and Research.

Towards the hydraulics of the hydroinformatics era

L'hydraulique à l'ère de l'hydroinformatique

M.B. ABBOTT, Knowledge Engineering, *Avenue François Folie 28, B 28, 1180 Brussels, Belgium*

V.M. BABOVIC, DHI, *Agern Alle 5/11, 2970 Hørsholm, Denmark*

J.A. CUNGE, IHE, *Westvest 7, P.O.Box 3015, 2601 DA, The Netherlands*

SUMMARY

Hydroinformatics is the study of the flows of knowledge and data related to the flow of water and all that it transports, together with interactions with both natural and man-made, or artificial, environments (Abbott, 1991). Hydraulics, understood as the study of flows of water, more recently extended to include the transport of matter in all its forms with these flows, is accordingly central to hydroinformatics. Without hydraulics, no hydroinformatics! From this situation it may at first appear as though hydroinformatics provides only a new periphery to hydraulics: a new way of transmitting hydraulics knowledge and data to society. In practice, however, the way in which hydraulics is viewed and practised is itself now changing as a result of its incorporation into the new paradigm that hydroinformatics provides. The first purpose of the present paper is to introduce some of the changes that are currently proceeding in hydraulics under the influence of developments occurring in hydroinformatics. The second purpose is to indicate the consequences of these changes for the application of hydraulics within society, and thus for the future direction of hydraulics and hydroinformatics themselves.

RÉSUMÉ

L'hydroinformatique est l'étude des flux de connaissances et de données concernant l'écoulement de l'eau et de tout ce qu'elle transporte, ainsi que les interactions avec les environnements naturels ou artificiels (Abbott 1991). L'hydraulique, comprise comme l'étude des écoulements d'eau, plus récemment étendue au transport de matière sous toutes ses formes, dans ces écoulements, est en conséquence au centre de l'hydroinformatique. Sans hydraulique, pas d'hydroinformatique!

A partir de là, tout se passe comme si l'hydroinformatique constituait seulement un nouveau domaine périphérique de l'hydraulique, une nouvelle façon de transmettre à la société le savoir et les données hydrauliques. Dans la pratique, cependant, la manière dont l'hydraulique est vue et pratiquée est elle-même en train de changer, du fait de son insertion dans les nouveaux paradigmes que fournit l'hydroinformatique. Le premier objectif du présent article est d'introduire quelques uns de ces changements qui apparaissent couramment en hydraulique sous l'influence des développements de l'hydroinformatique. Le second objectif est d'indiquer les conséquences de ces changements pour les applications de l'hydraulique dans la société, et donc pour la direction future de l'hydraulique et de l'hydroinformatique elles-mêmes.

The hydraulic engineer in the post-symbolic era

The first and most obvious change that has occurred in hydraulic engineering is in the way in which the hydraulic engineer works. Like most other engineers nowadays, the hydraulic engineer works for a large part, and in many cases for the most part, through the graphical user interface of a computer. The era in which the engineer worked with symbols, making calculations directly from equations and the curves of graphs of such equations, is mostly over. Today, the engineer works for the most part with signs. Whereas the symbols of an earlier era *replaced* the world in the mind of the engineer, at least while he or she manipulated these symbols, the signs of the graphical user interface (the buttons, the pull-down menus, etc) *point towards* the world in the mind of the engineer. More basically, while in the earlier era the engineer was a repository of knowledge made expressible in symbols, that is, a *knower*, the engineer is now primarily a repository of the sum of all the means to access knowledge, so that he or she is primarily a *consumer of knowledge* made expressible in signs (Baudrillard, 1963; Lyotard, 1979//1982 and 1998, Appignanesi and Garatt, 1995). Correspondingly, the device that was previously a *computer*, as a means of making computations, now becomes a *knowledge processor*, as a means of manipulating encapsulated knowledge. Similarly, what was previously a *data network*, as a means merely of accessing data, now becomes an intranet, or even an extranet, or more generally an *internet* (in the generic sense and so with a lower-case 'i') as a device for communicating knowledge in the first place and data only in the second place. The new era, in which the engineer no longer works with symbols in the capacity of a knower, but instead works with signs in the capacity of a consumer of knowledge, is called quite generally the post-symbolic era (Abbott, 1998 a and b; Jonoski and Abbott, 1998; and, especially, Abbott, 1999 b).

Tool builders and tool users

Corresponding to this change in the status of engineer, and indeed in the status of all knowledge users, a division opens up between those who encapsulate knowledge for consumption, on the one hand, and, on the other hand, those who access and use this knowledge. The equipment whereby knowledge is transformed and transferred belongs to the category of tools, so that we identify here a division between *toolmakers* and *tool users*. Correspondingly the most immediate and direct consequence of this division is that the most advanced hydraulics knowledge has become useable to far more persons than hitherto. From their inception in 1986, fourth generation modelling systems alone have increased the number of persons who are able to draw upon and use

Revision received June 28, 2000. Open for discussion till February 28, 2002.

this knowledge by an order of magnitude every five years. As of 2000, well over 6.000 organisations were using tools of this class, representing around 15.000 users at any one time in more than 100 countries. This development has in its turn only been made possible through the radical changes that have occurred within modelling tools of this class and through the new social-institutional arrangements that have been set in place for their dissemination and support. This first stage in the more widespread application of hydraulics knowledge is illustrated in Fig 1 together with the later stages which will be introduced later in this paper. In principle, a numerical-hydraulic modelling tool is one that encapsulates generic hydraulic and related knowledge in such way that this may operate on site-specific data in order to provide site-specific knowledge. The extent to which this knowledge comes to presence in the mind of the tool user naturally depends upon the ability of that user to interpret the (mostly graphical) output of the system, and so upon that user's particular *knowledge frame* (Abbott, 1993). A considerable part of the effort put into fourth generation modelling systems has been directed to increasing the rate of knowledge processing by their users by the provision of more appropriate interfaces and supporting facilities, while most of the rest of this effort has been expended upon extending the range of application of these systems and automating their instantiations. Thus, whereas in 1985 some eighty percent of investment was still associated with the numerics of modelling tools, by 1998 at least seventy percent was being used for enhancing knowledge transmission facilities. The emphasis has passed correspondingly from number organisation and application, or *numerics*, to sign production and organisation, or *semiotics*. It has thus passed over to the study of sign production, distribution and consumption, which belongs to the subject area of the *theory of semiotics* (e.g. Eco, 1976; Klinkenberg, 1996). In the simplest terms, we observe the transformation:

Numerics → semiotics

This transformation is naturally reflected in the composition of the design and production teams of the tool producers, where quite new skills have had to be introduced alongside the existing skills. The highly visible developments in the latest versions of modelling tools – common-run-time environments, installation and license systems, hypertext-extended help facilities, seamless coupling to standard GIS and CAD packages and data management systems, video and multimedia presentation environments, etc. – are themselves enabled by a wide variety of hidden features, such as utility file standards, including data file standards and parameter file standards for text files. These developments are supported in their turn by a variety of computer-aided software engineering (CASE) tools and other advances in software engineering, continuing into such advances in communication technologies as those that are enabled by XML, ActiveX/Com, Java and CGI technologies. This is, however, only the beginning of a process that has continued to accelerate ever more rapidly as hydroinformatics has progressed beyond fourth-generation modelling, as is also introduced towards the end of this paper.

Experiences with the applications of fourth-generation systems

The very great success of modelling and simulation tools working within the post-modern paradigm of knowledge consumption in this area naturally also brings many problems with it. Although

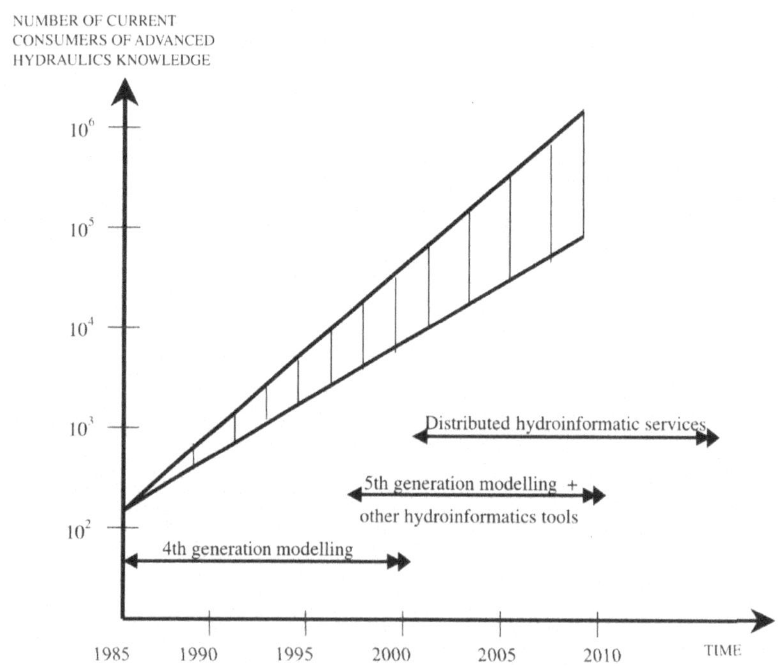

Fig. 1.Historical and predicted growths in the numbers of consumers of high-level knowledge in hydraulics, hydrology and water resources

not strictly speaking 'new', these problems have become exacerbated by inadequacies in the knowledge frames of many of the users of such systems. The most notable class of problems – and one that in fact dates back to the era of physical-hydraulic models – is associated with the process usually known as 'calibration', and this will be used here in order to exemplify the more general problem. It is well known how the results of models of fluid flow can nearly always be fitted to measurements of hydraulic behaviour recorded in a few places by varying the values of certain parameters, and most commonly the roughness coefficients, in the model. This process can however easily hide gross inaccuracies in the description of the modelled domain, such as in topography, in local head losses and in the dimensions of structures. On the other hand, the range of variation of, for example, Manning coefficient values is rather well known. In practice, it is almost certainly safer to make an error in estimating these values on the basis of a visual inspection of the terrain than to try to 'calibrate' a model by attributing unrealistic values to these coefficients in an attempt to compensate for unknown information or, worse, for physical phenomena not represented by the model laws, as expressed in the equations used in the models. In the former case one knows the range and the interval of possible imprecision. The latter is truly a black box. It does not mean that this kind of calibration should be forbidden altogether, but is should take place, within this example, only within a very strictly, physically-known range of Manning coefficient values and only where the type of bed or flood plain cover is also well known.

In general, it is acceptable in practice to calibrate roughnesses when small differences within a known range of variations cause, for example, computed and observed hydrographs to coincide. It is absurd however to do the same when differences cannot be explained or reduced by varying known and experimentally established coefficients, because this means that one has 'forgotten' something, such as singular head loss (which in fact provides the least disadvantage when one is trying to replace its influence by a longitudinally-distributed Manning coefficient of an unlikely value). Only too often the 'forgotten' part is a missing physical law or a process. In the former, acceptable situation, the model will be predictive even without calibration. In the latter, unacceptable situation, it will not be predictive even with an apparently excellent calibration.

Illustrative examples

A few examples, selected because they are simple and because they all correspond to cases which have actually occurred, may serve to illustrate the above thesis on the role of calibration in good modelling practice.

Example 1: Calibrating models within the context of a decision-making problem

A very detailed unsteady flow model was built for a several-hundred- kilometre long reach of a topographically very-well-documented European river. Water stages during recent floods with discharges culminating at some 3500 m^3/s were systematically recorded, as is usually the case, only at a limited number of stations, in this case some 20 kilometres apart. The model was carefully calibrated by varying the roughness coefficients in order to obtain the best possible coincidence between the computed hydrographs and those recorded at the stations. Higher floods occurred in the past (some 100 to 150 years previously) but their influences were not properly recorded and, moreover, the river bed had since been seriously modified, not only by natural processes but also by dikes. The dikes were intended to protect the valley from catastrophic discharges so long as these discharges were lower than about 7000 m^3/s. The decision-making process was concerned with crisis management. If the flood forecast gives a certain discharge propagating along the river, such that water may be expected to flow over the dikes, the authorities should have some 24 hours in which to evacuate the population and transportable assets from the threatened areas. Costs and damages resulting from a wrong decision are very considerable and, in particular, would still be considerable in the case of a false alarm. A flood peak of some 7000 m^3/s was accordingly simulated with the calibrated model. Maximum computed water stages were then found to be some 15 cm lower than the dike crests. The calibration of the model having been based upon the difference between observed and computed hydrographs for a 3500 m^3/s flood being within 10 cm, the authorities could have concluded that for an upstream 7000 $m^3/_S$ flood there was no reason to evacuate the population. Many hydraulic engineers, however, would conclude otherwise. They would do so simply because the model was calibrated with respect to water stage recording stations some 20 kilometres apart. Because of this, the calibrated roughness coefficients had to compensate for a considerable range of missing hydraulic and topographical data, with all manner of errors and assumption being implicitly made: The ranges of variation in all these phenomena were simply not known. It was pointless to run the usual 'sensitive range computations', corresponding to changing all Manning n values by +/–15%, because the real life situation may well have been very different from the resulting 'black box'-encapsulated 'reality' and flows under such circumstances may not at all necessarily vary linearly with *calibrated n* values. The predictive capacity of the model for this specific decision-making purpose was accordingly dubious, and the social consequences of its use could have been catastrophic.

Example 2: celerity model calibration and topographical data

This difficulty was encountered many more years ago when modelling a large South American river. The topography of the river was supplied by the client for whom a surveying company had done the work. The flood, simulated with one-dimensional modelling software, propagated much quicker in the model than in reality. There was no way to obtain a satisfactory coincidence between computed and observed hydrographs except by assuming values of Manning n values which could only possibly be justified if the flood plain was covered by jungle. This was quite simply impossible at that latitude. It was nevertheless quite possible by using these values to deliver a calibrated model to the client, whose requirements were limited to reasonably coinciding hydrographs. What actually happened was that the valley flood plain was generally very wide with a number of narrower locations. The surveyor made large savings by carrying out valley flood plain cross-section surveys only in the narrow locations. Thus the width of the modelled cross-sections was taken as much smaller than it was on average in reality and the computed celerity was seriously affected accordingly. The predictivity of the model for higher discharges was also jeopardised because of wrongly simulated storage volumes. However, the model had been 'calibrated'. Once the reason for the discrepancies had been found and a realistic topography introduced, calibration led to very limited changes in the roughness coefficient values that could have been estimated from the bed material and land cover description in the first place.

Example 3: Backwater calibration problem and topographical data

A similar problem was experienced recently when using a two-dimensional model to calculate a steady flow in a river for which observed backwater curves were available. The observed and computed backwater curves did not fit. Here again the data supplier (this time the client) was unaware of the mechanism and requirements of the model. He decided by himself that the available data (on cross-sectional geometry) were not all necessary and that the use of all the available cross-sections would unduly increase the cost of the modelling, of the data processing, and even of computer time! Hence he supplied the modeller with half of the available cross-sections, simply eliminating every second one. It was *possible* to calibrate the Manning coefficients in the model and obtain the required coincidence. But what predictivity could such a model have? Actually there was no need to calibrate at all: once all the cross-sections had been taken into account, the coincidence was very good when using 'standard' coefficients without calibration. Playing with the data was even more deleterious to realism in this example because two-dimensional modelling was used.

Prediction without calibration

The thesis that presents itself here is that deterministic models which are based on a set of laws representing all features that are essential and of interest to the problem should, in principle, *not* need to be calibrated at all. In most cases, calibration is based on insufficient data and justified by an imaginary and often false concept of improvement, and this malpractice usually only reduces the model's predictive capacities and usefulness for engineering purposes. However, this thesis is not sustainable unless at the same time one is sure that the equations or other representations of laws that one is using do in fact correspond to the behaviour of fluids in the modelled nature. The avoidance of abuses in calibration when using established laws of flow shows us the one side of the story, but this must be complemented by new methods to establish what these laws really are. For example, many hydraulics studies now deal with wetland developments, where a wide variety of kinds of vegetation is encountered, and for calculating the resistance to flow in such cases Manning's n values are often not known, and nor is the adequacy of the Manning formulation at all well established generally. Indeed, as vegetation is deflected to one side by the flow and ultimately flattened, so the resistance must vary in a way that departs markedly from the Manning formulation (Kutija and Hong, 1999 see also Baboric and Keijzer, 1999). Moreover, even in cases where an existing formulation of a resistance law such as that of Manning is adequate, the nature of the bed is not at all well known and may change over time, such as with the season of the year or as a result of the dumping of dredging spill. It is accordingly a first rule of hydroinformatics, corresponding closely to its sociotechnical endeavours generally, that no such change in application practice is possible without the simultaneous provision of new technical means to support this practice. This thesis corresponds to a more general change of paradigm in which a new interpretation and a new light is placed upon 'a set of laws representing all features that are essential and of interest to the problem'.

The data mining paradigm

The formative period of modern science that defined the hydraulics of the twentieth century covered the period between the late fifteenth century and the late eighteenth century. The new foundations were based on the utilisation of the concept of a *physical experiment* and the applications of a *mathematical apparatus* in order to describe these experiments. The works of Brahe, Kepler, Newton, Leibniz, Euler and Lagrange clearly personify such an approach. Prior to these developments, scientific work primarily consisted only of collecting the observables, or recording the '*readings of the book of nature itself*'.

This modern-scientific approach was principally characterised by two stages: a first one in which a set of observations of the physical system were collected, and a second one in which inductive assertion about the behaviour of the system – a hypothesis – was generated. Observational data represent *time- and space-specific knowledge*, whereas a hypothesis represents a *time- and space-generalisation* of this knowledge which *implies* and *characterises* all such observational data. One may argue that, through this process of hypothesis generation, it becomes possible to economise human thought, since more compact ways of describing observations are thereby provided.

As far as water and its movements were concerned, the mathematical formulation of physical phenomena led, along the way, to the theory of ideal fluid motion. But, in parallel, and especially at the beginning of the nineteenth and during the twentieth centuries, the experimental-scientific search for an understanding of nature brought to light the deviations between the behaviours of the observed real world and its proposed conceptualisations. Thus it was impossible to explain such problems as those involving frictional drag or fluid resistance using ideal fluid formulations. Thus, even within this field, two separate domains appeared, theoretical hydrodynamics and hydraulics, with the latter then being defined by Cole in 1962 as "*accumulated semi-empirical data about real fluids in real practical situations, compiled primarily for the engineer*". Most persons involved in hydraulics today would probably consider this definition unacceptable. However it constituted already at that time the dividing line between the engineer, who wished to solve 'real-world' problems, and those mathematicians who held out the prospect of a hydraulics that would die out when all the equations and their solution methods had been developed. Today, at the turning of the twenty first century, we are experiencing yet another change in the scientific process as just outlined. This latest scientific approach is one in which information technology is employed to assist the human analyst in the process of hypothesis generation. This computer-assisted analysis, usually of large, multi-dimensional data sets, is sometimes referred to as a process of *Data Mining for Knowledge Discovery*. The subject of Data Mining for Knowledge Discovery aims at providing tools to facilitate the conversion of data into a number of forms that provide a better understanding of the physical, biological and other processes that generated or produced these data. These new models, when combined with the already available understanding of the physical processes – 'the theory' – result in an improved understanding and novel formulations of physical and other laws, and so provide an improved predictive capability.

As we enter the communicational stage of the digital information era, one of the greatest challenges facing organisations and individuals is how to turn their rapidly expanding data stores into actionable knowledge (Fayed *et al*, 1996). Means for data collection, storage, retrieval and distribution have never been so advanced as they are today. While advances in data storage and retrieval continue at breakneck pace, the same cannot be said about advances in knowledge extraction from large data sets. Without such advances, however, there is a substantial risk of missing what the data has most to offer. The question is thus posed with ever increasing urgency: *what is to be done with all this data?* Ignoring whatever cannot be immediately analysed is wasteful and unwise. This is particularly unacceptable in scientific endeavours, where data usually represents observations carefully collected at considerable expense.

Knowledge Discovery in Databases (KDD) is concerned with extracting useful information from data stores. *Data mining (DM)* is one step (be it fully automated or human-assisted) in this larger KDD process. The broad KDD process includes: retrieving the data from a large data warehouse (or some other source); selecting the appropriate subset with which to work; deciding on the appropriate sampling strategy; selection of target data; dimensionality reduction; cleansing; data mining; model selection (or combination), evaluation and interpretation; and finally the consolidation and the putting to practical use of the extracted 'knowledge'. The data-mining step then fits models to, or extracts patterns from, the pre-processed data.

However, mining the data *alone* is still not the entire story, at least in scientific domains. Scientific theories have long encouraged the acquisition of new data and this data in turn has long led to the generation of new theories. Thus a great deal of 'data mining' has already been done, albeit in a quite other way. When revisiting the earlier methodologies above, we have observed how the traditional process usually began with experimental observations, after which generalisations were postulated, as a theory, and commonly expressed in the form of equations. Thus, traditionally, the emphasis has been on a theory, which demands that appropriate data be obtained through observation or experiment. In such an approach, the discovery process is what we may now refer to as *theory-driven*. Especially when a theory is expressed in mathematical form, *theory-driven discovery* may make extensive use of 'strong' methods associated with mathematics or with the subject matter of the theory itself. The converse view, that is now being so strongly advanced through data mining technologies, takes a body of data as its starting point and searches, using 'weaker' methods, for a set of generalisations, or a theory, to describe the data parsimoniously, and even, possibly but most desirably, to explain it. Usually such a theory takes the form of a precise mathematical statement of the relations existing among the data. This is the *data-driven discovery* process.

We strongly believe that the most appropriate way forward is to combine the best of the two approaches: theory-driven, understanding-rich, processes with data-driven discovery processes.

Model Induction

Histories of science, and especially those of mathematics, draw a particular attention to the development of a physical symbol systems, such as a scheme of notation in mathematics, interactively with the evolution of more refined representations of physical and conceptual processes in the form of equations in the corresponding symbols. It is then a common experience that one and the same physical symbol system, may serve for the expression of a great number of different equations. To the extent that each equation can be regarded not only as a string of symbols that can be manipulated mathematically, but also interpreted as a collection of signs which serves as a sign that points towards a particular physical object, process or event, so it constitutes a *model* of that object, process or event (Abbott, 1992; 1993; see also Klinkenberg, 1996, p. 180). It is then usual to speak of a collection of 'indicative' signs that, through this collectivity, points towards an 'expressive' sign. Data, on the other hand, remain as 'mere' data just to the extent that the set of data constitutes a collection of indicative signs that does not serve as an expressive sign, so that it does not point immediately to anything meaningful. From this point of view, the evolution of an equation within a physical symbol system as a means of better conveying the 'meaning' or 'semantic content' that is encapsulated in the data corresponds to the evolution of another kind of sign which does express something to us, and thereby defines a model. Evidently the 'information content' is very little changed, or even unchanged, when a body of data is transformed into an equation derived from this data, but the 'expressivity' or 'meaning value' is commonly increased immensely. Since it is just this increase in 'meaning value' that justifies the whole activity of substituting equations for data, there is a natural interest in processes for further promoting such means for effecting what are again essentially 'economies of thought'.

Model induction is one particular mode of data mining. Inferring models from data is an activity of deducing a closed-form expression based solely on observations. Observations, however, always represent (and in principle only represent) a *limited source of in-*

formation. The question then has to be posed of how the corresponding limited flow of information from a physical system to an observer can result in the formation of a model that is complete in the sense that it can account for the *entire* range of phenomena encountered within the physical system in question – and so a model that can describe even the data that are outside the range of previously encountered observations (Babovic, 1996b). Now traditional model induction can often be usefully characterised as the search for a model that is capable of *acquiring semantics from syntax*. Clearly, every model has its own syntax. Artificial neural networks, for example, have the syntax of a network of interconnected neurons, whereas genetic programming has the syntax of tree-like networks of symbolic expressions in reverse Polish notation, or RPN. The question is whether a particular syntax can capture the semantics of the system that it attempts to model. Certain classes of model syntax will surely be inappropriate to the representation of some physical systems. One might try to choose the model whose representation is as complete as possible, in the sense that a sufficiently large model can capture the data's properties to a degree of error that decreases with an increase in the model size. Thus, to revert to the standard methods of the earlier practice, one might decide to expand in Taylor or Fourier series to a degree that will decrease the least-squares error to a certain, arbitrarily given degree. However, in most of the cases of interest here the semantics would almost certainly not be captured using the syntax of such methods (see, more generally, Klinkenberg, 1996, pp 143-154).

Genetic Programming

Genetic Programming (Koza, 1992) is one instance of the evolutionary algorithms family. In Genetic Programming (GP) the evolutionary force is directed towards the creation of representations, often called 'models' that take a symbolic form. In fact the strings of tokens, such as constitute equations for example, although usually treated as strings of symbols and employed as such, can also be regarded as sequences of sign in the strict sense. In this last situation they can be regarded as models whose expressive signs provide definite meanings. Thus, although the term 'symbolic model' is an oxymoron in the strict sense of the term of semiotics – and indeed even more so than the term 'data model' – it can still be employed with this special understanding. In this evolutionary paradigm, evolving entities are presented with a collection of data and the evolutionary process is directed towards the creation of such closed-form symbolic expressions describing the data. In its primitive form, GP lends itself quite naturally to the process of induction of mathematical models based on observations: GP is an efficient search algorithm that need not assume the functional form of the underlying relationship. Given an appropriate set of basic functions, GP discovers a (sometimes very surprising) mathematical model that approximates the data well. At the same time, GP-induced models come in a symbolic form that is familiar to many scientists so that their production can be more readily assimilated as 'knowledge' as soon as their 'symbols' are in fact regarded as signs. (see, for example, Babovic, 1995).

GP iteratively applies variation and selection on a population of evolving tree structures standing for symbolic expressions in RPN. Standard variation operators in genetic programming are subtree mutation (replace a randomly chosen subtree with a randomly generated subtree) and subtree crossover (replace a randomly chosen subtree from a formula with a randomly chosen subtree from another formula). For a detailed description, see, for example, Babovic and Abbott (1997a). The types of functions used in this tree structure are user-defined. This means that they can be algebraic operators, such as *sin, log*, +, -, etc., but they can also take the form of *if-then-else* rules, making use of logical operators such as *OR, AND*, etc. A number of applications of GP has been reported, such as studies by Babovic and Minns (1994) in which salt intrusion data were analysed, Babovic (1995) related to experimental data for bed concentration of suspended sediment, and Babovic (1997) related to rainfall runoff modelling. In all of the above-mentioned studies, GP-induced relationships provided more accurate descriptions of data than those obtained using more conventional methodologies. An extensive survey of the applications of GP in water resources is provided in Babovic and Abbott (1997b).while recent issues of the *Journal of Hydroinformatics* provide other examples and references.

However, the application of standard GP in a process of scientific discovery does not always guarantee satisfactory results. In certain cases, GP-induced relationships are too complicated and provide little new physical insight into the process that generated the data. One may argue that GP, in such situations, blindly fits parse trees to the data (in almost the same way as in Taylor or Fourier series expansions). It can be argued that GP then results in a model with accurate syntax, but with opaque, or even meaningless, semantics. Moreover, in these cases, the dimensions of the induced formulae often do not fit, pointing to the physical inadequacy of the induced relationships and the presence of further variables 'hidden' within the coefficients.

Dimensionally Aware Genetic Programming

Recently, an augmented version of GP has been proposed – dimensionally aware GP – which is arguably more useful in the process of scientific discovery (Keijzer and Babovic, 1999; Babovic and Keijzer, 1999).

In all scientific endeavours data representing carefully collected observations about particular phenomena that are under study are usually accompanied by their units of measurement. However, the traditional methods usually exploit this information only through the introduction of dimensionless ratios (with such well known examples as Froude and Reynolds numbers). Once the dimensionless numbers are used instead of the original dimensional values, the problem of dimensional homogeneity is conveniently avoided, as all analysed quantities are dimension-free. It can also be argued that dimensionless ratios reduce the dimensionality of the original search space, making it more compact and thus providing a more effective behaviour of algorithms that fit models to the data. At the same time, however, much of the information originally contained in the units of measurement of the data sets is ignored entirely.

Standard GP is ignorant of the dimensionality of its terminals and can safely be applied to problems composed of dimensionless numbers only. Given the symbolic nature of GP and its ability to manipulate the structure of functional relationships, it seems strange that information contained in units of measurement has not been used earlier as an aid in the search process. After all, the criterion of dimensional correctness as used in science acts as a syntactic constraint on any formula it induces. It was therefore to be expected that the introduction of dimensions into the GP paradigm ought to result in an improved search efficiency. The example of the Manning roughness coefficient n may again be used. Now it appears quite generally that the dimensions of some of the best known parameters used in hydraulics are anything but well understood and 'transparent'. In the case of the example of the Manning number, this has dimensions of $s^1\ m^{-1/3}$ but is still referred to as a roughness 'coefficient'! The fact is that Manning's n must accommodate a number of phenomena not explicitly taken into account in the well-known formula defining average flow velocity under steady conditions.:

$$u = \frac{1}{n} R^{2/3} I^{0.5} \qquad (1)$$

Since functional similarity to the natural system is supposed to be comprehended by equation (1) itself, it is this *calibration coefficient n* that must capture the exact correspondence between the model and the real world. As explained above, parameters of this sort serve in effect as *error compensation devices* that artificially adjust the model results to compensate for the fundamental discrepancies that exist between the real world and its representation within the model. Manning's n is much more than a coefficient that can be associated with roughness-induced forces only, and as such is not well defined in the natural world. Instead it exists only at the interface between nature and model. It has to accommodate both: a part related to physical processes, but also a part that has to do with our schematisation of nature within a model. One may even ask 'What is the physical meaning of such a parameter – how well is it grounded, and indeed is it properly grounded at all?' We may be able to read a certain 'physical meaning' into such calibration parameters, but they do not exist-as-such and are thus 'disconnected' in a fundamental way from the world that they are supposed to model (Minns and Babovic, 1996). At the same time, it is quite obvious that the dimensions of such calibration coefficients will then be chosen in such a way that, although the overall dimensional consistency of the model will be maintained, the coefficients themselves may have little physical meaning.

Thus, within the data -mining paradigm it is possible to induce a completely new definition of coefficients, such as roughness. Such new definitions must then however be supported not only by collected observations but also by physical insights. In this way a descriptive, semantic component is added to the data-mining algorithm. This is in addition to the functional semantics that define the manipulations on numbers. While functional semantics grounds formulae in mathematics, the dimensional semantics grounds them in the physical domain. Hopefully, the kinds of misunderstandings outlined in the case of roughness at the beginning of this paper can then be prevented.

It should be emphasised here that in the case of data mining for knowledge discovery in particular, the hydroinformatician must have *both* a sound insight into the functioning of the tools that are employed *and* a profound insight into the physical processes towards which the data mining is directed. Whereas in the symbolic era the most difficult part of the work was often at its beginning, in the formulation of the basic equations, in this new paradigm the greatest difficulties usually arise at the end of the process, in interpreting and 'making sense' of the productions of the data mining tools, so that these productions come to constitute true 'knowledge'. It is then necessary also that the hydroinformatician can communicate meaningfully with experts in the domain of application so that this new knowledge can be related to existing knowledge and specifically to existing theory.

Further Influences of Hydroinformatics

We have traced the beginning of the new, hydroinformatics, paradigm in hydraulics itself to two streams of development, the one occurring outside and the other within hydraulics. The first of these, the one that is imported from outside of hydraulics, is that of data mining for knowledge discovery. The second, which has arisen within hydraulics, is that of the accumulation of massive amounts of data, often collected at great expense, which data is eminently suited to data mining for knowledge discovery. However, since the hydroinformatics paradigm is also one of combining technologies in order to lever added value from out of their combination, calibration procedures are nowadays increasingly supported by field and laboratory measuring programmes, together with new technologies for mining the resulting field data specifically for calibration purposes. These practices are commonly complemented by the direct assimilation of data into numerical models (e.g. Canazares, 1998; see, more generally, Abbott, 1996). For example, acoustic-doppler records can now be mined to determine the nature of the bed material and forms, thus supporting the modelling process in a more satisfactory way. Insofar as unsupported calibration proceeds at all, it is now increasingly in its turn automated, so as to be so much less 'lumped' - or so much more 'distributed'- thereby highlighting areas in which the calibration procedure itself becomes unrealistic. Thus, the process of discovering new knowledge becomes 'data driven' in other ways again. It is in the combination and integration of such ways that the process of creating knowledge in hydraulics is expected to proceed for the greater part in 'the hydraulics of the hydroinformatics era'.

The sociotechnical dimension of hydraulics

As a technology, hydraulics has always been associated with activities in the 'outer', or 'material' world of applications with a social utility. For example, in 1786 the French military engineer, Pierre Louis Georges du Buat, published the second edition (the 1779 first edition was not complete) of his major work, *Principes d'Hydraulique*. He enumerated there what he considered to be the most urgent list of problems which hydraulics should solve:

"Our understanding of hydraulics has been extremely limited:.. We are.. in almost absolute ignorance of the true laws to which the movement of water is subject: .... All that concerns the uniform course of the waters of the surface of the earth is unknown to us; and to obtain an idea of how little we do know, it will suffice to cast a glance over what we do not.

"To estimate the velocity of a river of which one knows the width, the depth and the slope; to determine to what height it will rise if it receives another river in its bed;.. how much it will fall if one diverts water from it; to establish ... the proper capacity of the bed to deliver to a city at a given slope the quantity of water which will satisfy its needs; to lay out the contours of a river in such a manner that it will not work to change the bed in which one has confined it; to calculate the yield of a pipe of which the length, the diameter, and the head are given; to determine how much a bridge, a dam or a gate will raise the level of a river; to indicate to what distance backwater will be appreciable, and to foretell whether the country will be subject to inundation; to calculate the length of a canal to drain marshes long lost to agriculture; to assign the more effective form to the entrances of canals, and to the confluences or mouths of rivers; to determine the most advantageous shape to give to boats or ships ..; ... All these questions, and infinitely many others of the same sort, are still unsolvable; who would believe it?

".. for lack of principles, one adopts projects of which the cost is only too real but of which the success is ephemeral; one carries out projects for which the goal is not attained; one charges the state, the provinces, the communities with considerable costs, without gain, often with loss; or at least there is no proportion between the cost and the advantages which results therefrom". [Du Buat, 1786; quoted from Rouse and Ince, 1957]".

Not only are the meaning and language intelligible to any literate citizen but they also correspond to social demands. Moreover, the domain of hydraulics defined on this basis of social application is already very wide indeed. It was just the century which followed Du Buat's observations, however, which witnessed the explosion of the mathematical sciences and, in particular, all the various attempts to formulate real-world physical phenomena in terms of equations which might in principle be solved and thus might open the way to engineering applications - with all the restrictions that this approach also entailed.

It was essentially this mathematical-scientific development of hydraulics that led to the gradual separation of the hydraulician from his or her immediate social environment: it thereby became possible for many hydraulicians to concentrate their attention upon the one or the other technical, and even scientific, aspect of a problem, while leaving the social aspects and even the constructional aspects to 'other people' (Abbott, 1999). The hydraulician could retreat into his or her laboratory and office and follow the principle that 'what is good for hydraulics and flow efficiency must also be good for the mankind and nature'. Oscillating unconsciously between the 'beaver mentality' commonly attributed to the engineer and the arrogance of an ancient Egyptian priest, the hydraulician increasingly usurped the decision power over water resources in society while becoming at the same time increasingly detached from the social and ecological consequences. In effect, although it was supposed that the technical results of studies would somehow and at some time find social applications, the processes whereby such social applications occurred were not seen as the direct concern of the hydraulician and these processes were even less themselves regarded as subjects of hydraulics research. In the new paradigm, on the other hand, *it is practically impossible to research in hydraulics at all without simultaneously researching how the results of this research will enter into social applications*. Within this new paradigm, new technological means can scarcely be developed to any effect at all without a thorough research being made into the ways in which these new means can best be applied within society. Entirely consentaneously, no changes can be realised and sustained in society without the elaboration of new technical means to catalyse and guide these changes. Hydroinformatics is an essentially *sociotechnical* endeavour, and hydraulics is carried along with hydroinformatics in this endeavour (Thorkilsen and Dynesen, 2001).

Thus the era when the hydraulician could concern his or herself exclusively with the technical aspects of the project, leaving its social application aspects to 'other people' has largely come to an end. In the language of sociotechnical studies, the hydraulician is no longer a homogeneous engineer, but a heterogeneous one. Of course there have always been heterogeneous engineers in practice (Edison and Bell commonly serve as exemplars on the sociotechnical literature) but in earlier times these could be regarded as exceptions. Nowadays, heterogeneity is the rule (Abbott, 1996).

In the same vein, the hydraulician, in his or her role as a hydraulic engineer, might previously have travelled to a project site to establish 'the facts'. This person was not normally called upon to enter into the making of judgements about the social desirability of the project to which the facts applied and even less to make arrangements for such judgements to be expressed by local persons, and even less again to research, analyse, design and install the integrated social and technical, or sociotechnical, arrangements most appropriate for this purpose. Today, the hydraulician, as a hydraulic engineer working within the hydroinformatics paradigm, is in the first place one who *persuades* the rest of society to follow one line of action rather than others, whether individually, or (as is much more common nowadays) as a team with this same purpose, backed by an veritable arsenal of technical equipment. And, increasingly, this line of action involves the indivisible provision of the technical and the social, or sociotechnical, means for the ordinary, and most commonly non-technical, persons directly concerned to make this choice of the one line of action rather than other such lines by interacting with one another in further persuading activities, In effect, the task of the hydroinformatician becomes increasingly one of 'persuading people to persuade people' (Abbott, 1999 a and b; Abbott and Ionoski, 2001).

The decisive role of knowledge management

The most important single factor by far in the current development of hydroinformatics is the exploitation of the potential of the Internet and such electronic networks, or 'internets' with a lower case 'i', that are coming to succeed it. Indeed, it is ultimately through its application of internet-based technologies that hydroinformatics comes to be legitimised as an independent discipline. In terms of the post-modern distinction between 'knowers' and 'consumers of knowledge', the most immediate effect of such networks is to increase the number of consumers of the most advanced hydraulic knowledge by at least two orders of magnitude again. The development in the number of such users and its future projection is also followed in Fig. 1. However, sociotechnical studies show again that such a massive change or 'paradigm shift' in the social application of hydraulic and related knowledge cannot be realised without the introduction of new technical means. In particular, insofar as non-expert members of the general public wish to make use of such knowledge in order to assess the impact of various projects (ground water mining, river embankment construction, urbanisation etc.) upon their quality of life, so they need new classes of tools to assist them. They need tools that will enable them to relate their own personal and group interests, concerns and intentions to any proposed intervention in their environment if they are to participate at all effectively in the judgmental processes governing such interventions. Any tool that enables persons and groups, whether expert or non-expert, to make judgements on the basis of presented facts is called a *judgement engine* (Abbott, 1998b; Abbott, 1999; Huang at al, 1999). Thus a judgement engine is a computational device that facilitates the making of judgements in a decision-making environment. The decision making process is commonly supposed to entail strings of inferences, typically of the form:

$$(beliefs, facts\ (data)) \rightarrow attitudes \rightarrow positions \rightarrow judgements \rightarrow decisions \rightarrow actions \quad (2)$$

Here the inferential processes are constructed as mappings, denoted by arrows, from one specific state of conscious activity to another. Each such state is specified by its type, as described by a noun in a natural language in (2) above. As its own name suggests, a judgement engine is one that facilitates the process described in (2) reading up to the level of judgements from left-to-right, so that it constitutes one (of many) representations of a judgmental process.

In so far as a number of persons participate in a judgmental process, so a great variety of beliefs, attitudes, positions, etc will be presented, but there is still an interest in maintaining a community of structure in the overall processes exemplified in (2). This notion of communality of structure can be made more explicit by associating it with the requirement that all inferential strings such as (2), when taken over all persons participating in the inferential process, should constitute a category in a specific mathematical sense (Abbott & Dibike, 1998a and b; see also Abbott, 2000b, where the judgement engines must be instantiated automatically from 'user profiles' which characterise the requirements for advice of each individual user.).

The problem of promoting cooperation between persons participating in an inferential process can then be posed as one of inducing an understanding of the origins of the judgements of other participants. This implies, to take the example (2), that the category that is the dual of (2),

$$actions \rightarrow decisions \rightarrow judgements \rightarrow positions \rightarrow attitudes \rightarrow (beliefs, facts\ (data)) \quad (3)$$

should be made as explicit as possible when 'actions' and 'facts(data)' are observables. A judgement engine that facilitates this process is said to be *transparent*.

It should be observed that judgmental processes occur through the process described in (2) and (3) through the presence of many other processes than those represented explicitly by the mapping themselves, but that these belong to a different class of judgements - which we may call 'intermediate judgements' - than those identified explicitly in these strings of inferences (See, again Abbott, 1999, and Huang *et al* 1999; and then, originally, Husserl 1938/1938//1957)). Since such engines are designed explicitly for use over electronic networks, they can best be appreciated through their hands-on use over an electronic network. (See, again, Huang et al, 1999)

Conclusions

Hydroinformatics cannot be regard merely as a new way of applying an otherwise unchanged hydraulics within society. Instead, hydroinformatics, while building upon existing hydraulics and taking existing hydraulics for the most part as its central core, introduces its own paradigm into hydraulics itself. Some of the consequences of this introduction have been elaborated here, but there are increasingly many others. Hydroinformatics takes up many activities of hydraulics research and practice that were hitherto considered as rather separated and brings them together in new and innovative ways using the new information and communication technologies. By these means it increases the value of the individual activities, often markedly. Thus, for example, the bringing together and integration of field measurements, numerical models and remotely sensed data provide new and highly valuable new facilities that increase greatly the value of the component activities. Indeed, studies of such biological-geographical problems as those of following changing distributions of eel grass plains and mussel beds under the impacts of depositions of fine sediments are scarcely possible without such integrations (Abbott, 1997; Thorkilsen and Dynesen, 2001). Hydroinformatics is strongly engaged in these combining and integrating activities. Such procedures then lead to new ways of doing hydraulics itself, as exemplified in this paper.

Further to this again, hydroinformatics also introduces new techniques to encapsulate existing knowledge, commonly with the view to accelerating the rate of access to this knowledge. Artificial neural networks are already widely used for this purpose (e.g. Dibike et al., 1999). Such developments drive several new areas of applications of existing hydraulics knowledge, such as real-time control of urban drainage and irrigation systems, real-time diagnostic systems and on-line risk analysis support systems. Just as the information revolution has already led to profound changes in the ways of working of hydraulicians, so the ongoing communication revolution is changing the whole relation between the hydraulician and society. From being one who personally 'thinks about water', the hydraulician is drawn (or even dragged!) by degrees to become one who has also to 'think about how other persons think about water''. Although the more purely technical developments in hydroinformatics, as exemplified here by data mining for knowledge discovery, are the most readily acceptable to the more traditionally minded hydraulician, the sociotechnical developments promise ultimately to lead to the most profound changes in the practice of hydraulics itself.

References

ABBOTT, M. B., 1991, *Hydroinformatics: information technology and the aquatic environment*, Ashgate, Aldershot, UK, and Brookfield, USA.

ABBOTT, M. B., 1992, The theory of the hydrologic model, or: The struggle for the soul of hydrology, in *Advances in theoretical hydrology; a tribute to Jim Dooge* (ed. O'Kane, J. P.) pp 237-254, Elsevier, Amsterdam.

ABBOTT, M. B., 1993, The electronic encapsulation of knowledge in hydraulics, hydrology and water resources, *Adv. Wat. Resources*, **16**, pp 22-39.

ABBOTT, M. B., 1994, Hydroinformatics: a Copernican revolution in hydraulics, *J. Hyd. Res., Special Issue on Hydroinformatics*, pp 3-13.

ABBOTT, M. B., 1996, The sociotechnical dimension of hydroinformatics, in *Hydroinformatics '96* (ed. Mueller, A.) pp3-18, Balkema, Rotterdam.

ABBOTT, M. B., 1997, Range of tidal flow modelling, *Journal of Hydraulic Engineering*, ASCE, Vol 123 (4): pp. 257-277.

ABBOTT, M. B., 1998a, New business opportunities; new human responsibilities, in *Hydroinformatics '98* (eds. Babovic, V. M. and Larsen, L.-C.) pp 1171-1175, Balkema, Rotterdam.

ABBOTT, M. B., 1998b, Hydraulik im Zeitalter der Hydroinformatik, in Steinebach, G., *Mathematische Modelle in der Gewaesserkunde*, BFG, Koblenz, pp 126-129.

ABBOTT, M. B., 1999a, Introducing Hydroinformatics, *J. of Hydroinformatics*, **1**, 1, pp3-19.

ABBOTT, M. B., 1999b, Forchheimer and Schoklitsch: a postmodern retrospection. In memorial symposium P. Forchheimer and A. Schoklitsch, 1[ed.] Graff, W. H., Proc. 28th IAHR Congress, Graz.

ABBOTT, M. B., 2000a, Fides et Ratio, a critical appraisal (Book review: Faith and Reason: Evangelical letter *Fides et Ratio* of the supreme Pontiff John Paul II etc.) *Journal of Hydroinformatics* , **2**, 2, in the electronic version only, but with printed proofs available from the author.

ABBOTT, M. B., 2000b, The gender issue in hydroinformatics, *Journal of Hydroinformatics*, **2**, 2, pp 87-106.

ABBOTT, M. B. and IONOSKI, A., 2001, The democratisation od decision making processes in the water sector, *J. of Hydroinformatics*, 3 (1), pp 35-48.

ABBOTT, M. B., and DIBIKE, Y, 1998a, On the representation of processes in the postsymbolic era, in *Hydroinformatics '98* (eds.Babovic, V. M., and Larsen, L. C.) pp1177-1184, Balkema, Rotterdam.

ABBOTT, M. B., and DIBIKE, Y, B.,1998b, The symbolic representation of hydroinformatics processes using elements of category theory, in *Hydroinformatics '98* (eds. Babovic, V. M. and Larsen, L. C.) pp 1185-1192, Balkema, Rotterdam.

ABBOTT, M. B. and JONOSKI, A..,1998, Promoting collaborative decision making through using electronic networking, *Hydroinformatics 98*, ed. Babovic, V., and Larsen, L. C., pp911-918, Balkema, Rotterdam.

APPEGANESI, R., and GARRATT, C, 1995, *Postmodernism for beginners*, Icon, Cambridge.

BABOVIC, V., 1996a, *Emergence, Evolution, Intelligence; Hydroinformatics*, Balkema, Rotterdam

BABOVIC, V., 1996b., "Can water resources benefit from artificial intelligence?", in Köngeter,J., (ed.), *Internationales Wasserbau Symposium – Computational Fluid Dynamics; Bunte Bilde in der Praxis*, Aachen, Germany

BABOVIC, V., 1997, "On the modelling and forecasting of nonlinear systems", in Refsgård, J.C., and Karalis, E.A., (eds), *Operational Water Management*, Balkema Publishers, Rotterdam

BABOVIC, V., and ABBOTT, M.B., 1997a., "The evolution of equations from hydraulic data, Part I: Theory", *Journal of Hydraulic Research*, **35**, 3, pp. 1-14

BABOVIC, V., and ABBOTT, M.B., 1997b.,"The evolution of equations from hydraulic data, Part II: Applications", *Journal of Hydraulic Research*, **35**, 3, pp. 15-34

BABOVIC, V., LARSEN, L.C. and WU, Z, 1994., "Calibrating Hydrodynamic Models by Means of Simulated Evolution" in *Hydroinformatics '94*, Verwey, A., Minns, A.W., Babovic, V., and Maksimovic, C., (eds.), pp.193-200.

BABOVIC, V., and KEIJZER, M., 1999, Genetic Programming as a Model Induction Engine, Journal of Hydroinformatics, **2**, 1, pp 35-60.

BABOVIC, V., and MINNS, A.W., 1994, "Use of computational adaptive methodologies in hydroinformatics", in Verwey, A., Minns, A.W., Babovic, V., and Maksimovic, C., (eds.), *Proceedings of the 1st International Conference on Hydroinformatics, Hydroinformatics '94*, pp. 201-210

BAUDRILLARD, J., 1963, *Le systeme des objets: la consommation des signes*, Gallimard, Paris.

DIBIKE, Y. B., MINNS, A. W., and ABBOTT, M. B.,1999, Applications of artificial neural networks to the generation of wave equations from hydraulic data, *J. Hydraulic Res.* **37**, 1, pp81-97.

ECO, U., 1976, *A theory of semiotics*, Indiana Univ., Bloomington, USA.

HUANG, Y., SOLOMATINE, D., VELICKOV, S., and ABBOTT, M. B., Distributed environmental impact assessment using internet, *J. of Hydroinformatics*, **1**, 1, pp59-70.

HUSSERL, E, 1938//1957, *Experience and judgement; investigations in a geneology of logic* (trans. Churchill, J. S., and

Ameriks, K.) Routledge, London.

JONOSKI, A, and ABBOTT, M.B., 1998, Network distributed decision support systems as multi agent constructs, in *Hydroinformatics '98* (eds. Babovic, V. M., and Larsen, L.-C.) pp 1219-1226, Balkema, Rotterdam.

KEIJZER, M., and BABOVIC, V., 1999, Dimensionally aware genetic programming, in Banzhaf, W., Daida, J., Eiben, A. E., Garzon, M. H., Honavar, V., Jakiela, M., & Smith, R. E. (eds.)., GECCO-99: *Genetic and Evolutionary Computation Conference*, Morgan Kaufmann, Orlando and San Francisco.

KLINKENBERG, J.-M., *Précis de sémiotique générale*, De Boeck et Larcier, Paris.

KOZA, J., 1992., *Genetic programming: on the programming of computers by natural selection*, MIT Press, Cambridge, Mass.

KUTIJA, V., and HONG, 1996, A numerical model for assessing the additional resistance to flow induced by flexible vegetation, *J. Hyd. Res.*, 34 (1), pp 99-114

LYOTARD, J-F., 1979//1992, *The postmodern condition, a report on knowledge* (trans. Bennington, G., and Massumi, B.) Manchester Univ., UK.

LYOTARD, J-F, 1998, *Le postmoderne explique aus enfants, correspondence 1982-1985,* Galilee, Paris.

PIATETSKY-SHAPIRO, G., and SMYTH, P., 1996., From data mining to knowledge discovery: an overview, in *Advances in Knowledge Discovery and Data Mining*, Fayyad, U., Piatetsky-Shapiro, G., Smyth, P., and Uthurusamy, R., eds., MIT Press, Cambridge, Mass., pp.1-36.

ROUSE, H. and INCE S., 1957, *History of Hydraulics*, Dover, New York.

THORKILSEN, M., and DYNESEN, C., 2001, An owner's view of hydroinformatics: its rule in realising the bridge and tunnel connection between Denmark and Sweden, *J. Hydroinformatics*, 3 (2), pp 105-135.

The following article has been reprinted from *Journal of Hydroinformatics*, Vol. 16, No. 2, pp. 516–530, with permission from the copyright holders, IWA Publishing.

Towards a hydroinformatics praxis in the service of social justice

M. B. Abbott and Z. Vojinovic

ABSTRACT

This paper introduces a new role for hydroinformatics in its sociotechnical environment. It introduces a novel and modern approach for dealing with flooding and other such destructive phenomena which have been increasing ever more rapidly throughout the world. By far the greatest toll from floods and flood-related disasters, however, is in the so-called 'developing' world where the resources for dealing with such problems are either inadequate or non-existent, even as these are often subject to the interventions of corrupt behaviour and practices and their related dubious legal systems. The necessity of promoting the nature of the force that is required to overcome these negativities are identified as those provided by active stakeholder participation and this paper indicates how this force can be informed and motivated largely through projecting ultra-realistic dynamic and coloured illustrations within a collective social environment.

Key words | active stakeholder participation, colour, creative imaginations, social justice

M. B. Abbott (corresponding author)
European Institute for Industrial Leadership,
Château Latour de Freins,
Brussels, 1140,
Belgium
E-mail: *knowledge.engineering@skynet.be*

Z. Vojinovic
UNESCO-IHE Institute for Water Education,
Westvest 7 2611 AX Delft,
The Netherlands

INTRODUCTION

To date, our experiences of applications of what has become a new kind of *praxis*, as this involves public participation, have been largely positive when there was substantial backing from central and local governments and other such bodies, whereby public participation could proceed satisfactorily. Moreover, in other social environments, where corruption and social injustices have not yet taken such a hold as to have become endemic, essentially the same approaches and praxes have been employed, ultimately also to good effect. To date, however, we have not encountered situations where these social deficiencies have become so effectively institutionalised in size and scope over so many years that they remain unpunished and even become socially acceptable within the miasma of unknowing that these deficiencies themselves produce. This situation must be confronted as and when it arises, and other means again will certainly be needed to overcome it.

It is observed, however, that these kinds of negative forces continue to have ever-increasing destructive consequences, with ever more serious deprivations for the poorest in their society, and these events have their own influence on the future nature of our praxes. The most intransigent component of this kind of force as it currently proceeds was presented in an article published in the *Financial Times* of 27 September 2012, p. 13, as authored by George Soros and Fazia Hasan Abed, observing that:

> 'An estimated four billion people [amounting to half the world's population] live outside the protection of the law, mostly because they are poor ...'
>
> 'Without basic legal empowerment, the poor live an uncertain existence, in fear of deprivation, displacement and dispossession. A juvenile is wrongfully detained and loses time at school, village land is damaged by a mining company without compensation, an illiterate widow is denied the inheritance she is entitled to and is forced on to the streets with her children. By what means can individuals and communities protect their rights in daily life?'

doi: 10.2166/hydro.2013.198

This then appears as the principal challenge with which we are currently confronted when venturing into this territory.

It is especially in view of the increasing menace of these negative influences that we must now pose the questions of how our own field of hydroinformatics can intervene to expose and counteract these negative forces, which we identified in the first part of Flood Risk and Social Justice (Vojinovic & Abbott 2012) and to which we shall return in the second part of this paper. This paper, as a whole, is correspondingly partly devoted to a praxis that is being increasingly applied to defeat these destructive forces in the areas of flooding, water pollution and other water-related damage and devastation, while accepting that this approach is still in its infancy, even though its feasibility was first demonstrated some 15 years ago in describing the road and rail connection across the Øresund that separates Denmark and Sweden. Even then, attempts were made by certain political forces and parts of the media to subvert and derail this gigantic project (see Thorkilsen & Dynesen (2001) and Abbott (2007)). Although that project was not itself concerned with flooding but with countering environmental damage, it has continued to provide a paradigm case for the introduction of *active stakeholder participation*. When seen in retrospect, however, a corresponding contribution to this approach was its demonstration of the power of *coloured dynamic graphics* in reinforcing the formations of stakeholder movements and their consequent empowerment, as means for stimulating the *creative imaginations* and consequent actions of these stakeholders. Many illustrations of these graphics in a variety of applications appear in the book of Vojinovic & Abbott (2012) and four major projects are described in the conclusion there.

The first awakening

It has become increasingly clear that the only kind of activity that can combat and defeat the increasingly powerful forces that we shall call *transgressors*, in the sense that they do not tell the whole truth and cause harm correspondingly, is that which mobilises a combined effort by the persons being transgressed, many of whom may have already experienced flooding, pollution and other water-borne disasters. There is, in this process, a substantial gender issue in that women almost always play a much more active role than do men in such movements; see Abbott (2000) who explains this in terms of the Object and Value Theory of Alexius von Meinong (see Findlay 1963 [1995]) and the Category Theory of Freyd & Scedrov (1990). However, we do not have space to follow up on this aspect here, despite its importance.

When confined exclusively within the ambit of secular law, there is the question of the cost of mobilising a sufficient legal force to overcome the disclaimers and denigrations mobilised by the much greater legal resources of so many of these transgressors. There is a popular adage that 'British justice is the best that money can buy', but certainly the rest of Europe, most of the Europeanised Americas, and even some others places, are nowadays not very far behind in this respect (Abbott 2012).

However, even in societies with functioning legal systems, it is only through *collective action* by those persons capable of foreseeing the potential destructive forces of flooding, pollution and other such events, and those who listen to them, that it becomes possible to raise the financial means to support the basic costs of those lawyers who remain honest and decent persons. Experience has shown that some such persons and means are to be found almost everywhere from those who are themselves so shocked by injustices. The Mahatma Gandhi was a shining example of this practice, acquiring an iconic status correspondingly, even as he suffered so much from it, and there are other lawyers (barristers and solicitors) who are even today prepared, *in extremis*, to follow his example. However, we now have other means to release ourselves from the *legalised tyranny of language* that otherwise holds sway in this environment, as we must now explain.

The formation of the stakeholder group

The task of the hydroinformatician is to form a force from out of those who truly care about flooding and other water-associated damage in such a way that, whether allied with engaged and honest lawyers or not, they are capable of exposing and repulsing such transgressions that may be so destructive to their lives. Such a movement is usually motivated by earlier flood and other water-related events that have provided warning signals of future water-driven damage. This is the task of our *new kind of praxis*, one that is only now coming into its own as a *socio*technology. This praxis is concerned in the first place with providing realistic simulations of flood, pollution and other such water-related events, as provided by the hydroinformatician over the web, whereby these phenomena can be projected onto the minds of the present and potential stakeholders using their web facilities, such as are demonstrated by the websites of www.knowledge-engineering.org and www.urbanhydroinformatics.com, both of which include simulated dynamic coloured illustrations, primarily of flooding. This process corresponds to one of catalysing the movements of the minds of both the existing and the potential stakeholders in such a way that they experience the real-life appearances, and even the physical experiences, of the outer-world, water-driven events. This first stage initiates what we can describe as *social infrastructure*.

In taking this route we can largely avoid what Cheetham (2005, pp. 2–3) described as 'this shift in relations between the subject and the object [which] involves a 'withdrawal of participation'... Another disjunction, another loss of participation, accompanies the transition from oral to literate society.' To continue with Cheetham (2005, pp. 2–3):

> 'For European history the crucial transition occurs in Greece roughly between Homer and Plato. The techniques of alphabetical writing and reading forever changed the relations of humans to language and to the nonhuman world. Socrates was very concerned about this new technology, and was afraid that it signified the death of real thinking, and that education would suffer irreparably. In fact the great sweep of Western history as a whole has been read as a story of withdrawal and the progressive 'death of nature,' and the birth of a mechanistic cosmology based on abstract materialism.'

There are of course many other authors who have emphasised this descent. Correspondingly, the first lesson in our praxes is that we must avoid so far as possible all our speaking, writing and reading as means of transmitting meaning, and in our case necessarily falling back upon our envisioning of events.

In the conflict between the consequently better-informed stakeholders and the potential *transgressors*, it is in this way that the stakeholders come into possession of that most powerful of all weapons: *a more complete truth*.

This estimation is based upon the twin-definitions introduced from a sociotechnical standpoint by the first-named author:

> 'Reality is the name that we give to the interface between our inner and our outer worlds and a truth is an intimation of the oneness of these two worlds.'

Following the admonitions of Hugh of St Victor, as adumbrated by Illich (1993), such a definition should be sounded just as if it were being physically consumed, very much as we taste and consume our food, so that it is another experience than that of reading a naked text (see Cheetham (2002)).

The first task of the hydroinformatician within this environment is to create *a truthful oneness* from these two worlds, thereby taking the first steps in catalysing the processes leading to deeper understandings, and thereby to more profound truths concerning the actual and potential threats – and therewith identifying the means for repulsing them.

These are the first steps in creating an environment populated by active stakeholders, as it proceeds through deeper individual understandings, and from there *transmutes* into more collective understandings, and from there again *transmutes* into the worlds of the *creative imagination*, leading to *creative understandings*, that the process of active stakeholder participation comes, through processes of repetition, to fruition – and thus to *a transcendence in the understanding of the innermost reality of the water-related threat*. But this necessitates that we ask how this process is to proceed in terms of communication.

Cheetham (2005), among increasingly many others, ascribes the origins of the now clearly disintegrating world of discourse in European modernism as follows (p. 3, with further italics added):

> 'Henri Corbin was a French philosopher, theologian and scholar of Islamic thought, particularly Sufism and Iranian Shi'ism. It was Corbin's contention that European civilisation experienced a 'metaphysical catastrophe' as a result of what we might call *The Great Disjunction*.

> This was signalled by the final triumph of the Aristotelianism of Averroes over the Platonic and neo-Platonic cosmology championed by Avicenna. To the defeat of that cosmology is coupled the disappearance of the *anima mundi*, the Soul of the World. The catastrophic event that gave rise to modernity is the loss of the soul of the world.'

> 'The details of this history hinge on the fate of the Aristotelian *nous etikos*, which became the Agent or Active Intellect in medieval Western philosophy. This Active Intellect operating through us was something equated in Islamic thought with the Holy Spirit or Angel of Revelation, the Angel Gabriel. *The realm of being to which this intellection gives access is this place of vision*, which depth psychology calls the world of the psyche and the imagination. Corbin called it the *mundus imaginalis*, the *imaginal world*, to underscore the fact that it is not imaginary or unreal. Through the agency of the active imagination we have access to an intermediate realm of subtle bodies, of real presences, situated between the sensible world and the intelligible. This is the realm of the *anima mundi*.'

We may now proceed to introduce one of the most powerful – and nowadays least expected – of the post-modern (and thereby pre-modern) instruments for entering the realm of the *anima mundi*, so as to provide the means for the catalysing of our transmutations, *namely those of colour*.

The role of colour in catalysing the processes of understanding

Active stakeholder participation, if properly conducted, is in the last analysis a means of promoting states of social justice, defined as states that provide the possibilities for the individual participants to transcend the selves with which they entered into the participation, even as they collectively seek to establish wider states of social justice in society and its physical and emotional environments more generally. As already introduced, the way in which this has been realised and promoted is to provide web-based hydroinformatic environments which use dynamic, highly detailed and relevant illustrations, *almost always in colour*, of the objects that are of the greatest concern to the individual participants and to society as a whole, who are then represented by their active stakeholders. These illustrations are increasingly dynamic, so that the effects of proposed changes in the environment, for example, can be followed in detail and in their own experiential time by the active participants.

THE CORRESPONDING HISTORICAL BACKGROUND

We have already introduced the notion that, although the physical inputs to such hydroinformatic environments are for a large part definable within the ambit of modern science, the functioning of the *socio*technical hydroinformatic environment cannot be so described. Thus, quoting from Wikipedia, the free encyclopedia, under the heading of the *Condemnations of 1210–1277*, referring to the *Système du Monde* of Pierre Duhem (Duhem & Brenner 1997, p. 24):

> 'According to Duhem, 'if we must assign a date for the birth of modern science, we would, without doubt, choose the year 1277 when the bishop of Paris solemnly proclaimed that several worlds could exist, and that the whole of heavens could, without contradiction, be moved with a rectilinear motion…'

> 'Duhem believed that Tempier, with his insistence of God's absolute power, had liberated Christian thought from the dogmatic acceptance of Aristotelianism, and in this way marked the birth of modern science. The condemnations certainly had a positive effect on science, but scholars disagree over their relative influence. Historians in the field no longer fully endorse his view that modern science started in 1277.'

Being postmodern – that is, functioning in societies of *Consumers of Knowledge* rather than in the modern sense, the sense of *The Condemnations*, as *Knowers* – its science reverts back to the Premodern, and then in the first place to Alchemy, as described in Bruno Latour's 1991 book among many others. It is only within this context and its manners of expression that it can be described at all. Thus, as the imaginations of the individual stakeholders are mobilised, providing *active imaginations*, and as these stakeholders coalesce into larger groups and even come to cover the group in its entirety, we enter into the realm of the *creative imagination* which is at the centre of the succeeding decision-making processes. We observe that these processes are of an *essentially qualitative nature*, so that, once again, they can only be described alchemically, as *transmutations*, rather than modern-scientifically, as mere *transformations*. For the nature of these transmutations themselves we reach back in the first place to the twelfth to thirteenth century works of Ibn' Arabi on the creative imagination, as recalled into the twentieth century by Henri Corbin and into our twenty-first century by many other authors. In the spiritual sense of this movement, we may speak of a *theophonic imagination* (Corbin 1958/2006, trans. 1969, p. 99).

For our purposes, the knowledge-transmitting components of the hydroinformatics environment, such as those delineating houses, gardens, schools, shops, roads, children, etc., have to be directly, subjectively and even immediately *identifiable*. This necessitates the *emotional identifications of and attachments to objects that are of the deepest concern*. The role of colour in hydroinformatics environments is to facilitate the evocations of these emotional identifications as clearly and precisely as possible.

The Bortoft intervention

It is now 16 years since Henri Bortoft's book, following French- and German-language editions, entered the English language as *The Wholeness of Nature: Goethe's Way Toward a Science of Conscious Participation in Nature* (Bortoft 2010). This was introduced by John Barnes, starting as follows:

> 'Few recognise the depth of the existential crisis into which our modern scientific world view has led us. Through its analytical approach to material processes, it has come to focus on a molecular realm far removed from the world of our human experience. This narrow focus has evoked a countermovement calling for the recognition of personal experience and yearning for meaning and wholeness. The result is an unhealthy polarisation of our culture in which there is a yawning gap between objective, materialistic science on the one hand and the subjective culture of human experience on the other.'
>
> 'What is urgently needed today is a further step in the evolution of science, leading beyond material analysis to a deeper, holistic understanding of nature. In *The Wholeness of Nature*, Henri Bortoft describes how, already 200 years ago, Goethe, the great German poet and scientist, began to lay the groundwork for this new development in science. I know of no other book written in the English language that articulates the principles of Goethe's scientific approach as clearly as this work.'

For his own part, Bortoft provided a summary of his lecture on *Goethe's Phenomenology of Colour* when this was presented on the 17 October 2011 in London through the good offices of the *Temenos Academy* and attended by the present first-named author, as follows:

> 'Goethe's 'theory' of colours is wrongly named. It is not a theory in the conventional sense because it does not set out to explain colour, but to 'make a phenomenon visible'. It is the development of a new way of seeing through the practice of working with the senses and exact sensorial imagination, which keeps the phenomenon in the centre of attention instead of replacing it with a theory. Goethe's way of working is illuminated by phenomenology, so that we can understand better what he was doing on his own terms without having to rely on comparisons with more theory-centred approaches, This opens the door to the possibility of a phenomenological approach to nature which 'allows the phenomenon to be'.'

The background to this subject is very extensive and very varied, see Goethe (1810 and 1820). In the present case we start out from Josef Albers' *Interactions of Colour* that first appeared in 1963 from the Yale University Press as its most magnificent production ever. By following examples and their different modes of expression, Albers observed that:

> 'This discrepancy between physical fact and psychic effect, called in this case a *haptic illusion* – haptic as related to the sense of touch – is so 'in the haptic sense'. To begin the study of how colour deceives and how to make use of this, the first exercise is to make one and the same colour look different.'

The value of the psychic means of personalised perception as opposed to the modern scientific means of impersonal perception lies in this direction, whereby a colour enhances the impact of an emotionally charged 'surreal' object and is no longer associated with an emotionally neutral 'real' object – and it is only in this sense that it 'deceives'. Colour can 'deceive' in this way, *but it cannot lie.* Thus, when the user interface projects the streets along which the various families' children are walking to school with mobile telephones and/or other such communication devices, the blobs that represent the children may be simply 'children-coloured' when there is no danger from flooding, with the families correspondingly indifferent to any flood-related danger, but may appear flashing with an intensified 'children-colour' when these children are in danger and the parents may need to intervene. We may call this 'thinking in terms of situations', whether precedented, actual or anticipated, and, in this example, it is realised by bringing the danger into coincidence with the potential victims in the minds of their parents, thereby contextualising their deepest concerns, whether to assuage them or to support them. We have to do here with an emotional impact which is entirely qualitative, even as it depends upon the quantitative resources of web technology.

To this, however, Albers (1963) observes that 'when it comes to colour intensity (brightness) occasionally one may find agreement among a few people but hardly within a large group (such as a class)'. So this 'solution' is not so simple as it may at first seem. Of course, we are now in the world of multi-media, so that sound transmission over mobile telephones may be used to back-up the visual impressions. Albers identified this possibility even before the advent of mobile telephony:

> 'Though we were taught, only a few years ago, that there is no connection whatsoever between visual and auditory perception, we know now that a colour changes visually when a changing tone is heard simultaneously. This, of course, makes the relativity of colour still more obvious, just as tongue and eye perceptions interdepend when colours of food and of its containers increase or diminish our appetite.'

Albers first made a fundamental distinction between *The Factual* and *The Actual*, observing correspondingly that 'in dealing with colour relativity or colour illusion, it is practical to distinguish 'factual facts' from 'actual facts'. He continued:

> 'The data on wave length – the result of optical analysis of light spectra – we acknowledge as fact. This is a factual fact. It means something remaining what it is, something probably not undergoing change.'
>
> 'But when we see opaque colour as transparent or perceive opacity as translucent then the optical reception in our eye has changed in our mind to something different. The same is true when we see three colours as four or two, or four colours as three ...'
>
> 'Gestalt psychology has proved that 3-dimensionality is perceived earlier and more easily than 2-dimensionality. This explains why children do not begin – as most art-teachers still wish – with painting and drawing, which are lateral abstractions on a two-dimensional plane, but begin all by themselves with building, constructing in space, on a ground and upwards, in three dimensions.'
>
> 'We believe that art education is an essential part of general education, including so-called higher learning. We promote therefore, after a natural and easy laissez-faire as an initial challenge, an early shift from aimless play to directed study and work, which offers with basic training, a continuous excitement to growth.'
>
> 'To say this in psychological terms, it means a shift from a recognition of the primitive drive for being occupied, entertained – *Beschäftigungstrieb* – to a more advanced drive, or better need, for being productive, creative — *Gestaltungstrieb*.'

It is within this framework of Heidegger's *Ge-stell* of 1927 that we identify an antinomy between the so-called Laws of Modern Science and the Laws of Nature, which subsume the laws of our own beings as nature has formed us. The one speaks about Albers' *factual fact* and the other about his *actual fact*. Albers provides many examples of the differences between these two ways of looking at and more generally as experiencing the world, as examples of *a fundamental difference* in our ways of experiencing the world, our *Weltanschauung*, but now in a situation where we have come to live, to continue with Heidegger (1963), in 'a world deprived of its worldhood'. The movement that Albers describes is completely at one with the natural drive to transcendence that was re-established in the seventeenth century by Blaise Pascal, was raised again by Kierkegaard with even more effect in the nineteenth century, and has now been adopted as doctrine by the Catholic and some other Churches. This is at one with the ambition to restore a social justice that is now being expressed ever more clearly in so many parts of our present-day world, even as it necessitates a totally other way of experiencing this world if this ambition is ever to be realised.

As Albers explains, this difference between what enters the eye and what is observed in the mind was already identified by M.F. Chevreul in his work on *The Laws of Contrast of Colour* of 1839. He showed how an exponentially increasing (1×, 2×, 4×, 8×, 16×, ...) intensity of colour, when presented to the eye, transmutes into a linearly increasing (1×, 2×, 4×, ...) intensity of colour as it appears in the mind. Several other examples of this nature are given in Albers' text.

Taking our cue from Edmund Husserl's masterpiece (originally of 1900, 1901 and as partly reworked in 1913), that introduced the world to *phenomenology*, as *the means for identifying the things themselves*, rather than accepting whatever happened to be the current 'modern-scientific representations' of them, we can most conveniently return to Bortoft.

'Zu den Sachen Selbst!'//'To the things themselves!'

It is with the language of modern science as with any other language, that *although we can more or less understand how we come to use that language, we cannot really understand how that language comes to use us*. We have already introduced the notion that knowledge is not achieved by the senses alone. Thus, with italics added (Bortoft 2010, p. 68):

> 'There is always a nonsensory element in knowledge, and this must be so whether this element is verbal-intellectual or intuitive. The difference is that, whereas the verbal-intellectual mind withdraws from the sensory aspect of the phenomenon into abstraction and generality, the intuitive mind goes into and through the sensory surface of the phenomenon to perceive it in its own depth. It is by first going into the full richness and diversity of sensory detail that the intellectual mind is rendered ineffective, so that we can escape from its prison into the freedom of intuition.'

> 'Etymologically, 'intuition' means 'seeing into', which clearly expresses the fact that it is the experience of seeing the phenomenon in depth. But this depth is peculiar inasmuch as it is entirely within the phenomenon and not behind it – so it should be seen as an intensive dimension, and not in the manner of an extensive dimension of physical space. It is in fact the depth of the phenomenon itself. It is as if something which appears to be two-dimensional suddenly turns out to be three-dimensional, so that what had seemed flat is now seen in relief. This is the experience mentioned earlier, of seeing the phenomenon 'standing in its own depth'. It was said then that there is no intellectual equivalent to this experience, and the reason for this is now clearly because it is *an intuitive experience which depends on a change of consciousness*.'

It is this change of consciousness that initiates and activates the mobilisation of the imagination of the individual stakeholders, whereby they become *activated stakeholders*, and then as activated stakeholders within the hydroinformatics environment that the hydroinformatician provides with his or her dynamic and coloured, interactive-user interfaces. It is this *active imagination*, as it is distributed over the different stakeholders, each one commonly with an own environmental interest, that leads them to explore the *actual facts* of their own 'real worlds' in relation to the *actual facts* of the 'real worlds' of their fellows, as the preparation for their mutating into participants in the *creative imagination of an increasingly unified stakeholder group*. This is in turn the *conditio sine qua non* for the transmutation from a higher level of collective consciousness (*Bewußtsein*) into a higher level of collective conscientiousness (*Gewissenhaftigkeit*) that, through its recursions, has the capacity to establish states of social justice, including ecological justice, in our otherwise so dreadfully misused world (Vojinovic & Abbott 2012).

HOMAGE TO GOETHE

In the German language there is a word for what we otherwise call 'concrete vision' in the present context, and this is *Anschauung*. Bortoft (2010, p. 90) expressed the essence of this word as follows:

> 'Agnes Arber, who spent her long life studying plants, said that in this context it 'may be held to signify the intuitive knowledge gained through contemplation of the visible aspect'. This indicated very clearly that Goethe's approach to animal form follows the same pathway as we have discovered in his work on colour. The method, as described above, is active looking followed by exact sensorial imagination, plunging into the visible aspect to provide dishabituation from the verbal-intellectual mind and the analytical mode of consciousness. This exercise of redirecting attention to seeing, inwardly as well as outwardly, removes an obstacle to the holistic mode of consciousness. At the same time, the exercise of trying to see the visible aspect as a whole promotes the restructuring of consciousness into the holistic mode. This procedure therefore has the result of taking the *Naturschauer* into the phenomenon intuitively and not just sensorially, while escaping from the prison of abstraction that is the intellectual mind.'

In the conjunctive knowledge that joins together the social and the technological, so as to provide a *sociotechnology* as an entirely other species of knowledge than the knowledges that it binds together, we have to face issues that have not been much considered previously, either in the modern sciences or in technologies. As we move into the arena of active stakeholder participation in projects of all manner of kinds, so we enter into new territories of thought and intuition that were not previously addressed in our own field. In the present authors' 2012 book we come face to face with the challenges that confront us, even as they hold us together in our mission, the source of which is expressed within the body of the work as 'a fascination with technology is a fascination with truth', but then in such a way that this fascination should never degenerate into a one-sided 'idolatry of truth without feelings', such as would threaten the most basic spiritual values of benevolence (*Wohltätigkeit*), benignity (*Freundlichkeit*) and compassion (*Mitleid*), those timeless values which are the only certainties in our lives. As expressed already in the words of Kierkegaard (1844/1855/1960, p. 186, 1980, p. 139): 'He who has observed the present generation can hardly deny the discrepancy in it, and the reason for its anxiety and unrest is this, that in one direction truth increases in scope and in quantity, and partly also in abstract clarity, while in the opposite direction certainty constantly declines'.

...still familiar?

Active stakeholder participation, as introduced by the first-named author in 2007 and by the present authors together in their paper of 2010, and as developed further in their 2012 book, is increasingly seen as the only way in which an otherwise world-catastrophic process might be avoided, and this possibility has been taken up by several authors in more recent times. Thus, His Royal Highness The Prince of Wales has expressed his support for this kind of intervention (2010, p. 19) as follows:

> 'Now there are many examples where communities have replaced the short-term impulse with the long-term plan. But part of that strategy – to my mind at least at the heart of it – is the need for a new public- and private-sector partnership which includes NGO and community participation. To work effectively this will require governments to provide policies which support community participation.'

From the *UNESCO Water and Ethics Series* of 2004 we select a piece from the introduction by our late-lamented friend, James (Jim) Dooge, as follows:

> 'There is no life without water, and those to whom it is denied are denied life. Water for all and meeting minimum basic needs are vitally tied to the principle of human dignity. The ethical principle of association means that the person is social as well as sacred. The principle of participation means that individuals,

> especially the poor, must not be shut out from participating in those institutions which are necessary for human fulfillment. Both these ethical principles mirror a major theme: namely, that those who are impacted and who would benefit from water (which is vital to their fulfillment as humans) must have the opportunity to participate in its planning and management.'

Turning to the other side of the world, to China, we have the situation where extensive public participation has become an accepted means of preparation for decision-making. Thus, already in an editorial to the *Beijing Review* of 5 March 2008 and entitled 'Transparency: it's the law', its editor, Yao Bin, after castigating the many examples of social injustice that so much concern the peoples and the governments at all levels of that nation, observed that: 'Clearly, a complete change in the above-mentioned situation lies in the establishment of a strict and transparent legislative system that features the broadest public participation'. This has been strengthened by new legislation on property and water rights and responsibilities that had become increasingly urgent and which involved the active participation of many tens of thousands of persons, even though this participation itself necessarily led to considerable legislative delays in many cases. It depends of course on the omnipresence of the web, where China currently has much more than 500 million users, and the mobile internet, with more than a billion mobile telephones in circulation as of October 2012.

As explained in the second part of this paper, this kind of action depends upon the hydroinformatician's ability to create a virtual environment that conveys the menaces and possibilities of projected interventions in the *outer worlds* of the activated stakeholders through conveying the consequences of these works into the *inner worlds* of the minds of the individual stakeholders. This process, in turn, motivates the collectivity of stakeholders to exercise their combined creative imaginations. All such processes must then employ the colours that are most suited to motivating the active and the creative imaginations and the choice of these most appropriate colours is an essential part of the hydroinformatics-created environments that are at the heart of success in this enterprise.

THE HYDROINFORMATICS OF THE THREE ECONOMIES – WITH COLOUR REVISITED

The definitions of the three economies

It is generally understood that the human economy, which most persons in the West like to think of as a 'first economy', is 'supported by' or, figuratively speaking, 'floats upon' the natural economy, which then becomes a 'second economy', where this 'floating' metaphor introduces water, which is that which most binds these two economies together. For the greater part of humanity, however, this all-embracing natural economy is founded upon *another economy again*, which created and has continued to support the natural economy and with this the human economy in all its manifestations, but then in such a way that it makes its presence known to those humans within their first economy who understand that they are charged by this 'other economy' with sustaining and supporting the 'second economy'. This 'other economy', this *Third Economy*, this third support that has served to preserve the stability of the other two economies for as long as human existence can be traced, and continues to sustain some semblance of stability even in the West today, is commonly subsumed within the general category of *the spiritual* to the extent that it is concerned, under the heading of 'Economy', with *the divine governance of the world*, to use the *Shorter Oxford Dictionary* definition. In the *Greater Webster's Dictionary* it is defined, again under 'economy', as 'God's plan or system for the governance of the world'. The term 'economy' itself refers back to the Greek word for the governance of the household, as *husbandry*, implying as it does an attribution of values and therewith responsibilities. Another term that is widely employed is that of *stewardship*.

We may return very briefly here to the most basic modes of recognition of the various societies and individuals of the 'second' and the 'first' economies by referring again to the value, and indeed the very purpose, of colour, in that, in both economies, the most distinguishing features of these worlds are their uses of colour, the 'first' through such agencies as clothing, homes, flags and banners, and the second by such devices of nature as grasses, leaves, skin, fur and feathers, whereby colour is in the first place an attribute of the second economy, as that for which we, in our self-styled 'first' economy, are charged with sustaining.

It is from such examples as these that the worlds of the imagination and especially of the creative imagination now arise, whereby colour becomes the first creative agent in our imagination.

The subversion of technology

The reason for introducing this, in Western eyes, obscure branch of human knowledge, and then, following a Western-theological tradition, as an economy, is that it presents as great a challenge to the West, in particular, as does anything else occurring in the human and natural economies. Since almost all technology proceeds nowadays through our being challenged, we see at once that, to the extent that we experience that there is something highly irregular in our technology, we must conclude that *we are being challenged in 'the wrong way'*. But what is it that causes *us* to be challenged 'in the wrong way', this being a way that we otherwise experience, albeit subjectively, as not being a way of truth at all, but a way of deception, of untruth? Since modern technology must still remain, in Heidegger's words, as 'the place where *alētheia*, truth, happens', what we experience as happening now is that our present-day applications of technology *are telling us the truth about an untruth* that is not itself attributable to technology as such at all, but must be attributable to something *within our own selves*, something that comes to presence through the frame that *we* set around our worlds, in *our* action of Enframing that serves to negate our spiritual virtues.

But what is this 'something'? This is a question that has in fact persisted throughout history, but which became increasingly urgent in the Europe of the nineteenth century as the negative influences attributed to modern technology and the changing ways of life that it engendered became increasingly more evident. It became posed with increasing force again during the first half of the twentieth century, where this question demanded increasingly clear answers in response to the rise of the forces of fascism and Nazism, for much the greater part in Europe, with their glorification and prosecution of modern technology in its most terribly efficient, brutal and barbarous forms. Since that time, this negative force has taken on less obviously ugly forms in that part of the world that likes to call itself the First World, having transferred its nefarious attentions much more to the so-called Third World, but it still remains most active, and ever more dangerously so in Europe, and indeed just as much and perhaps even more so than elsewhere just because it is so much less evident. Thus in the words of Jacques Derrida (1993), responding, in his *Spectres of Marx*, to the empty-headed 'optimism' of Fukuyama's (1992) book entitled *The End of History*:

> 'For it must be cried out, at a time when some have the audacity to neo-evangelize in the name of the ideal of a liberal democracy that has finally realized itself as the ideal of human history: never have violence, inequality, exclusion, famine, and thus economic oppression affected as many human beings in the history of the earth and of humanity. Instead of singing the advent of the ideal of liberal democracy and of the capitalist market in the euphoria of the end of history, instead of celebrating the 'end of ideologies' and the end of the great emancipatory discourses, let us never neglect this obvious macroscopic fact, made up of innumerable singular sites of suffering: no degree of progress allows one to ignore that never before, in absolute figures, have so many men, women and children been subjugated, starved or exterminated on the earth.'

If only for these reasons alone, we have no alternative but to outline this response here already if we are to understand the essential features of modern technology in its *false enframing*, in its divorce from spiritual values, but we must do this for other reasons besides. The hydroinformatician cannot ignore, and indeed must take into proper account, the existence of this negative force in all that passes around that person. Our object is thus also to arm the hydroinformatician properly with an understanding of this force so that he or she can more quickly identify and subsequently defeat this adversary whenever it is encountered.

But what, we may ask, is the name of this adversary? It – for we have here to do with an 'it' – was introduced in Søren Aabye Kierkegaard's *Begrebet Angest* (1844/1855/1960, p. 179)//*The Concept of Anxiety* (1960) as *Intet*, which translates literally as 'nothing', and *Intethed*, which we can only translate as 'nothingness'. It correspondingly entered German as *das Nichtige*, and into French as *le néant* and into English, and more awkwardly so than ever, as *nothingness*. In an essentially atheistic philosophy, it was described as a *nihil* and its manifestations were described as *nihilism*. We now have, most unfortunately because it is an unpleasant matter, to introduce this 'it-which-can be-no-thing'. As we shall see, it is this 'nothing', this *nothingness*, which touches everything as it strives to destroy all that has been created, whether by 'The Absolute' or, by extrapolation, by humankind and the whole world of nature. Correspondingly, *almost all the much-vaunted gains of one relatively small part of the first economy are bought at the cost of the devastation of both the greater part of the first and the second economies*, whereby the third economy abandons its first economy on the grounds of its treachery, which has allowed nothingness to instil itself in it, leading to the *Deus absconditus* of our Western world, thereby setting it on the path of certain destruction.

From nihilism to nothingness

It was primarily the experiences of nineteenth-century European industrialisation, with all its negative manifestations, that gave rise to a new, more highly educated, so-called 'middle' class that made the conventional sophistries of the Western churches untenable. In the incisive words of the foremost among such critics, in the words of Nietzsche (1882 [1969]) written with his usual biting irony:

> 'The decline of belief in the Christian God, the victory of scientific atheism – this is a combined European achievement (*ein gesamteuropäisches Ereignis*) for which all races will claim their own share of merit and honour.'

And to this he elsewhere added, in deadly seriousness:

> 'The greatest recent event – that God is dead, that the belief in the Christian God has become unbelievable – is already beginning to cast its first shadows over Europe.'

To Nietzsche's rhetorical question that arose immediately from this situation: 'Are we not then straying through an infinite nothing?' Heidegger, in his own time, of the third quarter of the twentieth century, responded (1967, pp. 60–61):

> 'The pronouncement that 'God is dead' contains the confirmation that this Nothing is spreading out. 'Nothing' means here: absence of a suprasensory, obligatory world. Nihilism, 'the most uncanny of all guests', is standing at the door …'

> 'Nihilism is a historical movement, and not just any view or doctrine advocated by someone or other. Nihilism moves history after the manner of a fundamental ongoing event that is scarcely recognised in the destiny of the Western peoples. Hence nihilism is also not simply one historical phenomenon among others – not simply one intellectual current that along with others, with Christendom, with humanism, and with Enlightenment – that comes to the fore within Western history.'

> 'Nihilism, thought of in its essence, is rather the fundamental movement of the history of the West. It shows such great profundity that its unfolding can have nothing but world catastrophes as its consequence. Nihilism is the world-historical movement of the peoples of the Earth who have been drawn into the power realm of the modern age.'

This is to say that, for Heidegger, nihilism was not only a product of the nineteenth century, even though it was primarily in that century that its presence became more clearly recognised and its name more firmly established, but its origins proceeded much further back, back through all the histories and pre-histories of the European peoples.

It is in a spiritual reply to Nietzsche's *Sieg des wissenschaftlichen Atheismus*, that we nowadays speak of this loss in terms of an *absconded God*, this *Deus absconditus* (See Cheetham (2005), pp. 19, 55–57, 74–75, 99, 117).

This view of the mysterious nature but only too manifest influence of 'nihilism' was, however, essentially a philosophical one, and it was by no means the view of those with a more theological foundation who persisted in speaking of 'nothing' and 'nothingness'. Thus, even though Heidegger, for example, did have a substantial theological education, he was not himself a theologian: his knowledge of theology provided him with a certain theoretical apparatus for his philosophical work, but this work was itself in no way theological. The same could be said of Sartre, who also entered into this analysis. It is accordingly first necessary to explain this difference between the philosophical and the theological in order to follow why it is that nothingness, in its essence, can only be understood from a theological, and indeed from a 'dogmatic-theological', standpoint. We emphasise that it cannot be at all expected nowadays that everyone interested in such matters is Christian, or indeed has any explicit faith at all, but nonetheless all persons concerned with such matters as the present one can and should learn from theological studies. As concerns the most obvious world-catastrophic consequences of nothingness – the two world wars of the last century – both had their main focus within Europe. As Barth observed (1960, Vol. 3, Part 3, p. 345) in his masterly commentaries on the disparate views of Heidegger and Sartre (1960, Vol. 3, Part 3, p. 345):

> 'We experience nothingness, and in so doing we experience ourselves and all other things as well. Heidegger's astonishment [as expressed in his masterly *Der europäische Nihilismus*] is no less eloquent than Sartre's defiance [as in his equally famous *L'être et le néant*], nor does the latter bear lesser witness. Their thought is determined in and by their real encounter with nothingness. Their thought and expression are determined in and by the considerable though not total upheaval of Western thought and expressions occasioned by the world wars.'

In the words of the present first-named author, as written in *Hydroinformatics* (Abbott 1991):

> 'In his *Church Dogmatics*, Barth explained how that motion in the depths of the collective unconscious that is the innermost expression of nothingness has continued to take on ever more concrete names and forms in our outer world. Today we see in the destruction of the natural environment and an ever-increasing poverty of the majority of our populations, and so in the undoing of the creation, the further allegorical representation of nothingness and one which gives expression to 'the kingdom of nothingness' in the most concrete and material way possible.'

> 'Now of course the manifestations of an ever-encroaching nothingness are by no means a feature that is peculiar to the European peoples. By no means is it this! It is however centred upon Europe; it has a focus, in the same way as does a physical sickness, and this focus is situated in the collective psyche of the European peoples. Then, in so far as hydroinformatics constitutes in its essence one part of the total field of contest of nothingness, so it follows that *hydroinformatics begins as a European possibility*.'

This is to say that the *Hydroinformatics* of 1991 was conceived as something directed to a specifically European situation, so that it proceeded for the most part through the actions of the orderable, the numerable, the countable and the computable: it was for much the greater part *a mathematical hydroinformatics of the quantities*. Now, however, as the pendulum of socioeconomic development moves increasingly, inexorably and ever more dramatically towards Asia, and so away from the European *Zeitgeist*, this hydroinformatics has to be rethought and reformulated within new and essentially different social, cultural and increasingly religious-cultural contexts, whereby it is described more by *a mathematics of the qualities*. For this purpose we have used or – as probably some at least will say – misused, some fragments of *category theory* in our earlier works and most recently in the 2012 book, so as to provide an appropriate descriptive apparatus.

Reassessing technology

In this latest book (Vojinovic & Abbott 2012), as in some earlier works, we have observed that *a fascination with authentic technology is a fascination with truth*. Technology is grounded in a seeking after truth in the world of human creativity, and in our present case it seeks and finds its truths in acts of industrial creativity. Industrial technology is only one, but still one, of the many, many ways of seeking after truth and experiencing truth that gives us hope, cohesion and guidance as seekers after truth in an otherwise so troubled world. In Christian-theological terms, all such searchings and strivings are borne by that covenant of the spirit that we commonly call 'love', and the most profound of these movements of the intellect is the love of wisdom itself.

For well over 2,000 years now, it has been normal within the so-called Western tradition to distinguish between a love of wisdom translated into works without the explicit supposition of any agent external to humankind itself, and so by reason alone, which is called *philosophy*, and a love of truth as the issue of a wisdom that includes but also necessarily transcends the powers of unaided human reason, which is called *theology*. Thus philosophy requires human understanding, as mediated by reason, while theology requires not only this human understanding but also something more again, which is called *faith*. Thus, within the Christian tradition, as in other religious traditions of the book besides, faith is that which surpasses human understanding, which transcends reason, so that, following Kierkegaard [(Sygdommen til Døden 1849//The Sickness unto Death, 1983, p. 39]),]'faith is a miracle, otherwise it is not faith'.

The theologian Karl Barth correspondingly introduced our present subversive element as follows (Vol. 3, Part 3. p. 289):

> 'There is amongst the objects of God's providence an alien factor. It cannot escape God's providence, but is comprehended by it. The manner, however, in which this is done is highly peculiar in accordance with the particular nature of this factor. It is distinct from that in which God's providence rules the creature and creaturely occurrence. The result is that the alien factor can never be considered or mentioned together in the same context as other objects of God's providence. Thus the whole doctrine of God's providence must be investigated afresh. This opposition and resistance, this stubborn element and alien factor, may be provisionally defined as nothingness.'

All theology involves a struggle to express in words the many experiences, impressions and feelings which our present-day languages were never developed to describe and for the expression of which these languages remain always inadequate. This is nowhere more evident than in the present case. In Christian-theological terms, the relation between God and man becomes broken by this alien element, with the consequence that every attempt to describe nothingness, even theologically, must itself be broken in thought and in utterance. This is to say, however, that *this description cannot form a system*.

FOR WHAT THEN IS THE MODELLER SEARCHING?

That for which the hydroinformatician is constantly searching, as the *initiating modeller* in active stakeholder participation processes, is *cause and causality*. For example: what is it that is causing the issues that are of concern to the active stakeholders to lead to a failure to agree on a mutually acceptable arrangement, or verdict? From this standpoint it is clear that *he or she is looking for the cause of failure*, so that for the initiating modeller, just as for the psychoanalyst described by Lacan (1973/[1977], pp. 9 and 22): 'In short, there is cause only in something that does not work'. When it can be brought 'to work', through the processes of repetition and transcendence, then we arrive at 'The Great Work', the *magnum opus* of our latter-day alchemists, as both an individual transcendence of the Self and as a group transcendence of the community of Selves of the active stakeholders. Clearly, no such process is realisable in the case of passive stakeholders. In the same vein, only in the case that the initiating modeller can overcome the temptations of exercising only 'deficient modes of solicitude' in the now-classical Heideggerian sense (whereby the stakeholders are maintained in a passive state, such as happens when these stakeholders are only being 'consulted', 'informed' and 'directed', rather than being challenged-out to exercise and develop their own inherent knowledges, imaginations and judgments, and to exercise these both independently and interactively) can 'The Great Work' succeed.

The second immediate consequence is that the function of the modeller changes, indeed drastically, in such situations, as corresponds to this reversion from the chemical back to the alchemical. Whereas in the pre-internet era of hydroinformatics the emphasis was on the model, in its new web-connected and mobile era the emphasis of hydroinformatics is much more on the modeller: *the emphasis changes from the operand to the operator*. In the words of Lacan (1973/[1977] p. 9):

> 'What is it that makes us say that, despite the dazzling character of the stories... from ages past, alchemy, when all is said and done, is not a science? Something, in my view, is decisive, namely, that the purity of the soul of the operator was, as such, and in a specific way, an essential element in the matter.'

In hydroinformatics similarly, the *purity of the soul* of the operator, which we can better here describe as the *quality of the character* of the modeller, becomes inseparable from the quality of the model within the quality of the total production. Thus, in modern science, the outcome of an experiment should be independent of the nature of the experimenter, whereas in a pre-modern (and thereby, as in the present case, post-modern) science this is not the case. Correspondingly, the central question now comes to be posed of 'what is the modeller's *desire*?' It is this in turn that establishes the economy of the modeller's libidinal (in the original sense of emotional or psychic energy) resources. Thus the proper employment of the web-based *Software as a Service* (*SaaS*) paradigm in hydroinformatics depends vitally upon an authentic answer to this question, and thus, so far as is possible, an authentic understanding. Our question then becomes: with what kind of human activity, and correspondingly with what kind of participating Selves, are we concerned here? To put this question in the terms appropriated by Lacan for his field of psycho-analysis: 'What grounds it as praxis?' But then we must pose the question of what we now mean by *praxis*? We follow Lacan, again adding italics, when he says that 'It is the broadest term to designate a concerted human action, whatever it may be, which *places man in a position to treat the real by the symbolic*'. What we then do is to take our praxis with us, as interconnected applications in our outer world, and we let our observations of the applications 'direct us at once towards some fairly well-located, specifiable points of practice'.

Clearly the hydroinformatician must be able to identify nothingness, but he or she must also be able to combat it. For the 'non-believer', Barth's observation on present-day man and woman's chances in this combat must then appear at first sight as decidedly discouraging: 'God alone can summon, empower and arm the creature to resist and even to conquer this adversary...The creature as such would be no match for nothingness and certainly unable to overcome it.' From the Christian-theological point of view that is adopted here, however, it is not of much consequence whether the person who is summoned, empowered and armed to combat nothingness is immediately aware at all of the source of that person's strengths, and this person may well be a match for nothingness without the slightest inkling that she or he is such a match because she or he is *by no means standing alone in this confrontation*. From a theological point of view of course, both would be better prepared again if they were aware and alert to this support, but it is not the most immediately essential issue. The essential point is that the person so 'chosen' is prepared to face her or his responsibilities in such a combat – and after that it is our present responsibility to prepare him or her for this combat with such weapons as theology and other means have prepared.

It follows that the weapons necessary to combat nothingness are those of expressing and communicating and inculcating truth at this most exalted level. Thus (Barth (1950) *loc. cit.* p. 529):

> 'That the lie should be exposed is what is most appropriate to the lie itself and most helpful to those who are threatened, oppressed and tormented...And as it is done the lie loses the vital breath which enables it to threaten, oppress and torment. It is vanquished and driven from the field.'

We conclude, again with the words of His Royal Highness The Prince of Wales (2001, p. 14):

> 'I believe that if we are to achieve genuinely sustainable development we will have to rediscover, or re-acknowledge, a sense of the sacred in our dealings with the natural world and with each other. If nothing is held sacred any more – because it is considered synonymous with superstition, or in some other way 'irrational' – what is there to prevent us treating our entire world as some 'great laboratory of life', with potentially disastrous long-term consequences?'

ACKNOWLEDGEMENTS

The authors thank the referees of this paper for their engagement in providing well-founded criticisms and several corrections that have been of great value in finalising its form (see also www.knowledge-engineering.org).

REFERENCES

Abbott, M. B. 1991 *Hydroinformatics: Information Technology and the Aquatic Environment*. Avebury Technical, Aldershot, UK and Brookfield, USA.

Abbott, M. B. 1996 The sociotechnical dimensions of hydroinformatics, *Proc. Second International Conference on Hydroinformatics*, Balkema, Rotterdam, pp. 3–18.

Abbott, M. B. 2000 The gender issue in hydroinformatics, or Orpheus in the Underworld. *Journal of Hydroinformatics* **3** (2), 87–104.

Abbott, M. B. 2007 Stakeholder participation in creating infrastructure. *New Civil Engineer* **160** (1), 26–32.

Abbott, M. B. 2012 Rich will always decide their own tax. *Financial Times* 10 August, 8.

Albers, J. 1963 *Interactions of Colour*. Yale University Press, New Haven, CT.

Barth, K. 1950 Die Kirchliche Dogmatic, III.3, Evangelischer Verlag, Zollikon-Zűrich // 1961/1983 Church Dogmatics, III. 3 Transl. G. W. Bromiley and T. F. Torrance, Clark, Edinburgh.

Bortoft, H. 2010 *The Wholeness of Nature, Goethe's Way of Science*, 4th edition. Lindisfarne Books and Floris Books, Edinburgh.

Cheetham, T. 2002 Consuming passions: the feast, the stars and the science of the balance. *Temenos Academy Review* **5**, 117–137.

Cheetham, T. 2005 *Green Man, Earth Angel: the Prophetic Tradition and the Battle for the Soul of the World*. State University of New York Press, Albany, New York.

Chevreul, M. E. 1839 De la loi du contraste simultané des couleurs. Pitois-Levrault, Paris // 1854/55, The Principles of Harmony and Contrast of Colours, and their Applications to the Arts. Trans. C. Martel. Longman, Brown and Green, London.

Corbin, H. 1958/2006 L'imagination créatrice dans le soufisme d'Ibn Arabi, In: *Introduction by Gilbert Durand, Flammarion, 1958 / Editions Médicis-Entrelacs, 2006//1969/ 1997 (Alone with the Alone) Creative Imagination in the Sūfism of Ibn 'Arabi, Trans* (R. Manheim & H. Bloom, eds). Princeton University Press, Princeton, New Jersey.

Derrida, J. 1993 Spectres de Marx: l'état de la dette, le travail du deuil et la nouvelle Internationale, Éditions Galileé, Paris// 1994, Specters of Marx: The State of the Debt, the Work of Mourning & the New International, Routledge, London.

Dooge, J. 2004 *UNESCO Water and Ethics Series (No. 1. Water and Ethics Overview, of J.D. Priscala. J. Dooge, J, and R. Llamas)*. IHP Publications (July 2004 –August 2007), p. 95.

Duhem, P. & Brenner, A. 1997 *L'Aube de Savoir: Épitomé du Système du Monde*. Hermann, Paris.

Findlay, J. N. 1963 [1995] *Meinong's Theory of Objects and Values*. Oxford University Press Gregg Revivals and Ashgate, Aldershot, UK and Brookfield, USA.

Freyd, P. J. & Scedrov, A. 1990 *Categories, Allegories*. North Holland, Amsterdam.

Fukuyama, F. 1992 *The End of History and the Last Man*. Free Press, London.

Goethe, J. W. v. 1810 [1820] *Zur Farbenlehre*. Cotta, Tübingen.

Heidegger, M. 1927 Sein und Zeit. Niemeyer In: *Tübingen//1962 Being and Time* (Transl. J. Macquarie & E. Robinson). Blackwell, London.

Heidegger, M. 1963 [2002] *Die Technik und Die Kehre*, 10th edition. Cotta, Tübingen, Germany //1977, The Question Concerning Technology and Other Essays. Transl. W. Lovitt. Harper, New York.

Heidegger, M. 1967 Der europäische Nihilismus. Neske, Pfullingen.

HRH The Prince of Wales 2001 A reflection on the Reith lectures. *Temenos Academy Review* **4**, 13–18.

HRH The Prince of Wales 2010 Facing the future. *Temenos Academy Review* **13**, 9–22.

Husserl, E., 1900 Logische Untersuchungen, 2nd edition: M. Niemeyer, Halle; 1913; Vol 1 and Part 2 of Vol. 2 // Logical Investigations, 1982 (trans. Findley, J. N.), Routledge and Kegan Paul, London, and Humanities Press Inc., USA.

Illich, I. 1993 *In the Vineyard of the Text: A Commentary on Hugh's Didascalicon*. University of Chicago Press, Chicago.

Kierkegaard, S. Aa. 1844/1855/1960, p. 186, Begrebet Angest, reprinted in 1960 from the Danish Fifth Edition of the Collected Works, Introduction and Notes by Sørensen, V., Gyldendal, Copenhagen//1980, p. 139, The Concept of Anxiety, Transl. by Thomte, R., and Anderson, B., Princeton University Press, Princeton, NJ.

Kierkegaard, S. Aa. 1849 Sygdommen til Døden, reprinted in 1960 from the Danish Fifth Edition of the Collected Works, Introduction and Notes by Sørensen, V., Gyldendal, Copenhagen//The Sickness unto Death, 1983, Edited and translated by Howard V. Hong and Edna H. Hong, Princeton University Press, Princeton, NJ.

Lacan, J. 1973/[1977] reprinted 2004: translations of pp. 22 and 9 in the text are taken from Lacan, J., The four fundamental concepts of psycho-analysis, translation by Sheridan, A., KARNAC, London and New York.

Latour, B. 1991 Nous n"avons jamais été modernes, La Découverte, Paris // We have never been modern. Trans. C. Porter, Harvard University, Cambridge, MA, and Harvester, London.

Nietzsche, F. 1882 Die fröhliche Wissenschaft, Verlag von Ernst Schmeltzmer, Chemnitz/1969, Ullstein, Frankfurt/M.

Thorkilsen, M. & Dynesen, C. 2001 An owner's view of hydroinformatics: its role in realising the bridge and tunnel connection between Denmark and Sweden. *Journal of Hydroinformatics* **3** (2), 105–136.

Vojinovic, Z. & Abbott, M. B. 2012 *Flood Risk and Social Justice*. IWA Publishing, London.

Yao, Bin 2008 'Transparency: it's the law'. *Beijing Review* 5 March, 14.

First received 4 November 2012; accepted in revised form 26 March 2013. Available online 15 May 2013

IWA Publishing's authorised EU representative for General Product Safety Regulations is Diane D'Arras, 15 rue Duret, 75116 Paris, France, e-mail: safety@iwap.co.uk.

Printed and bound by CPI Group (UK) Ltd, Croydon, CR0 4YY
16/06/2026
02140374-0001